FLOTATION

Theory, Reagents and Ore Testing

Related Pergamon Titles of Interest

Books

DOBBY and RAO
Processing of Complex Ores

FINCH and DOBBY
Column Flotation

GILCHRIST
Extraction Metallurgy

WILLS
Mineral Processing Technology

Journals

Acta Metallurgica et Materialia

Canadian Metallurgical Quarterly

International Journal of Rock Mechanics and Mining Sciences & Geomechanics Abstracts

Minerals Engineering

The Physics of Metals & Metallography

Scripta Metallurgica et Materialia

CONTENTS

Introduction 1

Chapter 1. Flotation Fundamentals 5
Definition of terms 5
Reagents employed in flotation 9

Chapter 2. The Mechanism of Flotation 11
Coursing bubble flotation 12
Contact angle and Hallimond tube measurements 13
Nascent bubble flotation 17
Electron transfer to mineral surface 18

Chapter 3. Sulphide Mineral Flotation 25
Stereochemistry of xanthates and dithiophosphates 25
Six coordination complexes 25
Seven coordination complexes 26
Dithiolate chemistry and surface compounds 29
Thiol collector bonding with sulphide minerals 36
The nature of the bubble-particle bond 40
Frother-collector surface complexes 44

Chapter 4. Thiol Collector Chemistry 46
Properties of xanthates 47
Manufacture of alkali metal alkyl xanthates 49
Manufacture of thiophosphates 55
Manufacture of xanthogen formates 58
Manufacture of dialkyl thionocarbamates 62
Miscellaneous collector structures 64

Chapter 5. Commercial Sulphhydryl Collectors 66

Chapter 6. Properties of Flotation Froths 85
Flotation frothers and froth hydrodynamics 85
Dynamic froth stability 85
The effect of the collector on froth properties 86
The influence of the frother on the rate of flotation 88
Over oiling 92

Chapter 7. Chemical Properties of Frothers 94
Partially soluble frothers 94
Completely water miscible frothers 98

Chapter 8. Flotation Modifiers 101
Activating agents 102
Inorganic depressants 103
Organic depressants 105

Chapter 9. Sulphides - Depression and Activation 107
Sodium sulphide 107
Chemical properties of sulphides 107
Molybdenum concentrate cleaning 111
Copper depression in the molybdenite separation 116
Sodium sulphide as a depressant 117
The chemistry of the Nokes reagents 120
Sulphidisation of refractory ores 122

Chapter 10. Mill Tests - Case Histories 127
San Manuel - a non-sulphide moly separation process 127
CODELCO Andina - slime sulphidisation 136
CODELCO El Teniente - molybdenite recovery 147

Chapter 11. Sulphide Mill Practice 174
Reagent consumption in the flotation industry 174
Copper mill data 176
Flotation process design 200
Copper mill flotation practice 205
Non-sulphide copper ores 208
Flotation of complex ores 209

Chapter 12. Non-metallic Mineral Flotation 212
Non-sulphide mineral flotation 212
The nature of oily collector flotation 213
Structure of mineral surfaces 216
Mechanism of collector adhesion 217
Collectors for industrial minerals 220
Fatty acids 221
Anionic collector chemistry 224
Industrially important fatty acid collectors 225
Cationic collector chemistry 225
Factors affecting selective flotation 228
Metallic oxide minerals mill practice 229
Industrial minerals mill practice 230

Chapter 13. Flotation Testing 258
Planning a test 258
Experimental design 260
Testing procedures 276
Equipping the flotation laboratory 278
Comparison of laboratory flotation machines 285
Sampling and sample preparation 287
Sampling working concentrators 291
Testing routines 295
Test procedures used by a reagent supplier 297
Recommendations on setting test conditions 303
Batch flotation tests 306
Standard test procedure used by Sherex on coal 310
Column testing 313
Flotation column testing examples 317

Bibliography 324

Subject Index 336

Author Index 343

LIST OF TABLES

Table 1.- Classification of polar minerals 6
Table 2.- Collector structures 7
Table 3.- Frother structures 8
Table 4.- Modifiers 10
Table 5.- Xanthate contact angles for selected sulphide mineral 16
Table 6.- Properties of dixanthogens and xanthic anhydrides 31
Table 7.- Dithiolate reagents 33
Table 8.- Reduction potentials for dithiolate/thiol couples 34
Table 9.- Products extracted from thiolated surfaces 35
Table 10.- Mineral electrochemical properties 38
Table 11.- Band-gap energies of sulphide mineral semiconductors 40
Table 12.- Decomposition of Na ethyl xanthate on drying at 50°C 51
Table 13.- Properties of alkyl xanthates 51
Table 14.- Xanthate assays, slurry suspension process 52
Table 15.- Commercial stability data on xanthate solutions 55
Table 16.- Density and viscosity of Aerofloat collectors 58
Table 17.- Properties of alkyl xanthogen ethyl formates 61
Table 18.- Structures, and suppliers of alkyl xanthates 67
Table 19.- Structures, and suppliers of xanthic esters 68
Table 20.- Structures, and suppliers of xanthogen formates 69
Table 21.- Structures, and suppliers of thionocarbamates 70
Table 22.- Structures, and suppliers of thiophosphate collectors 70
Table 23.- Structures, and suppliers of miscellaneous collectors 72
Table 24.- Properties of alcohols used as frothers 94
Table 25.- Properties of terpenes and ketones used as frothers 96
Table 26.- Properties of water soluble frothers 99
Table 27.- Critical pH for flotation 101
Table 28.- Effect of collector on critical pH for flotation 102
Table 29.- The effect of sodium cyanide on critical pH 102
Table 30.- Sodium sulphide saturation solubility 108
Table 31.- Effect of pH on the equilibrium values of sulphide ions 110
Table 32.- Effect of temperature on the pH of NaSH solutions 111
Table 33.- Reagent practice, moly recovery from porphyry ores, Cu circuit 112

Table 34.- Reagent practice, moly recovery from porphyry ores, moly circuit 113
Table 35.- Regional mine performance 115
Table 36.- Moly plants using the sulphide process - Shirley (1979) 119
Table 37.- San Manuel copper and moly assays 129
Table 38.- Molybdenite plant reagent consumption 135
Table 39.- Microscopic analysis of the moly plant streams 138
Table 40.- Lab results of sand/slime tests, without added reagents 138
Table 41.- Lab results of sand/slime tests, with split after 6 minutes 140
Table 42.- Lab results of sand/slime tests, with split after 12 minutes 142
Table 43.- Plant test of the effectiveness of the sand/slime separation 144
Table 44.- Screen analysis and copper distribution - Andina 145
Table 45.- The effect of NaSH conditioning time on copper recovery 146
Table 46.- Andina - results of test for addition point of NaSH 148
Table 47.- Material balances for El Teniente 152
Table 48.- Floatability of samples from different mine sectors 157
Table 49.- Material balance and screen analysis retreat regrind 167
Table 50.- Effect of Nokes reagent on rougher recovery 171
Table 51.- Effect of Nokes reagent on first cleaner recoveries 171
Table 52.- Moly/copper separation cleaner tests with columns 172
Table 53.- Froth flotation in the U.S.A.; source: U.S. Bureau of Mines 175
Table 54.- U.S. consumption of collectors in 1975 and 1985 176
Table 55.- Consumption of frothers in the total U.S. mining industry 177
Table 56.- Industrial copper mill flotation performance data 179/1
Table 57.- Industrial copper mill flotation reagent data 185/1
Table 58.- Average porphyry copper mine performance in 1975 192
Table 59.- Collector use frequency comparison for copper mills 194
Table 60.- Frother use frequency comparison for copper mills 196
Table 61.- Collector consumption in copper sulphide ore flotation 198
Table 62.- Power and water use in U.S. flotation mills 199
Table 63.- Screen size equivalents 201
Table 64.- Principal sulphide minerals 206
Table 65.- Secondary minerals in Pb-Zn complex ores 210
Table 66.- Collector reagents for non-metallic minerals 220
Table 67.- Fatty acid content of oils and fats (weight %) 222
Table 68.- The structure of some anionic collectors 222
Table 69.- Critical miscelle concentration for anionic collectors 224
Table 70.- Nitrogenous cationic flotation agents 226
Table 71.- Cationic amine collector structures 227
Table 72.- Molecular species solubility, pK_a, and polar head diameter 228
Table 73.- Commercially important boron minerals 232
Table 74.- Glass fiber colemanite specifications 233
Table 75.- Typical composition of spruce pine alaskite 235
Table 76.- Reagents employed in the flotation of feldspatic ores 237

Table 77.- Reagents for fluorspar, sulphide and barite flotation 240
Table 78.- Fluorspar grade specifications 242
Table 79.- Specification for commercial quartz and feldspar 242
Table 80.- Reagents employed in the flotation of glass sand in the USA 243
Table 81.- Experimental design and results - glass sands 245
Table 82.- Experimental design - glass sands flotation 246
Table 83.- Reagents employed in the flotation of phosphate rock in the USA 250
Table 84.- Reagents used in potash flotation 252
Table 85.- Reagents employed in the flotation of potash in the USA 253
Table 86.- Scheelite flotation results 253
Table 87.- Mineralogical analysis of Kings Mountain spudomene ore 255
Table 88.- Feldspar, quartz, spudomene minerals in N.C. pegmatites 256
Table 89.- Flotation testing variables 259
Table 90.- Influence of EtX and alpha-terpineol consumption on grade 268
Table 91.- Influence of EtX and octanol consumption on concentrate grade 269
Table 92.- Three-variable five-level Box-Wilson design layout 273
Table 93.- Flotation cell dimensions and standard flotation test conditions 285
Table 94.- Reproducibility of time-recovery curves for Leeds cells 287
Table 95.- Parameters to determine sample size for different minerals 288
Table 96.- Column flotation variables used in reground pulp cleaner study 318
Table 97.- Overall experimental plan for the RRC samples in the 2" column 320
Table 98.- Optimum flotation conditions for the 2" column 321

LIST OF FIGURES

Fig. 1.- Cavitation mechanism of bubble formation 18
Fig. 2.- The electrical double layer model 21
Fig. 3.- The structure of ice 22
Fig. 4.- Localized bond structure for water 23
Fig. 5.- Cation and anion hydration showing water orientation 24
Fig. 6.- General stereochemistry for $[M(bidentate)_3]^{x\pm}$ 26
Fig. 7.- $[Cu^{II}(xanthate)_2]$ or $[\{Cu(xanth)_2\}_x Cu^{II}{}_{x-2}(H_2O)_2]$ 27
Fig. 8.- Pentagonal bipyramid and capped octahedron geometry 27
Fig. 9.- $[Pb^{II}(ethyl\ xanthate)_3(lone\ pair)]$ 28
Fig. 10.- $[Pb^{II}(di\text{-}isopropylphosphorodithiolate)_2(lone\ pair)]$ 29
Fig. 11.- $[M(bidentate)_3(lone\ pair)]^{x\pm}$ 30
Fig. 12.- Structure of chalcopyrite and enargite 39
Fig. 13.- Wurtzite and sphalerite, from Pauling 39
Fig. 14.- Leja and Schulman (1954) interpenetration theory 43
Fig. 15.- Frother-collector interaction complex 44
Fig. 16.- Frother volume versus alcohol chain length 87
Fig. 17.- Effect of frother concentration 88
Fig. 18.- Iso-recovery curves, Hallimond tube 88
Fig. 19.- Iso-recovery curves for chalcocite 89
Fig. 20.- Rate of flotation of copper ores 90
Fig. 21.- Performance of frothers on recovery of coal/ash 91
Fig. 22.- Illustration of two laboratory time-recovery profiles 93
Fig. 23. Ionization of sodium sulphide solutions 109
Fig. 24.- Relative effect of the addition of NaSH on pH 111
Fig. 25.- Relationship between pH, NaSH depression and floatability 118
Fig. 26.- San Manuel molybdenite recovery circuit 128
Fig. 27.- Andina rougher circuit in 1976 136
Fig. 28.- Andina rougher circuit in 1977 137
Fig. 29.- Rougher circuit for Andina's sand/slime mill test 141
Fig. 30.- Overall molybdenum recovery, El Teniente 149
Fig. 31.- Simplified flotation schematic for El Teniente 151
Fig. 32.- Sewell ore CuT, CuNS and % oxide copper content 154
Fig. 33.- Sewell, Colon and combined ore non-sulphide assay 155

Fig. 34.- Mineral content of Teniente concentrate - 1984 155
Fig. 35.- Monthly moly assays and tonnage produced by sectors 156
Fig. 36.- Monthly moly assays for 1989 for Colon and Sewell 156
Fig. 37.- Floatability of samples from different sectors of the mine 158
Fig. 38.- The alkaline cleaner flotation circuit at Colon 160
Fig. 39.- Performance of the copper cleaner circuit 161
Fig. 41.- The effect of pH on pyrite and molybdenite recovery 163
Fig. 42.- Effect of pH on Cu and moly recovery from second cleaner 164
Fig. 43.- Molybdenum content of Teniente tailings water 164
Fig. 44.- Potential-pH diagram for moly-sulphur-water 165
Fig. 45.- First cleaners and copper circuit flow diagram 166
Fig. 46.- Effect of grind on copper and moly recovery 168
Fig. 47.- Moly content vs particle size - Colon 169
Fig. 48.- Moly content vs particle size - Sewell 170
Fig. 49.- Proposed new Colon cleaner circuit 170
Fig. 50.- Redox measurements of rougher flotation - Teniente 172
Fig. 51.- Overall mill performance, world survey 193
Fig. 52a.-Collector frequency graphs, world survey 195
Fig. 52b.-Frother frequency graphs, world survey 197
Fig. 53.- Complex ore differential flotation circuit 204
Fig. 54.- Effect of grind on copper recovery from ores 208
Fig. 55.- Forces in adsorption and chemisorption 216
Fig. 56.- Water clathrates 218
Fig. 57.- Effect of brine composition on ratio of oleate chemisorption 228
Fig. 58.- Feldspar, mica and quartz flotation, Spruce Pine, NC 236
Fig. 59.- Weathered decomposed pegmatite processing, N.C. 236
Fig. 60.- Fluorspar mill, Reynolds Mining Corp., Texas 238
Fig. 61.- Mill circuit for a fluorspar-zinc heavy media concentrate 239
Fig. 62.- Pilot plant flowsheet for sand beneficiation 244
Fig. 63.- Lock cycle tests and pilot plant - glass sands 244
Fig. 64.- Flow diagram for a phosphate rock plant 249
Fig. 65.- Potash scrubbing-desliming-reagent conditioning system 251
Fig. 66.- Foote spudomene flotation circuit 254
Fig. 67.- Spudomene and feldspar-quartz-mica flotation, Kings Mt, NC 255
Fig. 68.- Results of 2 kg laboratory flotation tests, single variable search 261
Fig. 69.- A two-variable factorial design 262
Fig. 70.- Experimental design based on a square pattern 264
Fig. 71.- Cubic experiment for a truncated 3 by 3 design 265
Fig. 72.- Effect of frother and collector on copper recovery 266
Fig. 73.- Collector/frother interaction on recovery and grade 267
Fig. 74.- Interaction of octanol and EtX on recovery and grade 270
Fig. 75.- Box-Wilson rotatable two-variable five-level design 272
Fig. 76.- Box-Wilson rotatable three-variable five-level design 273

Fig. 77.- False optimum in a one variable at a time search 274
Fig. 78.- Steepest ascent experimental procedure result 275
Fig. 79.- SSDEVOP: climbing a response surface, from Mular 276
Fig. 80.- Standard 5.5 liter (2 kg) flotation cell from ASTM 281
Fig. 81.- Flotation cell proposed by ISO committee on coal 282
Fig. 82.- Flotation cell impeller and diffuser 283
Fig. 83.- Comparison of the performance of flotation machines 286
Fig. 84.- Minimum representative sample size 289
Fig. 85a.-Standard 2 kg ball mill used in Chile (body) 301
Fig. 85b.-Standard 2 kg ball mill used in Chile (cover) 302
Fig. 86.- Schematic of a flotation column 314
Fig. 87.- CIMM's column flotation pilot plant 317
Fig. 88.- CIMM recommended flowsheet for cleaner column flotation 322

ACKNOWLEDGEMENTS

The author acknowledges with thanks the assistance given by the following companies and publishers in permitting the reproduction of illustrations from their publications:

John Wiley & Sons Inc., 605 Third Ave, New York, NY 10016
Stephen J. Lippard, Editor, Progress in inorganic chemistry, Volume 23, copyright - 1977; containing: D.L. Kepert, "Aspects of the stereochemistry of six-coordination", Figures 6. 7, 8 & 9; and Michael G. B. Drew, "Seven-coordination chemistry", for Figures 10 and 11.

Gordon and Breach Science Publishers, 270 Eight Avenue, New York, NY 10011,
Janusz S. Laskowski, Editor, "Frothing in flotation", copyright: 1989; containing Ronald D. Crozier, and Richard R. Klimpel, "Frothers: plant practice", Figures 15, 20, 21, and 22

Plenum Publishing Corporation, 233 Spring Street, New York, NY 10013
Jan Leja, "Surface chemistry of flotation", copyright: 1982, Figures 2, 55 and 57

Cornell University Press, 124 Robert Place, Ithaca, NY 14850
Linus Pauling, "The nature of the chemical bond", copyright: 1960, Figures 3, 12, and 13

Australasian Institute of Mining and Metallurgy (Inc), Clunies Ross House, 191 Royal Parade, Parkville, Victoria 3052, Australia
K.L. Sutherland, and I.W. Wark, "Principles of flotation", copyright: 1955, Figure 25

The Society of Mining Engineers, AIME-SME, Littleton, CO
Leja, J., & Schulman, J.H. (1954), Flotation theory: molecular interaction between frothers and collectors at solid-liquid-air interfaces, *Trans. AIME.*, 16, pp.221-8. Figure 14, depicting the interpenetration theory
Mular, A.L. (1989), Modelling, simulation and optimization of mineral processing circuits, *Challenges in mineral processing*, Eds., K.V.S. Sastry and M.C. Fuerstenau, SME-AIME, Littleton, CO, pp.323-349. Figure 77, 78, and 79
Mular, A.L. (1976), Optimization in flotation plants, *Flotation*, Ed. M.C. Fuerstenau, SME-AIME, Littleton, CO, pp. 895-936. Figures 75, and 76
Redeker, Immo H. and Bentzen, E.H., (1986),Plant and laboratory practice in non-

metallic flotation, *Chemical reagents in the mineral processing industry*, Editors, Deepak Malhotra and W.F. Riggs. SME of AIME, Littleton, CO, pp. 3-20 Figures 58, 59, 61, and 67

The Institution of Mining and Metallurgy, 44 Portland Place, London, W1N 4BR
Grainger-Allen, T.J.N. (1970), Bubble generation in froth flotation machines, *Trans. I.M.M.*,79, pp. C15-C22, figure 1
Lekki, J., & Laskowski, J. (1971), On the dynamic effect of frother-collector joint action in flotation, *Trans. I.M.M.*, Section C,80, C174-80. figures 16, 17, and 18
Mathieu, G.I. and Sirois, L.L. (1984), New processes to float feldspathic and ferrous minerals from quartz, *Reagents in the minerals industry*, M.J.Jones and R.Oblatt, Eds, (London: IMM), pp. 57-67; Figure 62, and 63

Mine and Quarry Engineering, 19 Fairlie Road, London, UK
Wrobel, S.A. (1953), Power and stability of flotation frothers, governing factors, *Mine and Quarry Engineering*, 19, pp. 363-7, Figures 16

Pergamon Press, Headington Hill Hall, Oxford OX3 0BW
Beas-Bustos, E., and Crozier, R.D. (1991) Moly/copper separation from concentrate of the combined acid and basic circuits at El Teniente, *Reagents in Minerals Engineering*, Camborne School of Mines, September 18-20. Figures 31-51
Crozier, R.D., (1991) Sulphide collector mineral bonding and the mechanism of flotation, *Minerals Engineering*, vol 4, Nos 7-11, pp.839-858, Figure 5

The author is also grateful to Mr. Farlow Davis for permission to use unpublished material on molybdenum depressants; to CODELCO-Chile division management for providing the data used in the case histories on the Andina and El Teniente mines; to the Director of Chile's Centro de Investigation Minera y Metalurgica for providing the column testing data; and to Sherex Chemical Company, Inc. which provided flotation testing procedures and details of the test equipment. Sabine Slotta is thanked for the many tedious hours spent copy editing the manuscript.

INTRODUCTION

This text is primarily addressed to the mill metallurgist and to technical service and development engineers working for reagent suppliers, rather than to flotation researchers or mill process designers. With this audience in mind, flotation theory is covered so as to emphasise the existence of distinctly different mechanisms in the flotation of simple metallic sulphides, complex sulphides, metallic non-sulphides (loosely known as oxides), industrial minerals, and coals.

As a primary target of flotation testing is correction of a chemical problem in a working mill, be it the flotation media (water) or the selection of a collector, frother or modifier to improve the recovery efficiency of a mill, a considerable amount of space has been devoted to attaching the proper chemical composition to commercial trade names, as well as providing relatively detailed information on the properties of the industrial collectors, frothers and modifiers in normal use. Reagents produced by former major collector suppliers such as Minerec Corporation (out of business), and the Dow Chemical Company, no longer in this niche business, are also identified, to aid in the interpretation of older published mill data and because many laboratories still have samples of Dow and Minerec reagents which can be used to broaden a particular flotation study. A brief description of the synthesis of collectors is given in case special samples need to be prepared, and their physico-chemical properties are summarised, as these data are generally not available in standard handbooks. Water quality, per se, is too vast a subject to try and summarise, but as the interaction between collectors and cations in solution is a common problem in mineral pulps, some of the coordination chemistry of collectors is touched on, as is the semi-conductor properties of the principal sulphide minerals.

The technology to treat low grade ores has evolved by trial and error in the operating mills, and very little is based on fundamental research. Most of the published fundamental research has been of relatively little value in resolving practical problems, primarily because it has been carried out using synthetic mixtures of pure minerals. In addition, test equipment normally used in micro-flotation laboratories usually mimics the obsolete pneumatic flotation machines of the turn of the century. Thus, the results of much of the published fundamental research cannot be applied to the design of sulphide mineral flotation circuits. Because of mechanistic incompatibility of agitation flotation and pneumatic flotation, some of the conclusions and mathematical relations proposed in the literature on the rate of flotation of pure minerals is particularly dangerous if it is used in the development of scale-up algorithms and in choosing the key process variables in the design of experiments. This criticism of simplified flotation testing equipment does not mean that there is not a crying need for an accurate micro-flotation apparatus capable of obtaining scale-up data and accurate processability responses from gram-sized samples obtained from core drilling programmes. BUT, just as the grade and mineralogical data must be inserted into a representative 3-D grid of an ore body to generate reliable geologic modelling of a mining project's reserves, the same broad analysis must be applied to the floatability response to provide useful recoveries and operating costs. The statistical confidence range of the floatability model based on bench flotation data is poor, and if based on micro-flotation experiments it is considerably poorer than the reserve estimate because chemical analysis of exploration cores is more accurate. Thus, designing a mill based on current theoretical knowledge using micro-flotation data could well be disastrously more expensive than investing in adits, tonnage samples, and pilot plant tests.

The development of an optimum metallurgical sequence to process a completely new ore is always expensive, therefore it must be carefully budgeted in all new projects. An improper geologic evaluation of the probable ore grade and reserves can have serious financial consequences for a mining company if the ore runs out before all the capital invested has been recovered, so valuation of an ore deposit justifies a large expenditure in drilling and exploration. Similarly, unexpected processing difficulties, which result in a recovery reduction of 3 or 4% on predicted efficiencies, will affect the cash flow values used to calculate the debt service, and can also have a crucial effect on corporate existence, considering the mammoth investments required for today's low grade ores. So the allocation of resources, human and financial, to testing budgets during the pre-investment decision period must be carefully weighed.

Much has been written on how to design and manage the proper exploration of an ore body and the design of an optimum drilling programme. Computer programmes are capable of processing drill, core and tunnelling data, and presenting these in terms of three dimensional topologies of the mineral species, and spewing out the proven, probable and inferred reserves of the future mine, including an indication of the confidence level of the quantity of ore that will be available. However, no similar computer capability has been standardised in the more mundane area of testing ores for processability, particularly if flotation is involved.

The chemistry of mineral surfaces is dealt with systematically, and is readily available in standard publications, though not necessarily for conditions applicable to flotation. The review of the mechanisms of flotation summarised in this text is slanted to emphasise factors that affect the selection or design of flotation reagents, not equipment and process circuits. The greater emphasis and breadth of information on sulphide reagents reflects the writer's expertise and not the relative size of the industrial minerals reagent market. The testing procedures designed for sulphide ores, whose behaviour is more complex, can all be easily adapted to non-sulphide ores and also to non-metallic flotation, cleaning or separation processes, so sulphide testing is generally covered more fully.

As will be apparent from the text, the selection of appropriate reagents for the optimum flotation of a particular ore is more an art than a science; but the use of a systematic approach to obtain laboratory flotation data can reduce the hit or miss aspects of this particular art. Information on the performance of flotation reagents is frequently contradictory because non-identical flotation circuits may be conforming to a different flotation regime, and therefore the reagents in the full-scale plants act differently from the case histories described in the literature. Even if we limit our review to currently fashionable flotation equipment, we must qualify any reagent recommendations, as frothers and collectors will not react in the same manner if the flotation is carried out in small subaeration machines, forced air units, modern giant cells or flotation columns. Quite obviously, there are also operational differences between the generically similar machine designs of different manufacturers, which can affect the choice of the reagents.

This mechanistic factor, i.e. that a particular reagent's flotation response is different, depending on whether the flotation occurs because of bubble capture, or mineral surface catalysed micro-bubble formation, is glossed over in most texts on flotation. In these notes it will be over-emphasised, because it is so important in diagnosing problems in cleaner or scavenger circuits, and in extrapolating applicability of a particular reagent from one ore to another or from one type of flotation equipment to another (i.e., from cell to column flotation, where only bubble capture occurs).

To add complexity to our information matrix, there are at least four different types of minerals that are upgraded by flotation: (1) metallic sulphide minerals, (2) metallic oxides or tarnished sulphide minerals, (3) non-metallic insoluble minerals, and (4) non-metallic soluble, or semi-soluble minerals. In each case the mechanism of flotation can be bubble capture or nascent bubble formation, or a combination of the two.

Sub-groups of these four minerals are ores whose mineral dissemination requires that they be ground to colloidal sizes to liberate the valuable species; minerals where differential flotation is required to separate complex ores that include mixtures of hydrophyllic minerals; and those that are naturally hydrophobic and amenable to flotation without the use of collectors. It is also obvious that the mill metallurgist, the flotation equipment designer, and the reagent designer have quite different goals, and therefore possibly conflicting points of views, on how to apply the technology used in flotation. Hardware designers, for example, should be aware that designations such as a "strong" or a "weak" collector do not describe the strength of adhesion between a mineral and a bubble; rather, it is an inverse description of selectivity, or the effect of a particular reagent on the rate of flotation of individual minerals, so a change to a 'stronger collector' in a formulation is unlikely to compensate for poor over-all recovery because of an overly deep cell. When industrially proven flotation machines are involved in a side-by-side mill test and one is not performing competitively, its inadequate performance cannot be compensated by a reagent suite change, as the effect of the reagent change could affect the operation of both machines identically, in which case the cause of the difference is more likely to be due to a physical operating variable, such as pulp density, percent solids, air feed rate, impeller rpm, mineral particle size distribution or some other operating detail that has not been fully controlled during the test period.

Froth flotation, the surface dependent process par excellence, consists of a deceptively simple procedure - the separation of valuable constituents of a two or three phase mixture by levitation - but, in fact, flotation embodies a very broad gamut of non-obvious sequential micro-processes, the combination of which have as a final result the levitation of a desired mineral species, leaving other minerals and gangue in the tailings. The same reagent and micro-process sequence does not act the same when applied to different minerals, i.e. non-metallics, metal oxides and metal sulphides can be depressed or promoted by a particular reagent. This behaviour results in what appear to be completely contradictory results reported in research published by different reputable metallurgists. The reason for this inconsistency is the great diversity in mineral morphology of commercially important raw materials. Thus, there is no single valid theory of the mechanism of flotation which can be used to design universally applicable testing procedures, and therefore no 'cookbook' mineral processability test recipes.

Chapter 1

FLOTATION FUNDAMENTALS

Definition of terms

Froth flotation is a process used to separate minerals, suspended in liquids, by attaching them to gas bubbles to provide selective levitation of the solid particles. It is the cheapest and most extensively used process for the separation of chemically similar minerals, and to concentrate ores for economical smelting.

Floatable minerals can be classified into non-polar and polar types, according to Wills (1988). The segregation into these two types of minerals is based on their surface bonding. The surfaces of non-polar minerals have relatively weak molecular bonds, difficult to hydrate, and in consequence such minerals are hydrophobic. Non-polar minerals include graphite, sulphur, molybdenite, diamond, coal, and talc, all naturally floatable in the pure state. The ores containing these minerals usually require the addition of non-specific collectors to their pulp to aid the natural hydrophobicity of the floatable fraction, i.e. oily collectors, such as fuel oil, kerosene, coal distillates, etc.

Polar minerals have strong covalent or ionic surface bonding, and exhibit high free energies at these surfaces. Therefore, surface hydration is rapid due to the strong reaction with water molecules which form multilayers on the mineral. Thus these species are hydrophyllic. The listing in Table 1 is adapted from Wills (1988). The minerals are listed in groups of increasing polarity, divided into classes dependent on the magnitude of the polarity. Minerals in group 3 have similar degrees of polarity, but this polarity can be changed easily in the case of the minerals grouped under 3 (a) because they are susceptible to sulphidisation.

The minerals in group 1 are all sulphides, with the exception of the pure metals. Their flotation characteristics are not uniform, and, based on Finkelstein and Poling (1977), must be further divided into three more sub-groups: those minerals that form a xanthate, dithiophosphate, mercaptan, thionocarbamate, or xanthogen formate dimer (dithiolate) on their surface on contact with the collector (covellite, chalcopyrite, stibnite, arsenopyrite, pyrite, sphalerite, orpiment, realgar, and the native Au, Pt, Ag and Cu); those that form metal xanthates (galena, bornite, chalcocite); and the others, where the analytical data is ambiguous on whether the collector bond is with the metal cation or with the surface sulphur atoms in the mineral.

Table 1.- CLASSIFICATION OF POLAR MINERALS

Group 1		Group 1 continued	
Galena	PbS	Orpiment	As_2S_3
Chalcopyrite	$CuFeS_2$	Pentlandite	$(Fe,Ni)_9S_8$
Covellite	CuS	Realgar	AsS
Bornite	Cu_5FeS_4	Cinnabar	HgS
Chalcocite	Cu_2S	Alabandite	MnS
Energite	$Cu_3AS,Sb)S_4$	Native Au, Ag, Pt, Cu	
Argentite	Ag_2S		
Millerite	NiS	**Group 2**	
Cobaltite	$CoAsS$		
Arsenopyrite	$FeAsS$	Barite	$BaSO_4$
Pyrite	FeS_2	Anhydrite	$CaSO_4$
Pyrrhotite	Fe_7S_8	Gypsum	$CaSO_4.2H_2O$
Sphalerite	ZnS	Anglesite	$PbSO_4$
Stibnite	Sb_2S_3		
Group 3a		**Group 3**	
Malachite	$Cu_2CO_3(OH)_2$	Flourite	CaF_2
Azurite	$2CuCO_3.Cu(OH)_2$	Witherite	$BaCO_3$
Chrysocolla	$CuSiO_3.2H_2O$	Magnesite	$MgCO_3$
Wulfenite	$PbMoO_4$	Dolomite	$CaMg(CO_3)_2$
Cerrusite	$PbCO_3$	Apatite	$Ca_5((F,Cl)(PO_4)_3$
		Scheelite	$CaWO_4$
		Smithsonite	Zn silicate clay
		Rhodochrosite	$MnCO_3$
		Siderite	$FeCO_3$
		Monazite	$(Ce,La,Di)PO_4$
Group 4		**Group 5**	
Hematite	Fe_2O_3	Zircon	$ZrSiO_4$
Goethite	$FeO(OH)$	Hemimorphite	$Zn_4Si_2O_7(OH)_2.2H_2O$
Chromite	$FeCr_2O_4$	Beryl	$Be_3Al_2Si6O_{18}$
Pyrolusite	MnO_2	Garnet	$Ca_3Al_2(SiO_4)_3$
Borax	$Na_2B_4O_7$		
Wolframite	$(Fe,Mn)WO_4$		
Columbite	$(Fe,Mn)(Nb,Ta)_2O_6$		
Tantalite	$FeTa_2O_6$		
Rutile	TiO_2		
Cassiterite	SnO_2		

source: Wills (1988)

Table 2.- COLLECTOR STRUCTURES

Type	Structure	Water Solubility
Soluble Collectors		
Alkyl xanthates or Alkyl dithiocarbonates	$R-O-C(=S)-SNa$	Soluble
Sodium O,O Dialkyl dithiophosphates	$(R-O)_2P(=S)S^-Na^+$	Soluble
Insoluble Oils Mercaptans	R-O-SH	very sltly soluble
2-Mercaptobenzothiazol	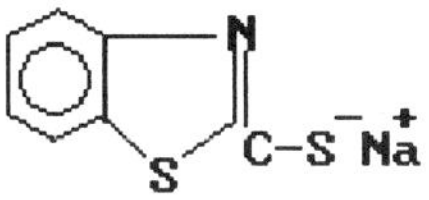 benzothiazole ring (N, S) $C-S^-Na^+$	Soluble
O,O-Aryl,O-aryl dithiophosphoric acids	$(Ph-O)_2P(=S)S^-Na^+$	Insoluble
Dialkyl xanthogen formates	$R-O-C(=S)-S-C(=O)-O-R'$	Insoluble
Dialkyl thiono-carbamates	$R-O-C(=S)-NHR'$	Insoluble
Xanthic Esters	$R-O-C(=S)-S-R'$	Insoluble
Metal oxide and non-sulphide mineral collectors		
Fatty acids	$R-C(=O)-OH$	Slightly soluble in water
Sulphonates	$R-S(=O)_2-O-OH$	Soluble
Alkyl amines	$R-HNH_2$	Soluble
Quaternary ammonium compounds	RR'R"R'"NCl	Soluble

The reason to make this differentiation is that the mechanism of flotation is unique when the collector reacts with the mineral surface to form a dithiolate. Nascent bubble flotation is the rate dominant mechanism, when dimers are formed, while coursing bubbles dominate when there is true oily collector flotation. With the exception of pentlandite, ambiguous minerals are less often encountered in the flotation laboratory.

Table 3.- FROTHER STRUCTURES

Type	Structure	Water Solubility
Aliphatic Alcohols R = Alkyl gp with 5 to 8 C MIBC	R-OH $CH_3CHCH_2CHCH_3$ (CH_3, OH)	slight
Pine Oils Terpineols	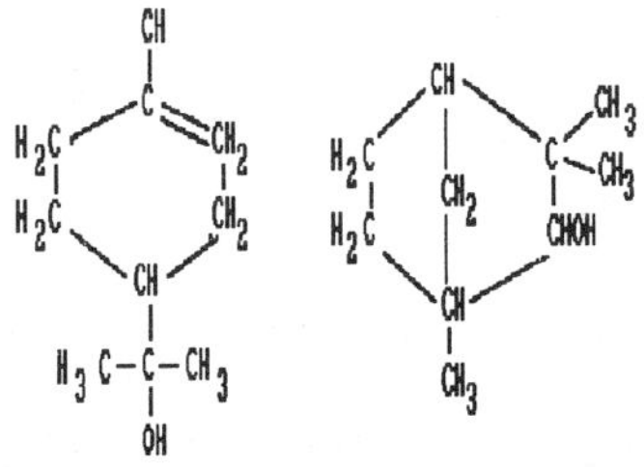 α - terpineol; fenchyl alcohol	slight
Cresylic acid	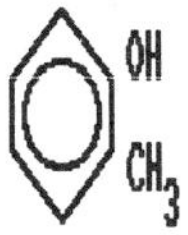 OH, CH_3	slight
Alkoxyparaffins 1,1,3-triethoxybutane or TEB	$CH_3CHCH_2C(OCH_2CH_3)(H)(OCH^2CH^3)$ with OCH_2CH_3	slight
Polyglycolethers Dowfroths	$R(OR')_xOH$	miscible at low 'x' to partially soluble at high 'x'
Aerofroths	$R(OR')_xOH$	

The largest volume flotation operations are those that involve the upgrading of commodities: iron ores, coal, phosphates, limestone, potash, copper... but the technically most interesting are the separation of similar metals from complex ore, such as mixed lead, zinc, silver and copper ores, or copper/molybdenum concentrates.

Bulk flotation is a somewhat imprecise term that covers nearly all the normal rougher or scavenger flotations, where a single mineral or a group of related minerals are separated from gangue and other low value minerals in a single flotation step. An example would be the recovery of a mixed copper sulphide concentrate from an ore containing pyrite and gangue. Others include the separation of KCl from sodium chloride, upgrading phosphate rock, desulphurising iron ores, etc.

Differential flotation is the term normally used to describe the separation of complex ores, and is generally restricted to describing the separation of similar minerals from each other (e.g. copper, lead, zinc, silver and gold from a single ore, or molybdenum from copper in concentrate, etc.) where the successful and economic recovery of each component involves the sophisticated use of collectors, depressants and flotation activators.

Reagents employed in flotation

The reagents employed in flotation are generally interfacial surface tension modifiers, surface chemistry modifiers, and/or flocculants. Usually they are classified under five headings: collectors (sometimes known as flotation promoters), frothers, modifiers, activators and depressants. The designation collector or promoter, interchangeably, reflects the opinion of the individual metallurgist, who may consider the primary function of the collector to be a flotation rate accelerator, or a fine particle agglomerator, in a particular flotation system.

Collectors are reagents that coat and/or react with mineral surfaces and make them water repellent or attachable to air bubbles. The chemical structures of most commonly employed collectors are shown in Table 2. Sulphide ore collectors all contain sulphur and are thiols or can hydrolyse to a thiol. Non-sulphide and non-metallic minerals are normally floated employing collectors such as fatty acids, amines, quaternary ammonium compounds, sulphonates or petroleum oils.

Frothers are surface active reagents that aid in the formation and stabilisation of air-induced flotation froths. The commonly employed frothing agents are alcohols which are only slightly soluble in water, or the more modern frothers, which are generally varieties of polyethers or polyglycol ethers that are completely miscible with water. Examples are listed in Table 3.

Modifiers, activators and depressants The boundaries between the functions of a specific inorganic flotation aid are rather fuzzy. For example, for pH control, an environment modifier such as lime may be used; but lime contains calcium, and the calcium cation is a known depressant for pyrites in copper flotation or for quartz and talc in the flotation of silver ores. So lime would need to be classified under both headings. The silicates and the phosphates act as modifiers when they are used to control the effect of cationic and anionic impurities in flotation water, but can also act as activators or depressants in different mineral systems. Another example of a dual function is that of the sulphhydryl anion which can be a depressant for sulphide copper, especially in the copper/moly separation, but, with very careful concentration control, this anion can be used as a sulphidiser to activate non-sulphide or oxide copper.

Table 4.- MODIFIERS

pH Modifiers	
Lime:	CaO
Soda ash:	Na_2CO_3
Caustic soda:	NaOH
Acids:	H_2SO_4, HCl
Resurfacing agents	
Cations:	Ba^{++}, Ca^{++}, Cu^{++}, Pb^{++}, Zn^{++}, Ag^{+}
Anions:	$SiO_3^{=}$, PO_4^{---}, CN^{-}, $CO_3^{=}$, $S^{=}$
Organic Colloids:	dextrin, starch, glue, etc.

Chapter 2

THE MECHANISM OF FLOTATION

The morphology of froth flotation is not well defined because full-scale systems are normally opaque. In macro terms, froth concentrates form by selective capture of mineral particles in the bubble generation zone, followed by a second zone where there is coagulation of the discrete bubble stream into a loose froth. These first two zones are normally identified as the "pulp" in a flotation cell, and can be fixed in height by a level control mechanism. The third region appears when the gas phase predominates over the liquid phase; here, the froth starts to condense and returns water to the pulp. Finally a structurally stable mineral-laden froth forms, and is removed over the lip of the flotation machine. The height (or thickness) of this stable froth can be controlled by the designer and operator by fixing the total froth height and the pulp height. The operating variables, that can be fixed by the foreman within the flotation bank, are: pulp mineral content (pulp density) and flow rate; aeration gas rate; in some cases agitation intensity (in conventional cells); pulp/froth interface height; and total overflow height, which fixes the froth drainage height and the froth removal rate.

A material balance algorithm for a single cell can quantify both the pulp mass flow rate in and out of the cell, and also the concentrate mass rate, if assays are available for the different streams. At a lower confidence level, the rate of return of water to the pulp, and the rate of flow of the different components of the concentrate, can also be calculated. Under steady conditions, the height of the interface between the concentrate froth and the draining froth can be inferred from froth physical property measurements, such as density or metal concentration profiles.

The key mechanistic variable that is undefined is exactly how a bubble selectively captures a mineral containing ore particles. Observation confirms that for selective capture the desired mineral must be naturally floatable or be preferentially coated with a collector. Once the desired particle is tagged, there appear to be two main, non-exclusive, bubble loading mechanisms in operation: a) multiple particle bubble collisions in the pulp which eventually result in particle adhesion, and b) micro-bubble formation directly on the mineral surface within the bubble formation zone of the pulp. There is consensus that the forces that join the mineral particle to the bubble, and that define the rate of capture of the mineral by the bubble, involve electron transfers when the mineral is conductive, and surface tension forces when the mineral is an insulator.

Coursing bubble flotation

The froth flotation literature labels these two particle capture processes as the coursing bubble process, and the air precipitation or nascent bubble process (the latter will be used in this text). Bubble capture, or coursing flotation, predominates when oily collectors are used to concentrate non-metallic minerals or non-sulphide metallic ores, or when collectors containing double bonded sulphur atoms are used with metallic sulphide ores which do not form dithiolates on the mineral surface, but ionise and form metal salts. As we see from the data in Table 53 (page 175), the bubble capture mechanism occurs mainly with minerals where the concentration ratio is less than 5:1, and selectivity is usually obtained by employing specific depressants rather than making collector adjustments.

In these cases, the flotation efficiency of different reagents can be measured by simple tests such as Hallimond tube experiments, though even with these simple ores one would hesitate to extrapolate the selectivity results of Hallimond tube flotation tests (which normally are carried out using only a collector) to a specific industrial flotation, as the concentrate obtained is primarily dependent on the rate of flotation. As is well known, the flotation rate depends more on the frother employed, rather than the collector.

With oily collectors, contact angle measurements, and the corresponding calculated adhesion force between a bubble and a mineral particle, may be a guide to the mineral carrying capacity of a bubble, at least when very dilute pulps are involved; but under normal industrial flotation conditions, where pulps contain at least 30% solids, the rheology of the inter-bubble fluid is the property that fixes the mineral-carrying capacity of the froth, not the strength of the particle to bubble bond, though the differential value of this force for two or more minerals and the gangue may help predict the selectivity of a reagent.

Klassen and Mokrousov (1963) describe the fundamental difference between the collectors for these two types of flotation by pointing out that: "In physical adsorption, the reagent maintains its chemical individuality, whereas, in chemisorption, [when electron transfers occur] it forms new compounds". They confirm other researchers, such as Gaudin (1957), that when the collector acts as a micro-bubble nucleator, the dosage must be such that only a small proportion of the mineral surface is covered by the collector, which means that the surface of a floatable sulphide mineral is NOT hydrophobic, i.e., in highly selective flotation with a concentration ratio greater than 7:1, for bubble capture to occur with a sulphide mineral, the bubble contact angle at the optimum reagent dosages must be zero or near zero.

Some of the effects of physical variables on bubble capture flotation can be deduced from the principal characteristics of oily reagents, which are: 1) a tendency to form miscelles when the concentration exceeds a critical concentration (CMC) and temperature (Krafft point); 2) a significant surface tension reduction in dilute solutions; and 3) dissociation, and, as a function of pH, ionisation through hydrolysis. The nuances of the application of non-sulphide collectors to flotation of minerals is covered extensively by Sutherland and Wark (1955). As in the case of thiol collectors, the oily collectors also show a depressing effect, due to "over-oiling", if the dosage is such as to produce a multi-molecular layer on the particle surface. Oily collectors are generally less selective than the thiols and have a serious drawback, which is the need to heat some flotation pulps, as their collector action is related to miscelle formation and for the longer chain alkyl groups their Krafft temperature is considerably above ambient.

Contact angle and Hallimond tube measurements and their meaning

Contact angles, as used by Sutherland and Wark (1957) in their studies to predict the effect of critical pH and depressant concentrations on floatability, provide go-no-go data on whether a particular collector will or will not form a monolayer under given pulp conditions. The authors recognise that they only measure whether this mono- to multi-layer of collector has or has not adhered to a polished sample of a pure mineral, as can be seen from the following passage from page 97 of their classic text:

> "No matter from what source they [the minerals] came, each gave a constant maximum contact angle with ethyl xanthate, and moreover, this angle was invariably 60° ... Metals gave the same angle, and it was immaterial whether the sodium or potassium salt was used. This angle of contact is therefore characteristic of ethyl xanthate; any surface that has responded fully to the collector exhibits it. There can only be one explanation for this fact, namely, that the mineral becomes coated with a film of xanthate, and that the effective surface is, in fact, the xanthate itself."

Other authors have pointed out the static nature of contact angle measurements and the danger of extrapolating contact angle results to flotation systems in general, be they sulphide or non-sulphide. For example, Laskowski (1974) cites Klassen and Mokrousov (1963) saying:

> "These authors point out that flotation is typically a non-equilibrium process. The phenomena that occur on the surface of particles and bubbles in agitated, aerated mineral pulps, when these particles strike bubbles, are dynamic-non-equilibrium processes which cannot be analysed based on contact angle values obtained at equilibrium".

Leja (1982, p.7) says:

> "in the opinion of the author, the contact angle is an indicator, but is not a measure of the hydrophilic character [of an ore]".

The seminal work on the significance of contact angles in flotation has been done by Gardner and Woods (1973, 1974, and 1977), and summarised by Woods (1976, 1977, 1981). The key to their work was the design of a cell which allowed the simultaneous measurement of contact angles and a surface potential on an electrode, and a micro-flotation device equipped with an electrode capable of transferring an electrical potential to a bed of mineral or metal particles which permitted dynamic measurement of electrochemical potentials and flotation recoveries. The initial work was with gold and platinum, which was then extended to pyrite and galena surfaces.

Based on contact angle measurements, their major conclusions were:

> "The presence of dixanthogen on the [platinum or gold] surface clearly renders the surface strongly hydrophobic.... when xanthate ions are adsorbed on platinum at potentials cathodic to dixanthogen formation, the contact angle in this region remained close to zero and hence specifically *adsorbed xanthate ions do not make the surface hydrophobic*".

In addition they found:

> "It is interesting to compare the results of investigations of the interaction of xanthate with a galena surface and with lead. Lead methyl, ethyl and butyl xanthates on the metal surface were found to be hydrophilic even at low coverage. Only at high potentials, where lead xanthate is oxidised to

> dixanthogen, were significant angles observed. If the potential was decreased after the dixanthogen was formed, the immediate contact angle remained that characteristic of dixanthogen. However, dixanthogen slowly reacts with the surface to reform lead xanthate, the contact angle decreased and the surface became hydrophilic".

And finally very pertinent to the thesis in this text:

> "It has been generally considered that chemisorbed xanthate will be attached to a metal atom in the sulfide surface. However, it has been pointed out by Winter (1975) that xanthates can form bonds with sulfur, compounds of the type ROCSS-S-SSCOR being quite stable. Also, such compounds cannot readily be distinguished from dixanthogens by spectroscopic analysis and could well be formed at sulfide surfaces. The possibility of involvement of such compounds, and the instability of surface species which give rise to the release of monothiocarbamates into solution, add complications to the general picture of the interaction of sulfides with collectors..."

Data on contact angles for xanthates from which many of the generalisations on the properties of collectors have been based are reported by Klassen and Mokrousov (1963), and summarised in Table 5. Extending the range of the data in Table 5, Taggart (1947) cites a series of contact angle measurements where the number of carbon atoms in the xanthate alkyl group went from 1 to 16, and the contact angle increased from 50° for methyl xanthate, steadily to 80° for amyl, 87° for hexyl, 90° for 7 carbons, 94° for 8, and 96° for 16 carbons (cetyl).

Statements in text books that longer chain alkyl group xanthates are "stronger collectors" are based on these contact angle values. In fact, beyond some minimum bond strength between the mineral and the bubble, persistence of adhesion is only a function of the froth rheology and turbulence. What does correlate with contact angles for xanthates is that longer chain alkyl groups coincide with lower selectivity, so amyl xanthates will result in the production of a larger volume, lower grade, rougher concentrate.

Xanthates with alkyl chains longer than hexyl provide lower flotation recoveries. Decyl and longer chain xanthates are good depressants. A well known, but generally unpublished, property of xanthates is that they may be used as emergency copper depressants in moly/copper circuits. Operators habitually correct for a momentary ineffectiveness of the inorganic Cu depressant by the addition of more primary copper collector, be it a xanthate or xanthate derivative. Curiously, the amount of additional collector that will act as a depressant with a particular moly circuit pulp corresponds to that amount which will generate a measurable bubble contact angle in the lab. An interesting experiment (which I believe has not been performed) would be to see if a collector concentration which generates any contact angle corresponds to an over oiling dose industrially.

We have commented on "over oiling", a term which was first used around the time of World War I to describe the phenomena that excessive amounts of collector in a flotation system will result in a decrease in recovery. This decrease in recovery with increased dosage tends to be ignored in published papers on flotation research.

The reason for this omission seems to be that Hallimond tube and other micro-flotation tests do not detect the drop in recovery with excess collector and/or frother, and for many of these papers the only flotation measurements carried out have been on a micro-scale.

Table 5. XANTHATE CONTACT ANGLES FOR SELECTED SULPHIDE MINERALS

N° of Carbon atoms in xanthate	Galena	Chalcopyrite	Bornite	Pyrite
1	0	0	0	0
2	59	60	60	60
3	68	69	68	67
4	74	78	73	74
5	88	90	86	82
6	100	94	95	95
Effect of Branching				
Butyl Xanthate				
Normal	74	78	73	74
Branched	77	78	78	78

The Hallimond tube has been very popular in research laboratories because it eliminates the cost of assaying head, tailings and concentrate. The need to assay is avoided by studying artificial ores made by mixing pure minerals. The most common system has been a mixture of galena and sand (or pure quartz). To perform a flotation experiment in a Hallimond tube, a very dilute pulp (usually about 1% solids) is placed in the bubble generating unit, which is the size of a test tube. The tube has a side arm welded to the upper region; to operate, the unit is tilted until the liquid is level with the side arm, and air is turned on to the bubble generating device (usually a glass frit). The concentrate overflows into the side arm, is collected for a fixed time, and weighed. As a frother normally sequesters collectors and interferes with the operation of Hallimond tubes, they are usually omitted. Thus we have the anachronism that a device without a controllable froth is used to simulate froth flotation.

Because Hallimond tube flotation does not require the use of a frother, the effect of frothers on flotation performance also tends to be ignored in articles reporting fundamental research. In addition, many of these articles only report the collector concentration in the water phase, and do not include the pulp density. Therefore, it is important to translate the published reagent dose shown in recovery curves into mill terms (grams or pounds per ton of dry mineral), using a best estimate of pulp density (1 - 30%?) if unspecified, when evaluating whether the information reported has any possibility of being pertinent to mill problems. Typical sulphide ore flotation dosage of collectors used industrially are about 0.02 to 0.2 pounds per short ton of ore, or 10 to 100 gm per metric tonne of ore. These ranges translate to approximately 5 mg/l to 50 mg/l or 0.029 to 0.29 mol/litre of the water phase, for the molecular weights of the typical xanthates studied.

Nascent bubble flotation

Klassen and Mokrousov, in their *"Introduction to the Theory of Flotation"*, postulate that the first bubble to adhere to a mineral particle is a micro-bubble (formed by desorption of dissolved gases) which serves as a bridge for the adherence of a bubble large enough to be capable of levitating the particle. Klassen proposed this mechanism based on a thermodynamic analysis showing that the bridging micro-bubble accelerates macro-bubble attachment rates. Taggart, already in 1930, had observed that it was probable that micro-bubbles were involved in flotation, and that they were formed on the mineral surface by precipitation from dissolved gases in the cavitation region of the pulp. Summarising work published over the past 20 or 30 years, flotation is thought to occur because bubbles form on the surface of a valuable mineral particle at points selectively covered by frother-collector molecules complexes. Simultaneously, secondary bubbles are formed at miscellaneous electrical discontinuities that do not provide strong bonding between the particle and its bubble; i.e. in the case of gangue particles. Possibly, if Klassen is right, the mineral particles first nucleate micro bubbles generated from dissolved gases, and these micro-bubbles then act as bridges for the formation of larger bubbles or the capture of larger bubbles that could be recirculating in the impeller zone. The important point is that, in the flotation process, there is a rate limiting step, which is the velocity of the formation, or attachment, of a micro- or macro-bubble to the mineral surface. Once the bubbles have attached themselves to ore particles, the bubbles rise in the pulp. First they form a lattice which appears like a loose sponge, then later the bubbles tighten into a froth. This tightening process results in the squeezing out of a great part of the occluded water, which returns to the pulp. This returning water mechanically cleans the froth and detaches the gangue. During this cleaning stage there may be some secondary capture of valuable mineral that has fallen from bubbles in the stable froth region, but this secondary capture is not a controlling step in the overall mineral recovery process.

In an agitation flotation machine, air is normally drawn into the impeller through a hollow shaft by mechanical suction induced by the rotating impeller, or with a positive air supply. After the air reaches the impeller region, the flotation bubbles form by a two-stage process. First, cavitation "sausages" form at the trailing edges of the impeller blades. These break up to form bubbles on the surface discontinuities of the mineral particles. These discontinuities can be chemisorbed frother-collector complexes adhered to the mineral surface, or naturally hydrophobic areas.

Grainger-Allen (1970) studied the formation of froths in a flotation machine equipped with flat rotor blades. He employed a cryolite pulp where the refractive index would allow photography of the bubble formation process. His high speed movies confirmed that the bubbles are formed from cavitation induced air sausages which break up into bubbles onto the mineral particles (shown schematically in Fig. 1).

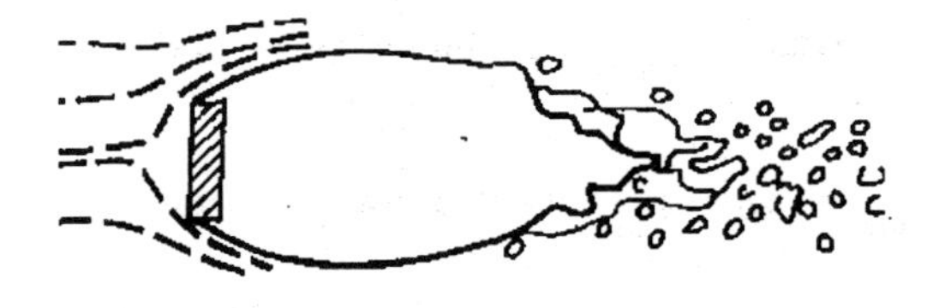

Fig. 1.- Cavitation mechanism of bubble formation

Electron transfer to mineral surface

Sulphide collector adsorption on to a sulphide mineral requires the transfer of two electrons to the surface of the mineral particle. This electron transfer mechanism is very similar to the electrochemical process of metal corrosion. During the corrosion process, metal goes into solution at a point on its surface in contact with water, where the electrolyte is lacking in dissolved oxygen; to go into solution the metal atom becomes a cation by losing the number of electrons corresponding to its normal valance. To maintain a constant surface potential, these electrons must travel through the conductive metal to an oxygen rich region where they take part in an oxidation reaction, the result of which is that metal cations in solution, that have diffused from the point of dissolution, are converted into insoluble metal oxide; so, in iron corrosion, the rust is deposited far from the point where a pit is being generated. In flotation, the electrochemical process is very similar. The collector molecule approaches the metal surface and is adsorbed only after it donates or has donated electrons to the mineral particle. So as not to generate a charge on the particle, the excess electrons migrate to an oxygen rich point on the particle surface where they are consumed by reduction of adsorbed oxygen to form OH^- ions, or H_2O, or H_2O_2, raising the pH, a phenomena very easily measured during the adsorption of collectors on minerals. Under most practical flotation environments, deaerated pulps will not float, and it is thought that some depressants (sodium sulphide, for example) may block flotation by scavenging oxygen. The reaction involved in xanthate adsorption on a mineral, see Woods (1976), are therefore:

ELECTRON TRANSFER TO A MINERAL SURFACE

step 1

$$RO\overset{S}{\overset{\|}{C}}\text{-}S^-_{soln} \longrightarrow (RO\overset{S}{\overset{\|}{C}}\text{-}S)_{ads} + e^-$$

step 2

$$2(RO\overset{S}{\overset{\|}{C}}\text{-}S^-)_{ads} \longrightarrow (RO\overset{S}{\overset{\|}{C}}\text{-}S)_2 + 2e^-$$

or

$$(RO\overset{S}{\overset{\|}{C}}\text{-}S^-)_{ads} + RO\overset{S}{\overset{\|}{C}}\text{-}S^-_{soln} \longrightarrow (RO\overset{S}{\overset{\|}{C}}\text{-}S)_2 + 2e^-$$

These reactions are balanced electrochemically by:

$$\tfrac{1}{2}O_2 + 2H^+ + 2e^- \longrightarrow H_2O$$

$$\tfrac{1}{2}O_2 + H_2O + 2e^- \longrightarrow 2OH^-$$

$$O_2 + 2H^+ + 2e^- \longrightarrow H_2O_2$$

Indirect evidence of this last reaction has been provided by Jones and Woodcock's (1978) discovery of perxanthates in tailings pulp from flotation mills in Australia. They postulate the reaction to form perxanthates in a flotation cell to be as follows:

$$2ROC(=S)-S^- + H_2O_2 \longrightarrow ROC(=S)-S-O^- + H_2O$$

and point out that the only source of hydrogen peroxide in a pulp would be the last reaction above.

The nature of the dixanthogen formed will depend on the metal cation involved. In the case of galena the infrared spectra show that the anodic reaction involved is

$$PbS + 2ROC(=S)-S + 1/2O_2 + 2H^+ \longrightarrow Pb(ROC(=S)-S)_2 + S^\circ + 2e$$

The *electrical double layer theory*, originally proposed by Helmholtz in 1879, evolved from a postulate that when a crystalline solid surface is generated by a fracture, the cleavage process severs the ionic bonds and leaves an electrical imbalance on the surface, and, if the solid is in contact with water, this imbalance is compensated by a charge gradient. This primitive Helmholtz model was refined by Perrin in 1904, and is shown in Fig. 2 as case (a), labelled as the compact layer theory; illustration (b) is based on Gouy (1910) and Chapman (1913) who refined the model to include dilute solutions, based on the assumption of the existence of a diffuse continuum of ions in a non-structured dielectric layer around the mineral particle; (c) adds ion interaction as proposed by Debye and Huckel (1923) and Stern (1924), which includes the premise that the individual ions can approach the mineral surface up to a minimum distance "d". This theory in fact is a triple layer theory, as can be seen from the illustration. While (d) corresponds to a case where a specific ion in solution has entered the double layer and resulted in a charge reversal.

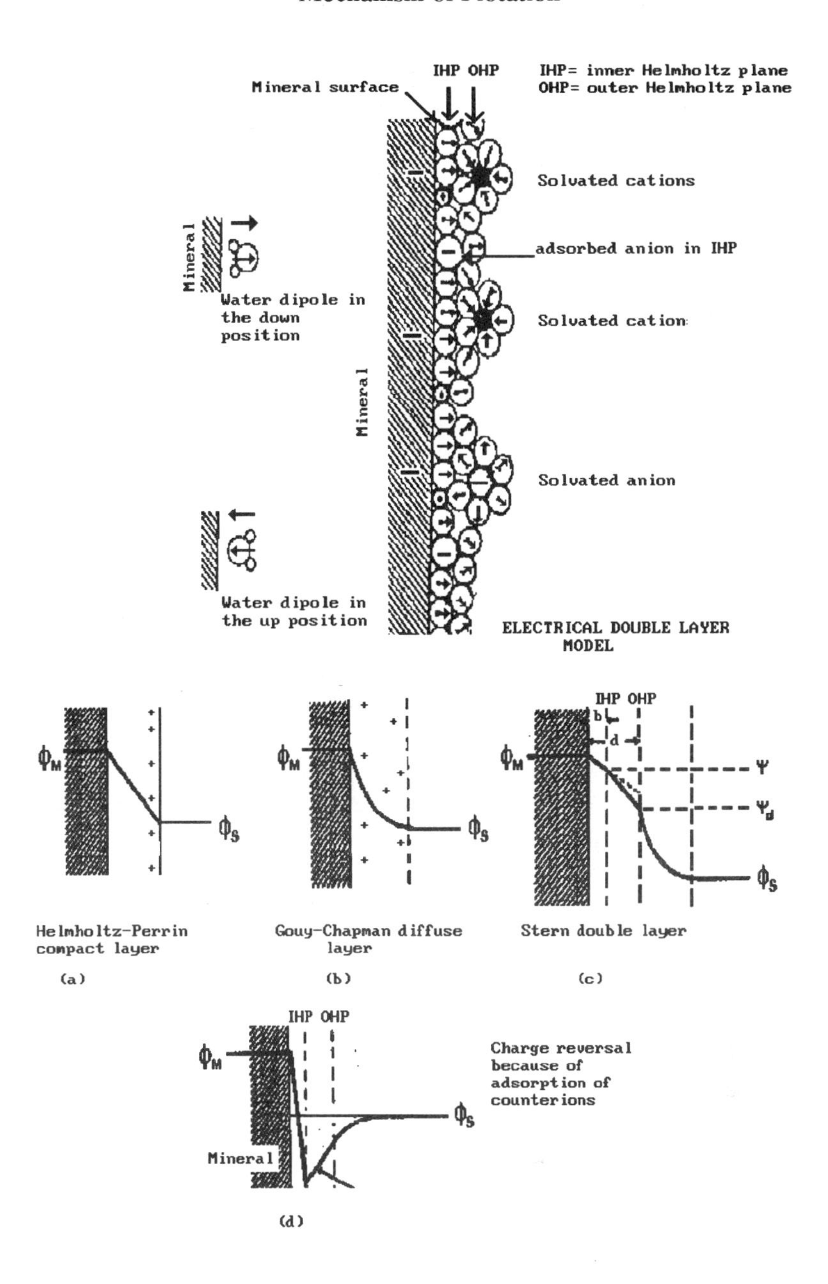

Fig. 2.- The electrical double layer model

The structure of water is also a factor. Most theories of the mechanism of flotation stick to the electrical double layer theory as an adequate description of the influence of the water phase on collector adhesion to a mineral surface. Possibly it is an over-simplification, as the structure of water is controversial. Linus Pauling suggests that for pure water, at ambient and lower temperatures there is considerable regularity to the structure; between 0 and 4° he suggests that the crystalline regularity of water molecules corresponds to a mixture of ice I crystals (Fig. 3), which has a density of 0.92, mixed with ice II, a structure similar to quartz, which has a density of 1.17. As the temperature approaches the maximum density of water, at 4°C, the relative proportion of ice II rises. Above 4°C, water becomes more amorphous, the crystalline regularity becoming more and more discontinuous as the temperature increases.

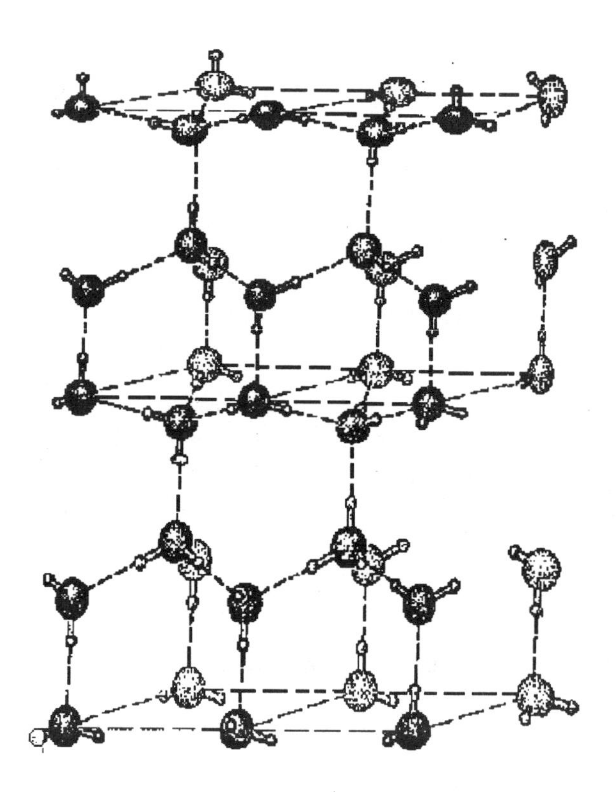

Fig. 3.- The structure of ice, showing the orientation of the hydrogen bonding, adapted from Pauling (1967).

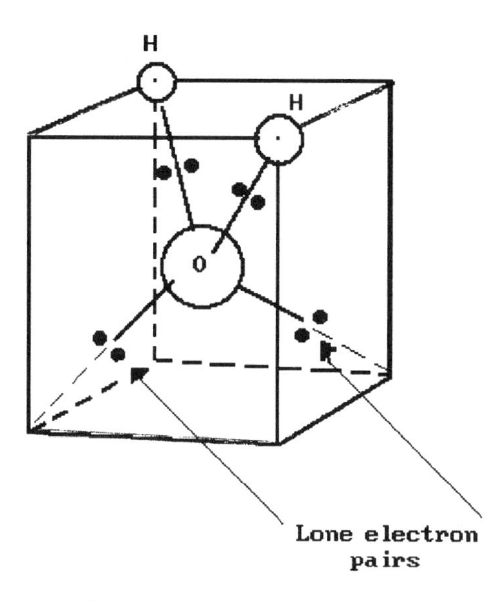

Fig. 4.- Localised-bond structure for H_2O with sp^3 orbitals for oxygen

Fig. 4 shows the spatial arrangement of the orbitals in water, and the explanation for the abnormally high dipole that this molecule exhibits. In mineral pulps, the aqueous phase is alkaline and contains common saline salts in solution, usually including calcium, magnesium, silicates, sodium chloride and excess hydroxyl groups. Schematically, the orientation of the water molecules in the hydrated anions and cations will be as shown in Fig. 5. Laskowski proposes that the hydration of a mineral surface in an impure aqueous solution will consist of multilayers of water molecules ordered into ice type crystals with stray water molecules and small hydrated ions captive within the clathrate structure. Presumably the value of the measured electrical double layer charges reflects the energy necessary to penetrate this hydration layer.

The topology of the layer may well be even more complicated than Laskowski's model, as concentrated transition metal salt solutions are known to have peculiar solubility characteristics which reflect a crystal-like regularity. For example, a 60.2 weight per cent zinc chloride solution, a concentration which corresponds to 5 molecules of water per molecule of zinc chloride, becomes a perfect solvent for polyacrylonitrile polymers, while a solution with less than a 45% zinc chloride (a nine to one mole ratio) is a non-solvent (see R.D. Crozier, et al, U.S. Patent 3,346,685, 1967). X-ray diffraction data on the concentrated solutions gives diffuse crystalline scattering, though the liquid ordering involved has not been determined.

The electrical behaviour of metal sulphide minerals is more complicated; most are semi-conductors, so their electro-potential properties can vary considerably from point to point on the particle surface as a function of the cation, the crystal structure, and the homogeneity of the ore. For these minerals the flotation process is complex and very difficult to predict from simple laboratory experiments.

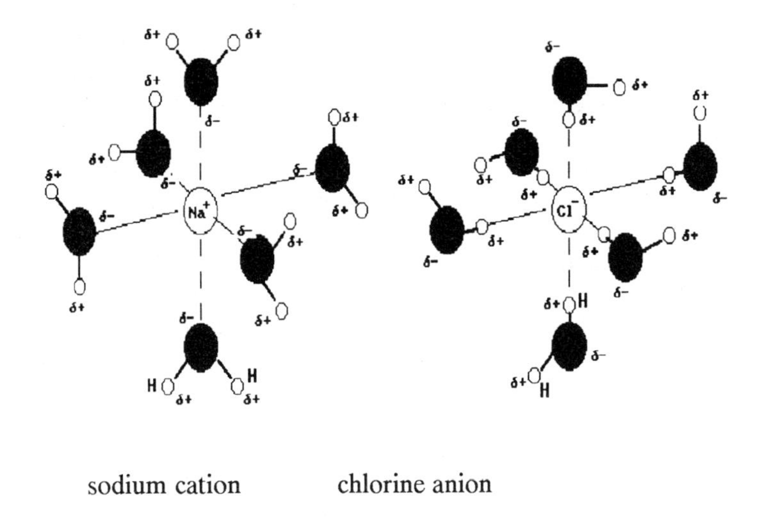

Fig. 5.- Cation and anion hydration showing water orientation

Jan Leja (1982) provides an extensive review of the mechanism of adsorption of fatty acids, sulphonates and amines on inorganic and metallic ores, and probably the most extensive, easily available reference on the adsorption mechanisms. In addition, there is considerable recent literature on zeta potential and other electrochemical sweep experiments, with interpretation of the results in terms of flotation conditions. Though, as virtually all of the data is based on pure, or at least selected mineral samples, the results of laboratory measurements are at best indicative rather than quantitative, if extrapolated to full scale equipment.

Chapter 3

SULPHIDE MINERAL FLOTATION

Stereochemistry of metal xanthates and dithiophosphates

An essential factor in the evaluation of spectroscopic data of the compounds formed by collectors on mineral surfaces is the stereo-chemistry of the complexes that the transition metals can form with thiolates in solution. These complexes are particularly important in the interpretation of the existence of dithiolates on the mineral surface, as the complexes themselves have sulphur-to-sulphur bonding, and many have active lone pairs.

Six coordination complexes

Iron, nickel, cobalt, chromium, Mn^{III}, and Cu^{II} can form six-coordination complexes with tris(bidentate ligands), with geometries as illustrated in Fig. 6 and 7. Structural parameters for xanthates and dithiophosphates are reported by Kepert (1977). For the ferric compound, the constants are b = 1.22 or 1.23 and Θ = 20.5° or 21.8°, according to three different investigators.

The reaction of the formation of the above bidentate complex is:

$$6EtXNa \; + \; Fe^{+++} \longrightarrow 2[Fe^{III}(EtX)_3] \; + \; 3Na^+$$

But the reaction is known to continue, through self reduction, to produce another, more massive, six-coordination complex and dixanthogen:

$$8[Fe^{III}(EtX)_3] \longrightarrow 2[Fe^{II}(EtX)_3]_3Fe^{III} \; + \; \underset{\text{dixanthogen}}{3EtX\text{-}XEt}$$

A pseudo four-coordination dithiolate complex is formed by Cu^{II}. In the case of copper xanthate, it forms a puckered square planar equatorial girdle of two xanthate molecules and either water or a bridging ligand in the axial position. It is considered pseudo four-coordination because the axial bond is

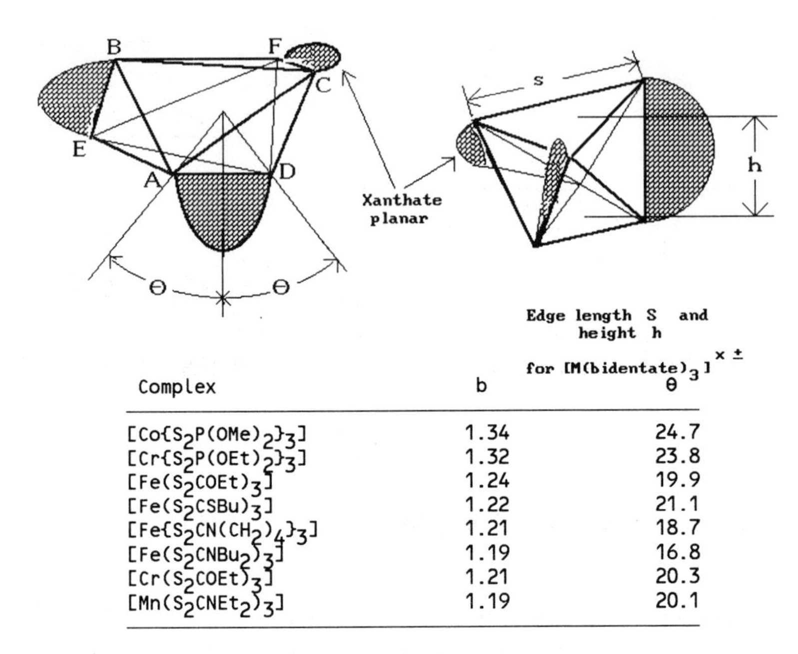

Complex	b	θ
$[Co\{S_2P(OMe)_2\}_3]$	1.34	24.7
$[Cr\{S_2P(OEt)_2\}_3]$	1.32	23.8
$[Fe(S_2COEt)_3]$	1.24	19.9
$[Fe(S_2CSBu)_3]$	1.22	21.1
$[Fe\{S_2CN(CH_2)_4\}_3]$	1.21	18.7
$[Fe(S_2CNBu_2)_3]$	1.19	16.8
$[Cr(S_2COEt)_3]$	1.21	20.3
$[Mn(S_2CNEt_2)_3]$	1.19	20.1

Fig. 6.- General stereochemistry for $[M(bidentate)_3]^{x\pm}$

much longer (3.5 A) and weaker than the sulphur bonds. The cupric reactions are:

$$2EtXK + Cu^{++} \longrightarrow [Cu^{II}(EtX)_2] + K^+$$

which react further, in solution:

$$3[Cu^{II}(EtX)_2] \longrightarrow [Cu^{I}(EtX)_2]_2Cu^{II} + \underset{\text{dixanthogen}}{EtX\text{-}XEt}$$

This complex will hydrate to form a bridging six-coordination cuprous complex. The water molecules, though, can be substituted by copper ions. For both the cupric and the cuprous bidentate complex, the geometry is shown in Fig. 7.

Seven coordination complexes

Seven coordination complexes, in all cases of interest in flotation, involve lone electron pairs in one of the coordination axis. These electrons can be transferred to the mineral surface or can be used to form polymeric complexes in the pulp. The most common geometry of the seven coordination complexes are the pentagonal bipyramid and the capped octahedron shown in Fig. 8.

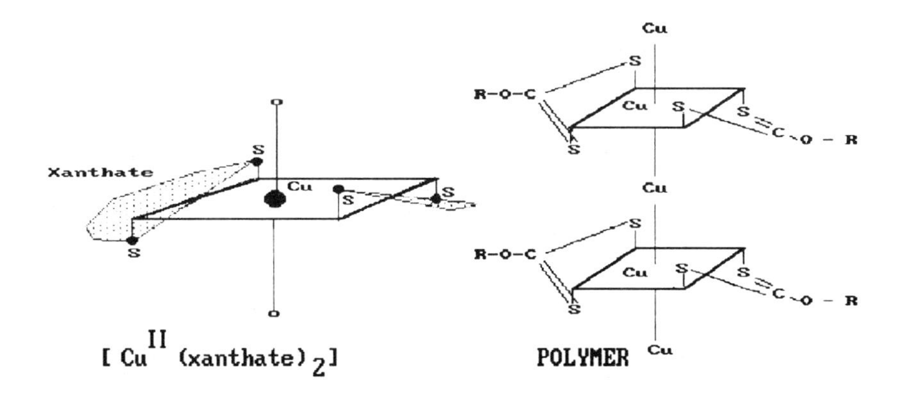

Fig. 7.- $[Cu^{II}(xanthate)_2]$ or $[\{Cu(xanth)_2\}_xCu^{II}{}_{x-2}(H_2O)_2]$

According to Mumme and Winter (1971), as reported by Drew (1977), bivalent lead forms a seven coordination complex with a lone electron pair which can polymerise. The opinion that polymer formation can occur is based on examination of samples where "close to the axial position presumably occupied by a lone pair an equatorial S atom of another molecule" was found. Drew describes the complex as a pentagonal bipyramid. The five positions in the equatorial plane surrounding the metal atom are at a shorter bond distance from the central metallic atom than the two axial locations. Certainly the existence of a complex will be sufficient to confuse surface infra-red spectra, and definitely it will show up as the lead

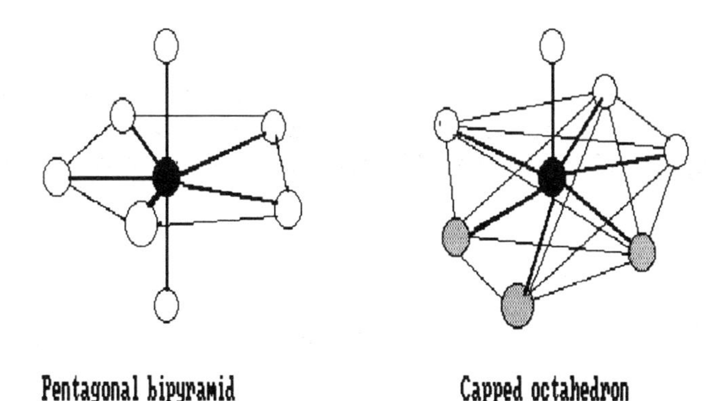

Fig. 8.- Pentagonal bipyramid and capped octahedron geometry

xanthate salt if leached from the concentrate by an organic solvent. Similar structures, with lone pair bridging, are produced for [Sb(dedtc)$_3$(lone pair)], [Bi(dedtc)$_3$(lone pair], and [Te(ethyl xanthate)$_3$(lone pair)], where dedtc is diethyl dithiocarbamate.

Dithiophosphates are more likely to produce a polymeric structure with lead, where the bridging bonds between the sulphur atoms in two complexes are considerably longer than the equatorial plane Pb ones (3.232 A, 3.175 A, compared to 2.761 A, 2.772 A, see Fig. 9). The sketch of a trimer (Fig. 10) shows that this complex will have more available electron pairs to form bonds with the mineral surface, which may explain the preference for dithiophosphates as collectors for galena.

Note that for the trivalent metals the geometry of the complexes are capped octahedrons. The evidence on polymer formation is tenuous, which probably means that xanthates reacting with lead cations in solution will usually only form dimers and trimers at the mineral surface. The complex will have available lone electron pairs to transfer to the mineral particle.

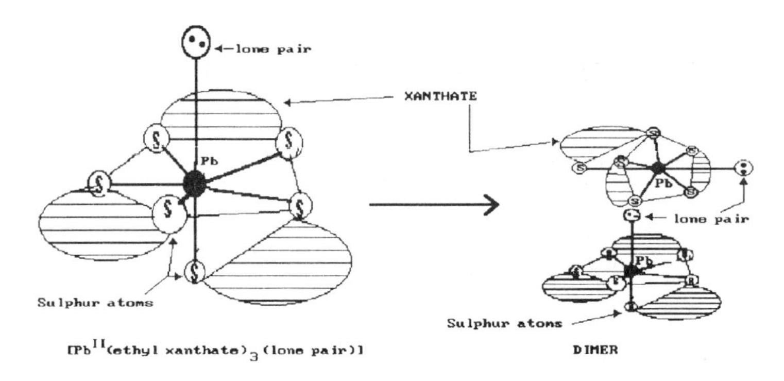

Fig. 9.-[Pb^{II}(ethyl xanthate)$_3$(lone pair)]

Trivalent arsenic, antimony, telurium, bismuth, and vanadium form a very similar complex with xanthates and dithiophosphates, except that they have seven coordination because there is a lone electron pair protruding out of one of the faces in the axial position. Fig. 11 shows the capped octahedron structure in a diagram where dimensions are easier to visualise, and some typical key dimcnsions for arsenic and antimony xanthates and a bismuth dithiophosphate reported by Kepert.

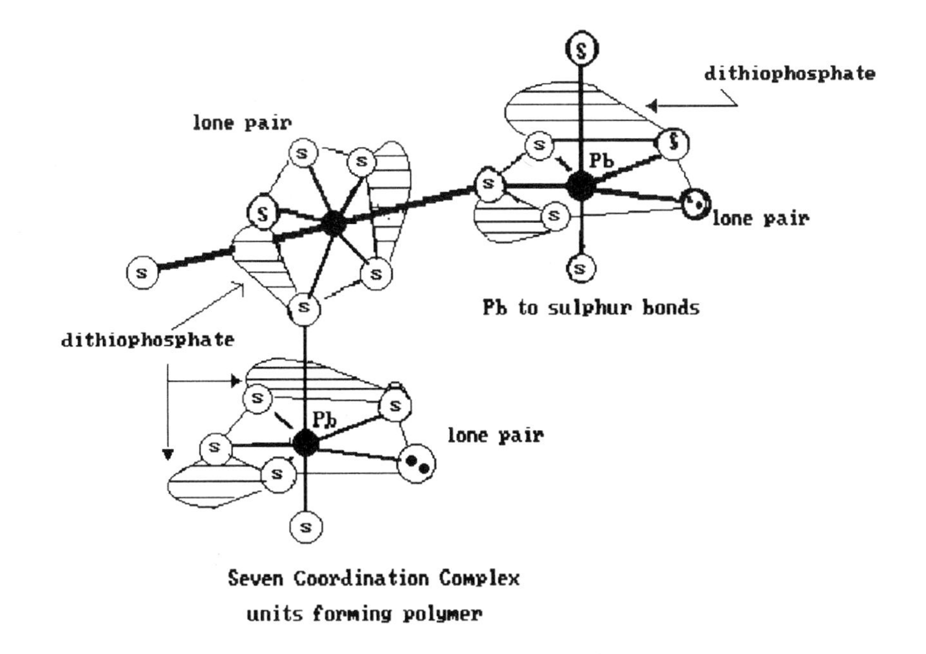

Fig. 10.- [Pb^{II}(di-isopropylphosphorodithiolate)$_2$(lone pair)]

Dithiolate chemistry and surface compounds

Dixanthogens: before going on to analyse the Finkelstein and Poling (1977) review of the nature of the compounds formed on the surface of minerals when they react with collectors, it is important to clarify the probable identity of the dixanthogens mentioned in the older flotation literature. Sixty years ago Cambron and Whitby (1930) were pointing out that reacting xanthates with iodine or chlorine produced impure liquid dixanthogens, while a more selective oxidiser, such as sodium tetrathionate, will produce solid dixanthogens without the need to recrystallise. More recent reviews, among them Reid (1962), have stressed that the melting point is variable, and an indicator of purity. Table 6 contains representative melting point and boiling point data on common dixanthogens, mostly from Reid, but rechecked against unpublished data taken by Professor J.C. Vega de Kuyper at the Catholic University of Santiago, Chile, in 1979/81.

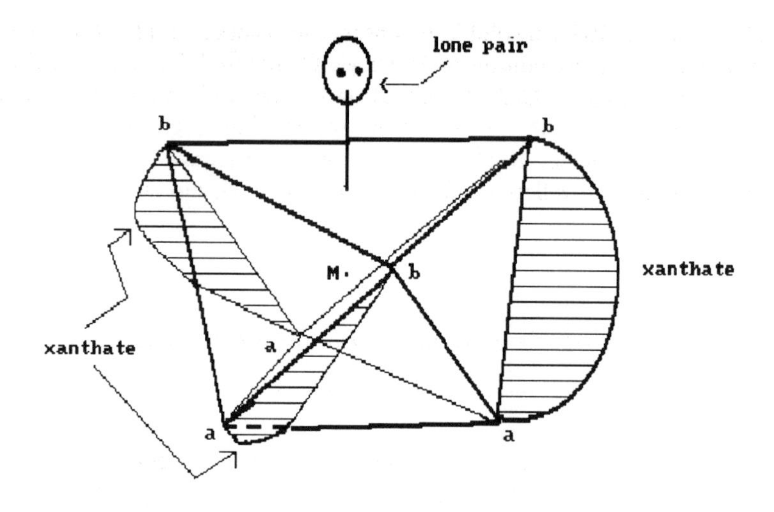

Structural Parameters of Tris(bidentate)(lone pair) Complexes

Complex	M - a	M - b	aMa	bMb	αa	αb
$[As(S_2CNEt_2)_3]$	2.35	2.85	90.1	106.0	54.8	67.2
$[As(S_2COEt)_3]$	2.28	2.94	92.0	107.6	56.2	68.7
$[Sb(S_2COEt)_3]$	2.52	3.00	87.5	113.1	53.0	74.5
$[Bi\{(S_2P(OPr)_2\}_3]$	2.70	2.87	91.2	99.9	55.6	62.1

Fig. 11.- $[M(bidentate)_3(lone\ pair)]^{x\pm}$

All the flotation literature surveyed, including recent compendia such as Leja (1982), always refer to dixanthogens as oily liquids. This suggests that impure products have been used in flotation experiments, because methyl, ethyl and isopropyl dixanthogens are solids at room temperature, as is the mercaptobenzothiazole dithiolate. The ethyl and mercaptobenzothiazole dithiolates, which are readily available commercially, are yellow solids, in all grades.

The 1983 Merck Index, under dixanthogen (compound 3402) lists as alternate names thioperoxydicarbonic acid diethyl ester, or dithiobis[thioformic acid] O,O-diethyl ester, and describes the product as yellow needles, mp 28-32°, onion like odour, solubility in alcohol 2g/100 ml, freely soluble in benzene, ether, petr. ether, oils. Almost insoluble in water. Use: insecticide formulations, herbicide, topical parasiticide. Under dithiobis[benzothiazole] (monograph 3389) it also lists the dithiolate of mercaptobenzothiazole as pale yellow needles, density 1.5, mp. 180°, (minimum commercial specification, mp 168°) insoluble in water, use: rubber accelerator. Under dixanthogen, the latest edition of Hawley's Condensed Chemical Dictionary lists ethyl dixanthogen (CAS: 502-55-6) under the trade name "Sulfasan", as yellow crystals, mp. 30°, insoluble in water, and used as a herbicide, insecticide, parasiticide. It also lists the mercaptobenzothiazole dithiolate with CAS: 120-78-5, yellow solid, properties, d=1.54, mp. 168°, use: rubber accelerator.

Table 6.- PROPERTIES OF DIXANTHOGENS AND XANTHIC ANHYDRIDES

		Xanthic anhydrides =	R-O-C(=S)-S-C(=S)-O-R		
			m.pt°C	b.pt°C	
Methyl	$R=CH_3-O$	yellow solid	54-55°		
Ethyl	$R=CH_3CH_2-O$	yellow solid	52-55°		
Isopropyl	$R=(CH_3)_2CH_2-O$	yellow solid	54-55°		
Propyl	$R=CH_3(CH_2)_2-O$	brown oil	55°?	136°?	
ter-Butyl	$R=(CH_3)_3C-O$	brown oil		?	
Isobutyl	$R=(CH_3)_2CH_2CH_2-O$	yellow oil		?	d=1.126
Amyl		yellow oil		?	
Hexyl		yellow oil		?	N_D=1.541
$(MBT)_2$		unknown			
		Dixanthogen =	R-O-C(=S)-S-S-C(=S)-O-R		
			m.pt°C	b.pt°	
Methyl	$R=CH_3-O$	yellow solid	23-23.5°	90°?	
Ethyl	$R=CH_3CH_2-O$	yellow solid	32°	107°?	d=1.2604
Isopropyl	$R=(CH_3)_2CH_2-O$	yellow solid	57.5-58.5		
Propyl	$R=CH_3(CH_2)_2-O$	yellow oil		117°	
ter-Butyl	$R=(CH_3)_3C-O$	yellow solid	58°		
Isobutyl	$R=(CH_3)_2CH_2CH_2-O$	yellow oil		165°	d=1.08
Amyl					
Hexyl		yellow oil			N_D=1.5569
$(MBT)_2$		yellow solid	180°		

A caveat in reviewing very old work on dithiolates is that in pre-WW II literature there is a confusion between diethyl dixanthogen and Minerec A. For example, Taggart, in his Handbook of Mineral Dressing (1945), says on page 12-11: "Minerec is principally dixanthogen, formed from treating xanthate with ethyl chlorocarbonate, which is a strong oxidiser"; but Minerec

A is *not* a dixanthogen, it is an impure xanthogen formate. Taggart's error arises from reading a rather deceptive U.S. patent of A.H.Fischer's (1928), that was a consequence of a patent priority problem between Guggenheim Brothers and du Pont. Guggenheim Brothers had hired Dr. A.H. Fischer to develop an improved collector for their El Teniente mine. Dr. W. A. Douglass (1927), working for du Pont, was simultaneously carrying out a general programme on testing xanthate derivatives as improved collectors. He obtained a clear priority on a xanthogen formate patent in the U.S.A., so the only U.S. patent issued to Guggenheim Bros. did not cover xanthogen formates, ergo the misinformation of Taggart and others. Minerec then licensed the du Pont patent, and later was the only commercial producer of xanthogen formates.

Physical or chemical adsorption of a collector to a solid surface involves an interaction of electrical forces; each type of bonding involves a reduction in free energy of the system, and the production of a definite amount of heat. The difference is that in physical adsorption the bond is amorphous, while in chemical adsorption there is a directional bonding involved. Metallurgically, the differences between the two bonds is that, in physical adsorption the bond is weak and easily reversible, so a reduction in the concentration of the absorbent in the liquid phase will desorb the species from the surface, while chemical adsorption is irreversible and normally localised on active sites of the mineral surface.

Experiments performed over half a century ago demonstrated that, when thiol collectors are employed to float sulphide minerals, the collection mechanism involves the transfer of electrons to the mineral particle and the formation of dithiolates. What has remained controversial has been whether dithiolates are always involved in the collection mechanism or whether in some cases the metal xanthate is formed and deposited on the mineral surface. Based on the premise that the active collector is a metal xanthate, a number of metallurgists have suggested that the effectiveness of a particular xanthate for the collection of a specific mineral could be deduced from the solubility of the metal xanthate.

The role of alkyl dithiolates in flotation systems, and the type of surface bonding involved, was extensively reviewed by Finkelstein and Poling (1977). The relevant dithiolates they studied are shown in Table 7.

They devoted much effort to evaluating the considerable volume of published information on the flotation of galena with xanthates. Among the more bizarre theories put forward that they evaluated was that the interaction of a xanthate with galena resulted in the formation of elemental sulphur on the

Table 7.- DITHIOLATE REAGENTS: NOMENCLATURE AND ABBREVIATIONS

Dithiolate			Corresponding Thiol	
Name	Formula	Abrv	Name	Abrv
Dialkyl dixanthogen	ROC(S)SS(S)COR	X_2	Alkyl xanthate	X
Bis (dialkoxyphosphono-thioyl) disulphide	$(RO)_2P(S)SS(S)P(OR)_2$	$(DTP)_2$	Dialkyl dithio-phosphate	DTP
2,2' Dithiobis benzothiazole	S S / \ / / CSSC / / \ / N N	$(MBT)_2$	Mercaptobenzo-thiazole	MBT

mineral surface, and that it was the natural floatability of the sulphur that caused the flotation of galena. More important was the controversy that, in all cases, for a xanthate to act as a collector, it must have reacted to form dixanthogen. The analysis is best summarised in their own words:

> "As Granville, Finkelstein and Allison (1972) noted, the case for dixanthogen as the essential hydrophobic entity rested primarily on the electrochemical evidence, in essence, that a bubble only adheres strongly to galena when the potential of the surface is high enough to allow dixanthogen to be formed. The same is true at the present time. .... Woods (1977) identified two stages in the reaction between the xanthate and galena. At potentials between -0.2 V and + 0.2 V on the hydrogen scale, a reaction that he identified as chemisorption took place. At potentials more anodic than 0.2 V, a reaction took place that increased in rate steadily as the potential was raised, until it was limited by the accumulation of product. This reaction is the formation of dixanthogen. ...."

After the evaluation of considerable contradictory data on the collector composition found on galena surfaces, including infra-red analysis of galena treated with dixanthogens, they concluded that the active species on galena is the lead xanthate. The work they reviewed on dixanthogen adsorption showed that galena reacts as follows:

$$PbS + X_2 \leftrightarrow PbX_2 + S^{\circ}$$

This explains why some of the early research suggested that elemental sulphur could be the active collector on galena after exposure to xanthates.

Finkelstein and Poling also reviewed the available literature on surface reactions of dithiophosphates and mercaptobenzothiazole with different minerals. They noted that dithiophosphates are weaker collectors than the corresponding xanthates, and that they are less readily oxidisable, as shown by their dithiolate/thiol reduction potentials (Table 8), and that their lead complexes are more soluble than the corresponding lead xanthates. They also concluded that the active collector on the galena surface is the lead thiolate, as is the case with the data they could examine for mercaptobenzothiazole. The reduction potential data for the xanthates and dithiophosphates is of interest because, at least theoretically, it can be compared with the rest potential of the mineral in water to see if the reaction will occur. The general conclusions of Finkelstein and Poling on the collector species present on the mineral surface at flotation is summarised in Table 9.

Table 8.- REDUCTION POTENTIALS FOR DITHIOLATE/THIOL COUPLES

Homologue	Xanthate	DTP	Homologue	Xanthate	DTP
methyl	-0.004	0.316	isobutyl	-0.127	0.158
ethyl	-0.060	0.255	amyl	-0.159	0.050
propyl	-0.091	0.187	isoamyl	n/d	0.086
isopropyl	-0.096	0.196	hexyl	n/d	-0.015
butyl	-0.127	0.122			

Finkelstein and Poling's most interesting conclusion in their overall survey of published data on infra-red surface analysis (which predates Woods review) is that the active species on galena is a lead xanthate. As there is considerable evidence that the lead atom in galena is labile, there should be Pb^{++} present in solution in flotation pulps, so the lead xanthate identified on the mineral surface is likely to be the seven coordination complexes depicted in Fig. 9. These complexes have lone pairs which can be donated to the mineral surface.

Woods' 1977 analysis, based on voltamograms for galena, and the concept that significant contact angles are essential to flotation, is as follows:

> "The more important question is whether the presence of dixanthogen is essential for flotation to occur. It is apparent that flotation takes place with a number of mineral-collector systems without disulfide formation. A problem to be

considered for the interaction of xanthates with galena lies in the fact that bulk lead xanthate is hydrophilic and multilayers of this compound on galena itself give rise to a low contact angle.

The electrochemical measurements indicate that the initial chemisorbed xanthate layer is hydrophobic and that flotation commences at potentials where this species is formed with only small quantities of dixanthogens being present. The chemisorbed layer on galena, therefore, has a different effect on the superficial tension from the lead xanthate and could be the important species in flotation of galena; this does not exclude the possibility that dixanthogen also plays an important supporting role."

Table 9.- PRODUCTS EXTRACTED FROM THIOLATED SURFACES (Finkelstein and Poling (1977))

Mineral		Product formed on mineral surface by collector						
Name	Formula	Ethyl	Propyl	Butyl	Amyl	Hexyl	Dithiophos	MBT
Orpiment	As_2S_3	☺	☺	☺	☺	♦	☺	☺
Stibnite	As_2S_3	☺	☺	☺	☺	♦	☺	☺
Realgar	AsS	☺	☺	☺	☺	♦	☺	☺
Cinnabar	HgS	☺	☺	☺	☺	♦		
Sphalerite	ZnS	☺	☺	☺	☺	♦	☺	☺
Antimonite	As_2S_2	☺	♦	♦	♦	♦	☺	☺
Alabandite	MnS	☺	♦	♦	♦	♦		
Bornite	Cu_5FeS_4	☺	♦	♦	♦	♦	$Cu(DTP)_2$	$Cu(MBP)_2$
Chalcocite	Cu_2S	☺	♦	♦	♦	♦	$Cu(DTP)_2$	$Cu(MBP)_2$
Galena	PbS	♦	♦	♦	♦	♦	$Pb(DTP)_2$	☺
Pyrrhotite	Fe_7S_8	◙	◙	◙	◙	◙	☺	☺
Arsenopyrite	FeAsS	◙	◙	◙	◙	◙	☺	☺
Pyrite	FeS_2	◙	◙	◙	◙	◙	$(DTP)_2$	$(MBT)_2$ +Fe(MBT)
Chalcopyrite	$CuFeS_2$	◙	◙	◙	◙	◙	$Cu(DTP)_2$	$Cu(MBP)_2$
Covellite	CuS	◙	◙ + ♦	◙ + ♦	◙ + ♦	◙+ ♦	$Cu(DTP)_2$	$Cu(MBP)_2$
Molibdenite	MoS_2	◙ +?	◙ + ?	◙ + ?	◙ + ?	◙	$Mo(DTP)_x$	$Mo(MBP)_x$

♦ = metal xanthate ◙ = dixanthogen ☺ = no positive identification

Thiol collector bonding with metallic sulphide minerals

To evaluate the validity of flotation theories, an important factor is to define the type of bonding that exists between the surface of a mineral and the thiol; i.e. is this a sulphur bond with the metal? The evidence is that the collector-mineral bond is a sulphur-to-sulphur bond and not a sulphur-to-metal bond, as is generally assumed. The following quote from Ronald Woods (1977) provides a corroborating opinion on this very controversial subject:

> "It has generally been considered that chemisorbed xanthate will be attached to a metal atom in the sulphide surface. However, it has been pointed out by Winter (G. Winter, (1975)"Xanthates of Sulfur: Their Possible Role in Flotation", Inorganic and Nuclear Chemistry Letters, Vol. 11, pp. 113-118) that xanthates can form bonds with sulphur, compounds of the type $ROCS_2$-S-S_2COR being quite stable. Also, such compounds cannot readily be distinguished from dixanthogens by spectroscopic analysis and could well be formed at sulphide surfaces."

Strictly speaking, from the point of view of differentiating between flotation mechanisms, it is more important to demonstrate that the chemisorption bond between the collector and the mineral surface is not a metal-sulphur bond, rather than that it is a sulphur-sulphur bond. For example, Clark (1974) points out that chemisorption on the surface of either an n or p type semiconductor results in a bond with hybrid covalent and ionic characteristics which makes it non-specific to a particular atom. The electron availability from the double bonded sulphur atom of the thiono group depends on the following equilibrium:

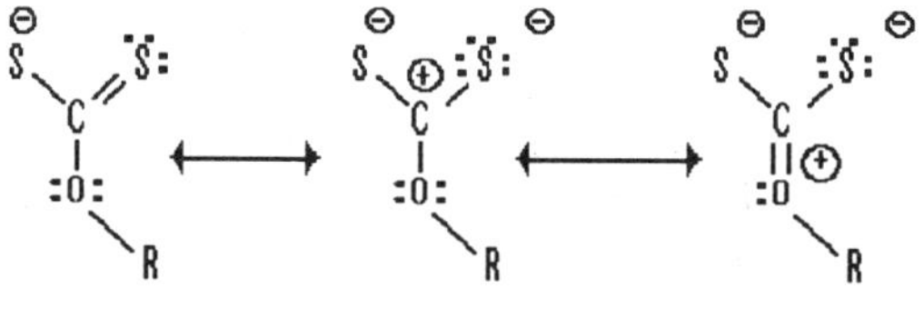

As the species is the anion of xanthic acid, it is obvious that the single bonded sulphur can provide an electron for an ionic bond. A very similar picture can be given for the dithiophosphates and the mercaptans, while the thionocarbamates, xanthogen formates and xanthic esters exhibit resonance in their sulphur-to-carbon bonding, as well as a similar resonance in the nitrogen or oxygen atom.

The authority for concluding that electrons donated by a collector to a sulphide mineral form a sulphur to sulphur bond is Linus Pauling. In his "The Nature of the Chemical Bond", pages 442/448, he says:

> "Many sulfide minerals have structures closely related to those of sphalerite and wurtzite. Chalcopyrite, $CuFeS_2$, is an example (Fig. 12). Its structure is a tetragonal superstructure of sphalerite, with the copper and iron atoms in the zinc positions of sphalerite. Energite, Cu_3AsS_4, has a structure that is a superstructure of the wurtzite arrangement. The sulfur atoms are in the same positions as in wurtzite, and the atoms of copper and arsenic replace those of zinc in an ordered way, so as to give discrete AsS_4 groups (Fig. 13). The observed As-S bond length, 2.22 A, agrees exactly with the calculated value for a single bond, 2.22 A (from the covalent radii, with the correction for electronegativity difference). The Cu-S bond length, 2.32 A, corresponds to a bond number about 0.7 (the appropriate single-bond radius of copper is 1.23 A). Approximately the same Cu-S bond length is found in other copper sulfide minerals. The copper-sulfur bonds have only a small amount of ionic character, and the conclusion may be drawn that the electric charge of the copper atom is negative, probably close to -1."

Pauling makes a similar analysis for a number of other sulphide minerals. In all cases he reaches the same conclusions, that in transition metal sulphide crystals the metal is negatively charged, and the sulphur atom is positive and an electron acceptor. In a 1970 article, with the title "Crystallography and Chemical bonding of Sulfide Minerals" (Mineral Soc. Amer. Spec. Paper 3, 125-131), he put forward a more controversial concept: bonding characteristics of "argononic or trans argononic sulfurs", i.e. sulphur atoms with the same number of electron pairs as argon, and trans argononic where the atom has more electron pairs than argon. In his conclusions in this paper he says:

> "The foregoing discussion of the structures of some sulfide minerals, especially the observed bond lengths and bond angles, leads to the conclusion that the sulfur atom may have either an argononic or a transargononic structure with formal charge 0, +1, or (argononic only) +2, sometimes with resonance between two structures."

Some specific calculated charges given in this paper are: for molybdenite, average charge for Mo, -0.74; for S, +0.37; for proustite (Ag_3AsS_3), formal charge for Ag -1; for sphalerite and wurzite, average charge for zinc -0.67; and for sulphur, +0.67; while for galena, which has a sodium chloride structure, the bonds are resonating, and the average charges are +0.43 for S, and -0.43 for Pb. These are listed in Table 10, and compared to the mineral's semi-conductor properties and flotation critical pH.

The general conclusion is that, for the majority of the sulphide minerals, the negative charge on the metal forces the collector- mineral electron transfer to be from the collector sulphur atom to the mineral sulphur atom.

Table 10.- MINERAL ELECTROCHEMICAL PROPERTIES

Mineral	Surface compound					Critical pH **		
Name	Formula	Ethyl	Amyl	DEDTP	Type	Ethyl	Amyl	DEDTP
Sphalerite	ZnS	?	MX	?	P	none	5.5	none
Pyrrhotite	Fe_7S_8	X_2	X_2	?	P	6.0		
Arsenopyrite	FeAsS	X_2	X_2	?	N	8.4		
Galena	PbS	MX	MX	MX	P	10.4	12.1	6.2
Pyrite	FeS_2	X_2	X_2	X_2	N	10.5	12.3	3.5
Chalcopyrite	$CuFeS_2$	X_2	X_2	MX	N	11.8	>13	9.4
Covellite	CuS	X_2	X_2+MX	MX		13.2		
Bornite	Cu_5FeS_4	?	MX	MX	P	13.8		
Chalcocite	Cu_2S	?	MX	MX		>14		
Molibdenite	MoS_2	X_2+?	X_2+?	MX				

Mineral	Charge on Atom*		Semiconductor	
Name	Cation	Sulphur	Type	Band-Gap eV
Sphalerite	-0.67	+0.67	P	3.8
Pyrrhotite			P	-
Arsenopyrite			N	0.16
Galena	-0.43	+0.43	P	0.38
Pyrite	- 1	+ 1	N	1.1
Chalcopyrite	- 1	+ 1	N	0.33
Covellite	- 1	+ 1		
Bornite			P	0.82
Chalcocite				
Molibdenite	-0.74	+0.37		

* = Charge on atom from Pauling
DEDTP = Diethyl dithiophosphate
** = critical pH for flotation above which mineral depressed (Sutherland and Wark)

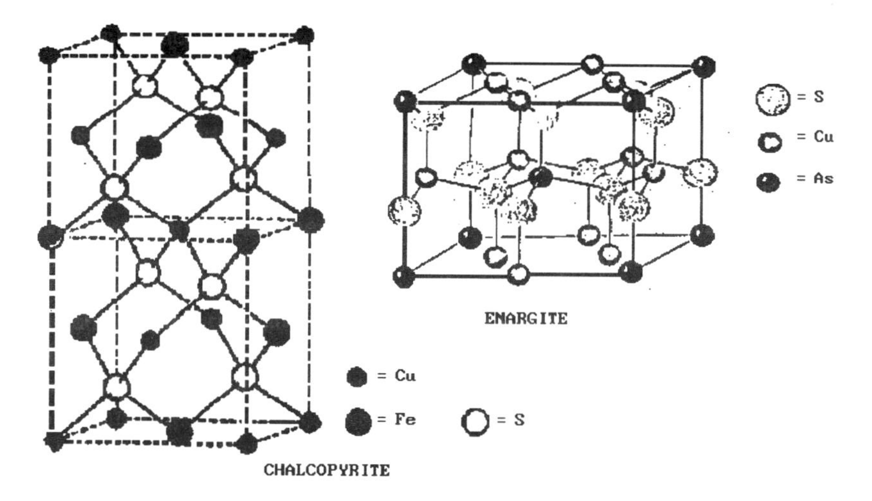

Fig. 12.- Tetragonal crystal chalcopyrite and orthorhombic enargite

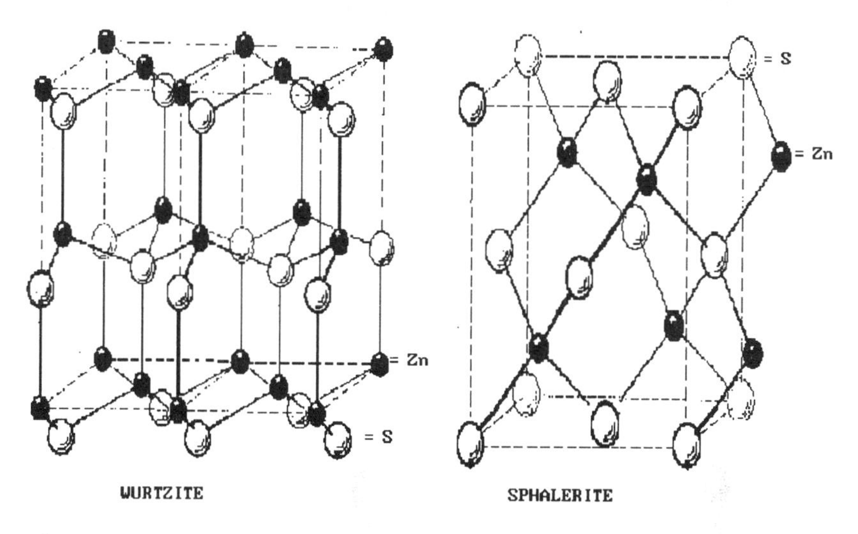

Fig. 13.- Wurtzite and sphalerite, from Pauling

The exact points to which collectors will attach themselves on mineral particles will depend on the micro semi-conductor topology of the surface. As will be recalled, the n-type semiconductor conducts through electrons in the conductance band, and thus is an electron donor, while the p-type conducts through the movement of vacancies in the valency band and thus is an electron acceptor. Ideally for rapid adsorption of collector molecules, the mineral particles should have n and p regions distributed on the surface, the p-type to provide a point of adherence for the collector, the n-type for the transfer of electrons to oxygen dissolved in the pulp to thus maintain electroneutrality for each mineral particle. The bulk semi-conductor properties (Baleshta et al, 1970) of the more important sulphide minerals are listed in Table 11.

Table 11.- BAND-GAP ENERGIES OF SULPHIDE MINERAL SEMICONDUCTORS

Mineral	Type	Formula	Band-Gap Energy eV
Arsenopyrite	N	FeAsS	0.16
Bornite	P	Cu_5FeS_4	0.82
Chalcopyrite	N	$CuFeS_2$	0.33
Digenite	P	$Cu_{1.8}S$	-
Galena	P	PbS	0.38
Marcasite	P	FeS_2	0.46
Pyrite	N	FeS_2	1.1
Pyrrhotite	P	Fe_7S_8	-
Zinc sulph.	P	ZnS	3.8
Germanium		Ge	0.75
Silicon		Si	1.2

The nature of the bubble-particle bond

We have looked at how collectors adhere to mineral particles but not at the nature of the collector-bubble bond, or for that matter the overall mineral particle-bubble bond. We know that mineral capture by bubbles is a reluctant phenomena. In all cases studied, an induction delay in gas-solid contact has been detected. One can visualise the time delay as the bubble girding its loins and battering its way past the defending forces to finally reach the mineral surface. In this case the defending forces resisting the bubble approach are the water of hydration around the particles and the oriented water layer at the liquid/gas interfaces.

The theory and mathematics of these surface tension phenomena and their effect on the dynamics of flotation are complex. Klassen and Mokrousov (1963) extensively reviewed the classic thermodynamic information available prior to 1960. Laskowski (1974) published "Particle-bubble attachment in flotation", a thorough up-date. His synopsis, which agrees with Klassen and Mokrousov, says:

> "The hydrophobicity, defined by the contact angle, is not enough to describe the flotation properties of a given solid, since the conditions connected with the speed of displacement of the liquid layer between the solid particle and the bubble must be also fulfilled."

The disqualification of the value of the contact angle in predicting flotation performance does not go far enough, though Laskowski puts his finger on the key factor that determines the flotation kinetics in bubble contact; i.e., that the rate controlling mechanism is the thickness and tenacity of the liquid layer surrounding the bubble and the particle, which must be parted before there is bubble contact and adhesion.

A very pertinent analysis of the alternative methods of obtaining bubble-particle adhesion is contained in Glembotskii's (1963) 1961 text, Flotation. Here the magnitude of the energy required for a bubble to attach to a mineral surface is compared to the energy necessary to generate a bubble in situ on the surface. His conclusion is that, particularly at large contact angles, the energy required to generate a new bubble on the surface is considerably lower than the energy expended in attaching a bubble to a mineral particle, postulating therefore that collection of mineral particles will preferably occur through micro-bubble formation on the mineral (nascent bubble flotation) rather than attachment of existing bubbles to a suspended particle (coursing bubble attachment). This is an obvious conclusion when one considers that the energy of adhesion is proportional to the surface tension and the area of liquid surface that must be displaced from the solid, and that the contact area of a micro-bubble is much smaller than that required to attach a preformed bubble. This relation also explains why the preferred frothers used in flotation barely affect the liquid surface tension, as otherwise, if they increased surface tension significantly, particle capture would become more difficult.

The forces holding a collector to the surface of a sulphide mineral are due to bonding through the transfer of lone electron pairs as well as ionic bonding by van der Waals forces. This bonding is obvious, but due to the use of contact angles to evaluate whether bubble attachment occurs at given conditions, an implicit subconscious assumption is present that the alkyl group of collectors is hydrophobic and therefore that a bubble attaches itself to this hydrocarbon layer through surface tension forces. In fact, the use of contact angle measurements to evaluate positive floatability is dangerous, as there is incontrovertible evidence that successful sulphide mineral recovery by flotation requires that the amount of collector present be considerably less than a monolayer of collector, while stable contact angles are measured when there are multilayers on the sample surface.

It would appear that to obtain a measurable contact angle on a sample, an area of the sample proportional in size to the bubble used to measure the angle must be covered with collector (or be hydrophobic), while, if a collector covers this much area of a sulphide mineral particle in a flotation cell, this results in depression rather than flotation. Therefore, if one detects a contact angle in a flotation system, the dosage of the flotation reagent is too great, and corresponds to the conditions at which over-oiling is present rather than those for optimum flotation.

Contact angle measurements are successfully used in industrial labs to screen depressants for any type of mineral. This may be significant in evaluating the mechanism of depression.

To determine what force, other than surface tension, accounts for collector-frother adhesion, one can analyse the bonding of each link in the chain joining the bubble to the mineral particle. First, what is the collector to mineral bond: it is obvious that as the collector has transferred electrons to the sulphide mineral, there is chemical adsorption, and a covalent or ionic bond between the collector and the mineral.

Second, what is the bond between the exposed side of the collector and the bubble: here the same reasoning as that used for the validity of the contact angle indicates that the surface tension forces that come into play with hydrophobic surfaces cannot be major contributors to bubble adhesion; but there is independent evidence that, in sulphide mineral systems, frothers adsorb onto collector coated minerals, so chemical adsorption between the frother and the collector is probably present. Supporting this is the fact that, to adsorb collectors, the sulphide mineral particle must maintain electroneutrality by oxidation of oxygen dissolved in the pulp; therefore, there must be a charge on the collector adhered to the mineral, and this electron deficit will be available to form a chemical bond between the frother and the collector.

Third, what is the bond between the collector-frother complex and the bubble: a simplified mechanism of nascent bubble formation on the mineral surface posits that this will occur at electrical discontinuities on the sulphide particle surface; this charge anisotropy will attract surface-active species in solution, which will then stabilise the growing bubble through normal surfactant action. In effect, an electron bond is formed at the same time as the bubble, and bubble stability is produced by independent frother molecules, with either the same or different chemical structures than the species forming the collector-frother complex.

Summarising, to obtain bubble-particle adhesion, a sulphide mineral particle must first become coated with a partial layer of collector molecules. The coated particle will then abstract frother molecules from solution. These will intermingle with the collector molecules already on the mineral surface; and then the nascent bubble (generated by hydrodynamic mechanisms) is nucleated by organic discontinuities on the mineral surface. This description has been limited to sulphide minerals because it only describes the flotation mechanism when starvation quantities of reagents are employed. Many oxide and salt systems require orders of magnitude greater quantities of reagents, due to mechanistic differences in the flotation process.

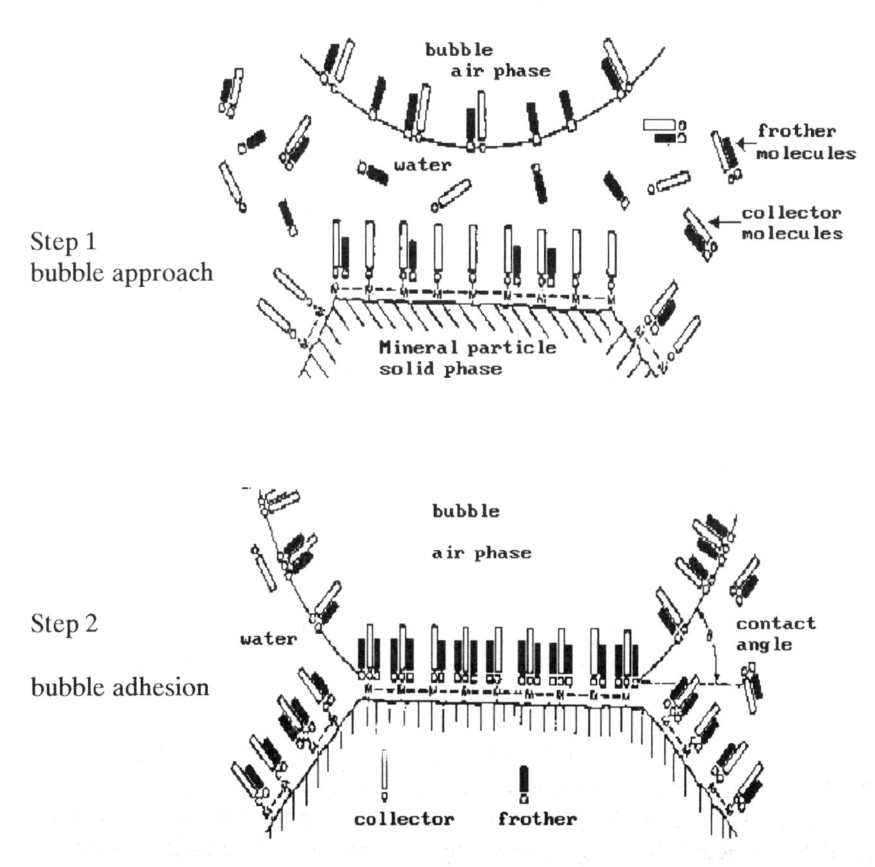

Fig. 14.- Leja and Schulman (1954) interpenetration theory

Leja and Schulman (1954) studied the interaction between collectors and frothers in 1954, and demonstrated that a collector coated mineral will remove frother from a solution. Their experiment was quite simple. They took a sample of ground galena, placed it in a xanthate solution, drained off the solution and then added water containing a known amount of a flotation frother. Analysis of the liquid phase demonstrated a reduction in the frother content and therefore the formation of some form of collector-frother adduct. The experiment was repeated with a gamut of collector-frother species, and gave the same results. As there were no bubbles present, the attraction between the collector and the frother molecules could not be due to surface tension. In this paper they also proposed their interpenetration theory; Fig. 14 reproduces their original diagrams. They assume that the attractive forces involved are ionic.

Frother-collector surface complexes

Fig. 15, taken from Crozier and Klimpel (1988) depicts a xanthate-alcoholic frother interaction complex on the surface of a sulphide mineral. It shows the chemisorption of the demised sulphur atoms and the physical adsorption of the double bonded sulphur atom. Note that these double bonded sulphur atoms are on the same plane as the single bonded atoms. The frother molecules within the collector matrix are oriented so that the hydroxyl groups are pointing away from the mineral surface, so, when bubbles adhere, these groups are located at the interface with the air water boundary to bond the bubble to the mineral surface.

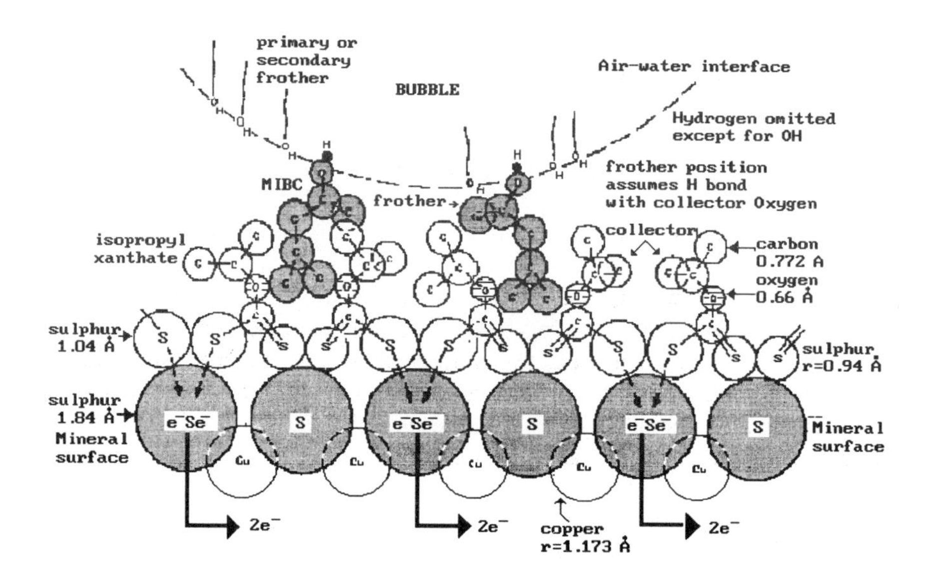

Fig. 15.- Frother-collector interaction complex

The bond between the frother alkyl group and the collector could be a hydrogen bond with the oxygen atom attached to the CS_2 group. See page 36 for the possible resonance structures for xanthate anions which would accommodate this sort of bonding. Note that the collector's single bonded sulphur atoms transfer electrons to form a bond with the mineral's sulphur atoms, not the metal atoms. Though the illustration assumes adsorption of the double bonded sulphur atom, it can chemisorb to the mineral sulphur atom because the double bond can resonate between the S=S and S=O positions in the collector molecule. This resonance can also provide the electrons for the frother-collector hydrogen bond. The atomic radii employed in the illustration are based on Pauling (1967).

There is evidence that the formation of collector-frother complexes is catalysed by minerals. The catalytic effect of the mineral in a flotation system is frequently overlooked when analysing the possible reactions occurring on the concentrate. The author once published the comment that thionocarbamates and xanthogen formates reacted vigorously, and received a letter from the chemist recognised as a world authority on thionocarbamates saying that he had tried the reaction and nothing happened, at any temperature, over a number of days of mixing. He was quite right, of course; in a glass beaker there was no reaction, but in a metal container there is a slow reaction, and in the presence of mineral there is a very vigorous reaction. At that time the author's company was marketing such a mixture as a collector. It was delivered to the sole client in tank cars, and the change in composition with time in the tank car had been monitored for a number of years trying to see if the active species that provided the unique, mineral specific, properties of the collector could be identified.

Sometimes industrial data can elucidate these types of theoretical problems, even when they are difficult to test experimentally in a laboratory with very sophisticated instruments. An interesting case was experienced over a number of years with the Anaconda, Butte concentrator. Because of changes forced on the reagent supplied, by competitive factors and periodic sharp mineralogical changes, we were able to deduce that for dialkyl thionocarbamates, the alkyl group location on either the C=S side or the N side did not affect metallurgical properties.

This was discovered because pressure for prompt supply forced the use of a temporary synthesis route, which on being modified into an economic industrial process reversed the location of the alkyl groups. Later bench flotation showed that, for that particular ore, a methyl group on the nitrogen side was as effective a collector as the traditional ethyl alkyl thionocarbamate, and much cheaper to manufacture. In the manufacture we ran into the problem that immiscible collector and the by-product brine densities were such as to make separation impossible, so the collector was extracted with a lower density alcohol - MIBC. The mixture was tested as a collector in the concentrator and was found to be as good or better than the pure thionocarbamate. Interestingly, Butte was employing MIBC as its frother at that time, at a dosage of about half that of the collector. It was found that, even though the MIBC contained in the collector mix was as much as the MIBC fed independently as a frother, it did not reduce the amount of frother required for good operation. The gross amount of collector added also remained unchanged, so in effect, concentrate and recovery grade was unchanged at half the collector consumption and double the frother dose.

The revival of the Leja and Schulman collector-frother interpenetration complex is the result of practical experience which has resulted in the greater (and more carefully chosen) use of collector/frother mixtures fed directly to the ball mill, plus the separate evaluation of frother structures to tailor these to specific hydrodynamic froth problems that are independent of the optimum collector from the mineral recovery standpoint. This choice of the second component should certainly be taken into account when a recommendation is made to dilute an insoluble oily collector with a surfactant to "emulsify" it and improve dispersion.

Chapter 4

THIOL COLLECTOR CHEMISTRY

Properties of xanthates

Xanthates are available commercially as solutions, powders or pellets. Pellets are the most desirable, as there are fewer dusting problems (xanthate dust is irritating and toxic) and better storage stability. All xanthates decompose in the presence of moisture, giving off carbon disulphide as one of the decomposition products, so that they should be treated as potentially highly flammable. There is a considerable range in the purity of commercially available xanthates. Quality control is somewhat of a problem, as accurate analysis requires the availability of an experienced chemist because of the rapidity of decomposition of solutions in contact with air. The commercial availability of xanthates, and their trade designations, are shown in Table 18, page 67.

Potassium xanthates crystallise as anhydrous salts, while the sodium salts have two molecules of water of crystallisation and are therefore less stable. Thus, commercially available sodium xanthates are rarely over 85% active. The stability of xanthate solutions cannot be predicted a priori, as it is a complex function of solution concentration, storage temperature, excess or deficiency of the alkali content (too much or too little will reduce xanthate stability). At a pH of below 3, the half life of all xanthates is reduced to minutes. With these caveats the decomposition of xanthates in solution, as a function of temperature and concentration, shown in Table 15, can be used as a practical guide to the stability of good quality xanthates.

The reaction employed in the production of xanthates is:

$$R\text{-}OH + NaOH + CS_2 = R\text{-}O\text{-}CSSNa + H_2O$$

As most sodium alkyl xanthates crystallise with two molecules of water, very pure xanthates can only be obtained by recrystallising from a non-aqueous solvent or by preparing the alcoholate with metallic sodium. Recrystallisation is done by dissolving the xanthate in alcohol and salting it out with ether. The anhydrous xanthate has a crystalline cream colour, and is easily soluble in water. If kept cool and dry (in a dessicator) the salt is relatively stable, but on exposure to moist air it will darken and acquire a disagreeable odour.

In the presence of water the following secondary reactions can occur, either during the synthesis reactions or the dissolution of the solid xanthate:

$$6R\text{-}O\text{-}\overset{\overset{S}{\|}}{C}\text{-}SNa + 3H_2O \longrightarrow 6\ R\text{-}OH + 2Na_2CS_3 + 3CS_2 + Na_2CO_3$$

$$6NaOH + 3CS_2 \longrightarrow 2Na_2CS_3 + Na_2CO_3 + 3H_2O$$

and in the presence of air the xanthate can be oxidised to dixanthates:

$$2R\text{-}O\text{-}CSSNa + \tfrac{1}{2}O_2 + CO_2 \longrightarrow R\text{-}O\text{-}CSSSSC\text{-}O\text{-}R + Na_2CO_3$$

The reaction will occur without CO_2 if there is a mineral surface present to catalyse it. With dissolved copper cations in the pulp the reaction is:

$$4EtXK + 2Cu^{++} \longrightarrow 2Cu(EtX)_2 + 4K^+ \longrightarrow Cu_2(EtX)_2 + EtX\text{-}XEt$$

The oxidation of a xanthate solution with iodine dissolved in an iodide solution is quantitative and is employed as an analytical method, using starch as the end point indicator.

$$2EtXK + KI_3 + 2I^- \longrightarrow EtX\text{-}XEt + 3KI3$$

Sodium peroxydisulphate ($Na_2S_2O_8$) reacts very similarly to iodine, as do ferric salts.

The trithiocarbonates can continue hydrolysing, so commercial xanthates may contain dixanthogens and free carbon disulphide, which make them highly inflammable, plus having the following inorganic impurities:

$$Na_2S_2,\ Na_2S,\ Na_2SO_3,\ Na_2S_2O_3,\ H_2S,\ \text{etc}$$

Another method for the synthesis of xanthates is based on reacting ethers with NaOH and carbon disulphide:

$$Et\text{-}O\text{-}Et + CS_2 + NaOH \longrightarrow 2Et\text{-}O\text{-}CSSNa + H_2O$$

Tertiary alcohols are difficult to react with caustic to form alcoholates, but do react readily with sodium metal, thus this is a good laboratory route to pure xanthates:

$$R\text{-}OH + Na \longrightarrow R\text{-}ONa + \tfrac{1}{2}H_2; \quad + CS_2 \longrightarrow R\text{-}O\text{-}CSSNa$$

Manufacture of alkali metal alkyl xanthates

The literature on the methods of xanthate manufacture consists almost exclusively of patents. All other published information on xanthate production appears to be descriptions of laboratory recipes and methods of purification. The author is not aware of any real patent on an industrial xanthate process, rather they all appear to have been written for defensive reasons and not to teach the art. Thus, descriptions of the processes in this section are based on the author's limited manufacturing experience, rather than on literature reviews.

An example is the sodium ethyl xanthate process described in the Kirk-Othmer "Encyclopedia of Chemical Technology". Dr. Guy Harris, at the time a chemist at the Dow Chemical company, a major xanthate manufacturer, wrote the xanthate article, but quite obviously could not reveal Dow proprietary information. He chose to quote from a German process described in the Department of Commerce, Office of Technical Service, P.B. Rept. 74736:

> "For the manufacture of potassium ethyl xanthate, one German plant used 400% excess alcohol and equimolar quantities of carbon disulphide and 50% aqueous potassium hydroxide. The reaction temperature was maintained at 40°C by cooling. This step took a half hour. The product was dried in a vacuum drier, and water was removed from the recovered alcohol by distillation. The xanthate was obtained in almost quantitative yield and assayed 95%."

From the author's experience, to make sodium ethyl xanthate using exactly the same process, it would be described as follows: in a steel kettle, equipped with agitation and cooling, a 300% molar quantity of recovered alcohol (a 96% azeotrope of alcohol and unreacted CS_2), plus an equimolar amount of fresh alcohol is charged. To this a slight excess of 50% caustic is slowly added. The rate of addition will depend on the cooling available, and should be adjusted to such a rate that the alcoholate reaction temperature does not exceed 60°C. The alcoholate is then cooled to about 35°C (but not less than 25°C) and a slight excess of CS_2 added through the bottom of the reactor at such a rate so that the cooling system maintains the reaction at, or below,

40°C. The reaction time will depend on the geometry of the reactor and its cooling capacity. Usually this is designed for about a 40 minute overall reaction time. Yields, in solution, are on the order of 95 - 98%, but at 40°C, in the reaction mix, xanthate loss is about 0.25% per hour. It is very unlikely that a commercial dried xanthate could be obtained by this method, because of decomposition and degradation during the drying step. The alcohol can be recovered by vacuum and/or steam distillation, and the solution used on site as a raw material or as a collector. From the description of the process, there are three critical steps involved: 1) the preparation of the alcoholate, 2) the xanthate reaction, and 3) the drying step. The sodium ethylate reaction is:

$$3EtOH + NaOH \longleftrightarrow NaOEt \cdot 2EtOH + H_2O$$

The sodium ethylate, or ethoxide, is a solid, very soluble in ethyl alcohol, which in the presence of water rapidly hydrolyses back to alcohol plus caustic. This is the reason for the 400% excess alcohol, as it is necessary to keep the equilibrium, in the above hydrolysis reaction, to the right. The yield can be improved by using solid caustic to limit the water present, but the alcohol excess must be increased to keep the reaction within bounds.

The second reaction also has side reactions that occur as a function of temperature and water content. The main reaction is:

$$NaOEt.2EtOH + CS_2 \longrightarrow EtO{-}\overset{\overset{S}{\|}}{C}{-}SNa + 2EtOH$$

The side reactions, in addition to the hydrolysis of the alcoholate, are:

$$6NaOH + 3CS_2 \longrightarrow 2Na_2CS_3 + Na_2CO_3 + H_2O$$

$$EtOH + Na_2CS_3 \longrightarrow EtO{-}CSSNa + NaHS$$

Higher reaction temperatures will favour both hydrolysis and the above reactions, reducing yields. The third, and always most critical step, is the recovery of the dry xanthate. In the presence of excess water, there is a significant decomposition of the xanthate to volatile components (CS_2 and alcohol) even at drying temperature as low as 50°C. Based on a 1959 Russian article, Harris, in the Encyclopedia article, reports the data in Table 12 on the decomposition of dry sodium ethyl xanthate.

In the presence of water the decomposition rate is much higher. The alkali metal alkoxides with alkyl group chain lengths of three or more carbon atoms, and particularly those prepared from secondary or tertiary alcohols, are considerably more unstable than ethyl alkoxide, as are the xanthates themselves. As they are considerably less soluble in their own alcohol than the ethyl xanthates (see Table 13), the excess alcohol process is only commercially interesting for the ethyl xanthate, and then only if it can be used as is. The economics of the solvent recovery and the drying step are questionable, when compared to suspension processes.

Table 12.- DECOMPOSITION OF Na ETHYL XANTHATE ON DRYING AT 50°C

Time of heating, hours	Xanthate content, %
0	79.32%
40	74.38
64	65.66
88	65.06

Table 13.- PROPERTIES OF ALKYL XANTHATES

xanthate		Mol. Wt.	Chain length angstr	Solubility g/100 g solvent In water 0°C	In water 30°C	In its own alcohol 0°C	In its own alcohol 30°C
K Methyl	CH_3OCSSK	146.3	7.1	-	-	-	-
K Ethyl	C_2H_5OCSSK	160.3	8.4	80.1	123.8	5.6	12.5
K Propyl	C_3H_7OCSSK	174.3	9.7	43	58	1.9	8.9
K Isoprop	C_3H_7OCSSK	174.3	-	16.6	37.1	-	-
K Butyl	C_4H_8OCSSK	188.4	10.9	32.4	47.9	-	-
K Isobut.	C_4H_8OCSSK	188.4	-	10.7	47.7	1.6	6.2
K Amyl	$C_5H_{11}OCSSK$	212.4	12.2	-	-	-	-
K Isoamyl	$C_5H_{11}OCSSK$	212.4	-	28.4	53.3	2.0	6.5
Na Ethyl	C_2H_5OCSSK	144.2	8.4	81.6	128.5	43.5	99.6
Na Propyl	C_3H_7OCSSK	158.2	9.7	17.6	43.3	10.2	22.5
Na Isoprop	C_3H_7OCSSK	158.2	-	12.1	37.9	-	19.0
Na Butyl	C_4H_8OCSSK	172.2	10.9	20.0	76.2	-	39.2
Na Isobut.	C_4H_8OCSSK	172.2	-	11.2	33.4	1.2	20.5
Na Isoamyl	$C_5H_{11}OCSSK$	186.3	-	24.7	43.5	10.9	15.5
Na Amyl	$C_5H_{11}OCSSK$						

Xanthates are commonly manufactured in a batch process, using solid caustic and an inert diluent to reduce the effect of the water on the alcoholate reaction. Solvent suspension simplifies agitation. The solvent chosen must be unreactive, and a solvent for alcohol and CS_2, but not for the water and xanthate. Its volatility should be such as to reduce the drying energy and temperature required. Most solvents commonly employed are petroleum ether, hexane, benzene, toluene or xylene.

In the suspension process a much larger reactor is charged with the recycled solvent and a moderate excess of alcohol (contained in the recycled solvent), and caustic (preferably beads, though flakes may be used) is dumped into the reactor at a rate compatible with the cooling system available. The peak temperature during the alcoholate reaction should be kept to less than 60°C. Adjusting the reactor pressure, this can be accomplished by refluxing the inert solvent. After the alcoholate reaction is complete, an equimolar amount of CS_2 is added, again at a rate dependent on the cooling system capacity. The reaction temperature depends on the alkyl chain length, but normally will be under 50°C. The product is a slurry of solid xanthate containing, hopefully, only a portion of the water of reaction. After the reaction is completed, the final xanthate slurry is filtered and the solid xanthate is vacuum dried. The recovered solvent is also dried, by any convenient method, and recycled to the reactor. At equilibrium conditions, an alcohol excess builds up in the recycle stream; thus, insufficient feed will reduce yields and excess will keep increasing the excess alcohol, so the optimum for each alkyl chain length must be determined by operating experience. The required excess usually is in the 10 to 40% range. For all the xanthates, a slight CS_2 excess is required to maximise yields, usually in the 2 - 5% range. Raw material specifications have dryness as the top priority, and purity second. Typical xanthate compositions obtained from the suspension reaction system are shown in Table 14. The higher moisture content of the sodium xanthates is due to their containing up to two molecules of water of crystallisation. Overall reaction yields will be somewhat lower than the dry basis xanthate purity, due to reactor losses and decomposition losses in the drying step.

Table 14.- XANTHATE ASSAYS, SLURRY SUSPENSION PROCESS

Alkyl group	Xanthate %	Moisture %	Dry basis %
K ethyl xanthate	88 - 92	0 - 1%	89 - 93
Na ethyl xanthate	82 - 86	3 - 5.5	87 - 91
K sec amyl xanthate	75 - 85	4 - 6	80 - 88
K amyl xanthate	88 - 93	0 - 2	88 - 94
Na isopropyl xanthate	75 - 80	12 - 18	88 - 94
Na sec butyl xanthate	70 - 77	10 - 15	80 - 88

Only the Dow Chemical Company is believed to have operated a continuous process for the manufacture of xanthates (Dow ceased collector manufacture in 1980). Their products were consistently of a higher xanthate purity and lower moisture content than competitive material, which suggests that all other producers employ batch processes. The Dow continuous process can only be surmised, as there is no public information on the process or even on whether it was non-batch. Their Na isopropyl xanthate averaged 90 - 93% purity, compared to a norm of 75 - 80% in the market, and moisture content ranged from 2.5 to 5% compared with a norm of 12 - 18%. For sodium sec-butyl xanthate, the figures are 80 - 85% purity, compared to 70 - 77%, and moisture 2 - 5%, compared to 10 - 15%.

The dry xanthate powder is unacceptably irritating, so nearly all commercial xanthates are pelletised before shipment.

Liquid xanthates: manufacturers consuming their xanthate directly, either in a flotation process or as a raw material for another process, will by-pass the drying step and will filter, then dissolve, the xanthate out of the inert solvent slurry with water and use the xanthate solution directly. The capital savings is considerable, as the vacuum drying equipment, plus pelletiser, can equal the cost of the rest of the plant. Alternately, a third process can be employed in which the xanthate is produced in a high intensity mixer vessel. In this case, the reactor is similar to a bread dough mixer, except that the container is equipped with a mechanically refrigerated cooling jacket. To manufacture a given xanthate, the reactor is charged with flake caustic and a modest excess of alcohol, 10 to 30%, depending on the alkyl group involved. After the alcoholate reaction is completed, using the same maximum temperatures as those stipulated in the suspension process, the paste is reacted with 1.02 to 1.05 molar ratio of CS_2, with the carbon disulphide fed at a rate compatible with the cooling available, but generally maintaining the temperature below 30°C (considerably lower if very high purity xanthate is required). After the reaction is completed, the paste is dissolved with water and the solution employed immediately in the mill or further process. The quality of the xanthate produced varies considerably as a function of the mixing intensity and the efficiency of the cooling system. The use of mechanical refrigeration in these units can be justified by the improved xanthate yields and the smaller equipment volumes per unit of production. The drawback of the process is that it is difficult to recover the excess alcohol fed, so if alcohol interferes with the subsequent reaction, the process cannot be used. The advantage of the process is that it can produce xanthates with alkyl groups of any chain length.

Stability of xanthate solutions: industrial experience on the decomposition rates of good quality xanthate solutions, in storage, are shown in Table 15. At ambient temperature, these rates normally are below 1% per day. In a pulp, the stability of the xanthate cannot be predicted. It is quite obviously a function of the pH of the environment, but the decomposition is very significantly affected by the catalytic effect of transition metal sulphides, as well as, to a lesser degree, all the different constituents of the gangue. In storage, the alkali content is crucial, as both an excess or a deficit reduce xanthate stability. Most published stability data is unreliable because it has been obtain in vitro. They must be employed with considerable caution when inferring or analysing results in flotation testing.

Commercial xanthates are normally supplied as pellets, which can be dissolved in water and then used in flotation mills in concentration of 5 to 25%. At normal concentrations, the solution viscosity is on the order of 1.3 cps.

Xanthate handling and safety: in powder form it is a strong irritant, so operators should employ chemical safety clothes, gloves, goggles, and a dust respirator, when handling. Its toxicity is moderate, with an LD_{50} of 500 to 2,000 mg/kg of body weight, dependent on the alkyl group. Xanthate solutions, particularly concentrated, are a severe eye irritant, so eye baths should be provided in the handling areas, as well as the use of the appropriate safety clothes.

Xanthate powder has been known to ignite spontaneously, though this does not seem to have been observed with pellets. Pelletised material must be kept dry and cool to maintain long term stability in storage. Protection from moisture is particularly important, so drums should be stored on their sides, if they are in the open. Because decomposition products are extremely flammable, storage precautions should be those usual for flammable materials. If stored in bulk, bins should be nitrogen purged.

Xanthate solutions may be handled in steel or plastic pipes without problems, but not in contact with copper, copper alloys, zinc or aluminium, etc., so care should be taken that valves do not have copper alloy components. Xanthate solutions must not be agitated with air sparges because oxidation can result in hydrolysis to CS_2, with the consequent fire hazard and ambient toxicity problems. Combustion of solid xanthates is similar to sulphur, so the same fire fighting methods apply.

When cleaning storage tanks that have contained a xanthate solution, toxic gas precautions should be taken as there is always the possibility of decomposition to CS_2, so suitable equipment for dealing with a poisonous explosive atmosphere must be used. The maximum worker exposure limit to CS_2 vapours in air is 20 ppm, average, during an 8 hour work day.

Xanthate analysis: the most generally accepted quality control method of analysis for xanthate pellets is the Dow Chemical Company procedure, described in its "Mining Chemical" trade publication. Summarised, the finely ground powder is dissolved in acetone (0.5 g in 35 ml of acetone), the solution is filtered and neutralised with about 40 ml of 0.1 N standard HCL solution and promptly back titrated with 0.1 N NaOH solution, using 3 to 5 drops of methyl red as the indicator. This method is preferable to titration of an aqueous solution, as, in water, impurities such as thiocarbonates and sulphides will report as xanthate. Numerous instrumental methods are recommended in the literature for pulp, or dilute solution, analysis.

Table 15.- COMMERCIAL STABILITY DATA ON XANTHATE SOLUTIONS

xanthate	Trade name Dow	Trade name Cyan-amid	Solution conc. % real	Specific gravity Hydrometer 60/60 15°C	20°C	25°C	Crystal Point °C	Stability Avg. loss daily % 20°C	30°C	40°C
K ethyl	Z-3	303	10%	1.048	1.046	1.045	-3°	1.1	2.7	4.6
			25%	1.122	1.120	1.118	-9°	1.4	2.1	4.4
			35%	1.174	1.172	1.170	-15°			
Na Ethyl	Z-4	325	10%	1.048	1.046	1.045	-3°	1.1	2.7	4.6
			25%	1.122	1.119	1.117	-11°	0.7	2.0	4.3
			35%	1.174	4.171	1.168	-12°			
Na iso-propyl	Z-11	343	10%	1.041	1.040	1.039	-4°	0.3	0.6	1.8
			25%	1.103	1.101	1.098	1°*	0.2	0.7	1.7
Na Iso-butyl	Z-14	317	10%	1.036	1.034	1.032	-3°	0.6	1.8	3.6
			25%	1.087	1.084	1.081	-9°	0.6	1.6	4.0
			35%	1.118	1.117	1.112	-13			
Na sec-butyl	Z-12	301	10%	1.039	1.038	1.036	-3°	0.2	0.6	1.2
			25%	1.099	1.096	1.094	-8°	0.1	0.5	1.4
			35%	1.136	1.133	1.130	12°*			
K iso-amyl		350	10%	1.037	1.036	1.035	-3°	0.8	2.1	4.2
			25%	1.088	1.086	1.084	9°*	0.7	2.0	4.8
			35%	1.122	1.119	1.116	19°*			

Thiophosphates

O,O-Dialkyl dithiophosphate acids: are prepared by reacting phosphorous pentasulphide (P_4S_{10}) with an alcohol in an inert media. The reaction is dangerous because of the production of H_2S, a gas which is deadlier than hydrogen cyanide at low concentration. The pentasulphide reaction has an irregular incubation period, so start-up is usually helped by leaving a heel from the previous synthesis in the reactor. The reaction by-products include dialkyl sulphide, trialkyl dithiophosphates, sulphur, and other minor substances. The acids are unstable, and must be converted into alkali metals salts for use as collectors.

$$P_4S_{10} + 8R\text{-}OH \longrightarrow 4\left[\begin{array}{ccc} RO & & S \\ & P & \\ RO & & SH \end{array}\right] + 2H_2S$$

The dithiophosphoric acids are clear liquids with a characteristic smell, soluble in organic solvents such as ether, chloroform, benzene, ethyl alcohol, etc., but insoluble in water. They react with alkalis to form their salts. Sodium and ammonium salts are soluble in water and are employed as collectors. Zinc salts are employed as lubricant additives or polymerisation stabilisers, while those of Hg, Pb, Co, and Ni can be used to purify the dithiophosphates, as these are insoluble in water. The acids can be purified by vacuum distillation. Boiling points of some acids are:

	B.P. °C	at mm Hg
O,O-dimethyldithiophosphoric acid	65 - 68	3
O,O-diethyldithiophosphoric acid	80 - 82	2
O,O-dipropyldithiophosphoric acid	8 - 10	2
O,O-diisopropyldithiophosphoric acid	82 - 85	1.5

These acids react, in many cases, in a like manner to xanthic acids, for example they form dithiols when treated with oxidising agents such as iodine:

$$2(RO)_2\overset{S}{\overset{\|}{P}}\text{-SH} + I_2 \longrightarrow (RO)_2\overset{S}{\overset{\|}{P}}\text{-S-S-}\overset{S}{\overset{\|}{P}}(RO)_2 + 2HI$$

bis(O,O-dialkyldithiophosphate)

With chloracetic acid (Houben-Weyl (1973)) they form esters:

$$2(RO)_2\overset{S}{\overset{\|}{P}}\text{-SH} + ClCO\cdot OC_2H_5 \longrightarrow (RO)_2\overset{S}{\overset{\|}{P}}\text{-S-CO}\cdot OC_2H_5 + NaCl$$

S-ethoxycarbonyl
O,O-dialkyldithiophosphate

With chloramine the reaction products are structures similar to the thionocarbamates:

$$2(RO)_2\overset{S}{\overset{\|}{P}}\text{-SH} + ClNR'_2 \longrightarrow (RO)_2\overset{S}{\overset{\|}{P}}\text{-S-NR'}_2 + HCl$$

O,O-Diaryl dithiophosphoric acids: differ from the dithiophosphates in that they are stable acids, generally insoluble in water, with most of the commercially available formulations resulting in collectors with frothing characteristics. The most common types are based on phenols or cresols, partially reacted with phosphorus pentasulphide. The reaction, as in the case of the alkyl acids, usually is carried out using an inert atmosphere and a heel in the reactor from the previous synthesis.

In flotation they behave similarly to the alkyl equivalents, with the exception that as the acids are stable, they need not be converted to the alkali salts prior to use as a collector. They will decompose in an alkaline media with the evolution of H_2S, so may act as mild sulphidising agents during the flotation process. Their collecting properties are very dependent on the cresylic acid purity used in their manufacture, so the technology of the manufacture of reliable collectors, based on dithiophosphates, includes having a reliable source of raw materials. Today's availability of synthetic cresylic acid has ameliorated the problem. As will be noted in the descriptions of the aryl dithiophosphate collectors marketed by Cyanamid, the usual manufacturing process consists in reacting an excess of cresylic acid with P_4S_{10}, so the exact composition of the collectors is unknown. The amount of unreacted cresylic acid fixes the frothing properties of the collector.

Aerofloat 15 consists of the reaction product of cresylic acid plus 15% by weight P_4S_{10}. Theoretically the mixture will contain a 42.9 % solution of dithiophosphoric acid in cresylic acid.

Aerofloat 25 similarly is produced by reacting cresylic acid with 25% by weight P_4S_{10}; it theoretically contains 72.6% collector dissolved in cresylic acid, and therefore has a lower frother strength than AF 15.

Aerofloat 31 consists of AF 25 mixed with 6% of thiocarbanilide.

Aerofloat 33 appears to be similar to AF 31.

Aerofloat 241 is an aqueous solution of AF 25, neutralised with ammonia.

Aerofloat 242 is AF 31 neutralised with ammonia; it is similar to AF 31, but a slightly stronger frother.

Manufacture of dithiophosphoric acids: according to M. Oktaweic (Chem. Abstr. 52, 12786 (1958)), the best yields (about 80%), are obtained when cresylic acid (8 moles) is reacted with P_4S_{10} while heating to 115°C, for about

15 minutes. He does not mention the need for a heel to start the reaction.

Table 16.- DENSITY AND VISCOSITY OF AEROFLOAT COLLECTORS

Aerofloat No	Density g/ml @ 20°C	Viscosity, cps 15°C	30°C
15	1.10	50.9	18.0
25	1.19	129.9	40.3
31	1.19	127.5	62.5
33	1.19	1004.7	209.5
242	1.13	157.0	57.5

The most common method of carrying out the reaction is to suspend the phosphorus pentasulphide in an inert solvent, heat to the right temperature and carefully add the alcohol to the agitated reaction mix. The most common inert medias employed are benzene, toluene or xylene; the choice depends on the reaction temperature.

An example of the preparation, from a patent, is: to 440 g of technical P_4S_{10} (75% grade) suspended in 500 ml of benzene, maintained at a temperature of 40°C, 380 g of absolute ethyl alcohol are added over a 2 hour period. Large quantities of H_2S are given off. The mixture is allowed to decant for an hour, and then heated to 70 - 75°C for 3 hours. The reaction mix is cooled, filtered, and the filtrate vacuum distilled. The yield is 80%. Using the same method, the yield on production of O,O-diisopropyl dithiophosphoric acid is 85%.

The aryl dithiophosphoric salts are used as flotation reagents. The solutions are corrosive and irritating, and must be handled in stainless steel piping and pumps, or cautiously selected plastics. They are toxic and should be handled using normal chemical safety precautions. They are more corrosive than the alkyl equivalents, and should be handled with greater care. They are especially corrosive to plastics, so hoses should be avoided unless previously tested for durability. They are highly flammable and toxic, especially for fish. They will produce strong allergenic reactions on handling, so complete chemical protection should be used by the operators. They are smelly, and can give off H_2S, which is highly poisonous, so should be stored in well ventilated places.

Xanthogen formates

These are the product of the reaction of a sodium alkyl xanthate with ethyl chloroformate:

$$\text{Et-O-}\overset{S}{\overset{\|}{C}}\text{SNa} + \text{Et-O-}\overset{S}{\overset{\|}{C}}\text{Cl} \longrightarrow \text{Et-O-}\overset{S}{\overset{\|}{C}}\text{-S-}\overset{O}{\overset{\|}{C}}\text{-O-Et} + \text{NaCl}$$

Dialkyl xanthogen formates are yellow, odorous, oily liquids, essentially insoluble in water. The diethyl xanthogen formate was used formerly as the collector for cement copper in Leach Precipitation Float processes, but in recent years has only been employed in the El Teniente acid circuit. The other xanthogen formates listed are used in the copper circuit at Lepanto in the Philippines, in Flin Flon in Canada, and Inspiration in the U.S.. The collector formula employed in El Teniente until recently contained 30% gasoline and 10% MIBC, which is said to have improved moly collection slightly and reduced collector costs significantly because the total dosage was equal to that used with pure XF, with no loss in recovery.

Commercial xanthogen formate contains only about 75% XF. The main impurities are active collectors: xanthic anhydride (10 to 20%), diethoxy carbonyl sulphide (less than 10%), while diethyl carbonate, the only inert, is present at the 2 to 5% level. The probable by-product reactions are:

$$\text{RO-}\overset{S}{\overset{\|}{C}}\text{-S-}\overset{O}{\overset{\|}{C}}\text{-OR'} + \text{RO-}\overset{S}{\overset{\|}{C}}\text{-S}^- \longrightarrow \underset{\text{xanthic anhydride}}{\text{RO-}\overset{S}{\overset{\|}{C}}\text{-S-}\overset{S}{\overset{\|}{C}}\text{-OR}} + \text{R'}\overset{O}{\overset{\|}{C}}\text{-S}^-$$

$$\text{R'O-}\overset{O}{\overset{\|}{C}}\text{-S}^- + \text{Cl-}\overset{O}{\overset{\|}{C}}\text{-OR'} \longrightarrow \underset{\text{diethoxy carbonyl sulphide}}{\text{R'O-}\overset{O}{\overset{\|}{C}}\text{-S-}\overset{O}{\overset{\|}{C}}\text{-OR'}}$$

$$\text{Cl-}\overset{O}{\overset{\|}{C}}\text{-OR'} + \text{R'-OH} \longrightarrow \underset{\text{diethyl carbonate}}{\text{R'O-}\overset{O}{\overset{\|}{C}}\text{-OR'}} + \text{HCl}$$

$$\text{Cl-}\overset{S}{\overset{\|}{C}}\text{-OR'} + H_2O \longrightarrow \text{R'-OH} + CO_2$$

$$\text{R'O-}\overset{S}{\overset{\|}{C}}\text{-S}^- + H_2O \rightleftharpoons \text{COS} + \text{R'-OH} + OH^-$$

Xanthogen formates hydrolyse slowly in the presence of water, giving off COS. Normal decomposition rates are less than 1% per year, except in the presence of catalysts, such as Zn, Cu, Fe, or mineral surfaces. In steel containers or in pulps, but not in glass, xanthogen formates will react with thionocarbamates and dithiophosphates. They react readily with alkyl amines to give thionocarbamates. With tertiary amines they react to quaternary ammonium salts,

$$2CH_3HO\overset{S}{\overset{\|}{C}}S\overset{O}{\overset{\|}{C}}OCH_2CH_3 \;+\; 2\underset{\underset{CH_3CH_2}{|}}{N}(CH3)3 \longrightarrow (CH_3)_3\overset{+}{N}\text{-}\overset{-}{S}\text{-}\overset{S}{\overset{\|}{C}}OCH_3 \;+\; (CH_3)_4\overset{+}{N}\text{-}\overset{-}{S}\text{-}\overset{S}{\overset{\|}{C}}_3$$

which could be interesting collectors.

Like the xanthates in solution, xanthogen formates can be handled in steel pipes, and plastics, but not in the presence of copper, or copper alloys, etc.. Plastics must be used with caution as the reagents are good solvents for PVC, but polyethylene and propylene stand up very well. If stored in galvanised drums, these should be vented at least once a year. Steel storage tanks that are agitated will show considerable corrosion where fluid velocities are significant, especially if the xanthogen formate has been diluted with an alcohol, such as MIBC. Unagitated standard steel storage tanks have been used for over 30 years without a failure.

Like all organic compounds, xanthogen formates are flammable; due to decomposition they may contain 1% CS_2, which will reduce their flash point into the red label range of less than 70°F. The pure product has a high flash point. Because the structure is pesticide-like, the reagent is toxic to fish and a human irritant, with occasional subjects showing high sensitivity to skin contact and nausea in its presence. Normal chemical safety equipment is recommended to handle it; untreated leather boots have a limited life in contact with reagent.

Manufacture of dialkyl xanthogen formates: the sole producer in the world of xanthogen formate collectors, between 1929 and 1977, was the Minerec Corporation, in Baltimore, U.S.A.. At this plant there was a dedicated plant for diethyl xanthogen formate, plus a number of small units for the production of the longer chain alcohol xanthogen formates reacted only with ethyl chloroformate.

The diethyl xanthogen formate was manufactured in the Baltimore plant by a batch process. First the sodium ethyl xanthate was made by the process described in the xanthate section. The xanthate, in solution in its excess alcohol, and the water remaining from the original 50% caustic solution, were loaded into the xanthogen formate reactor and cooled to less than 35°C with mechanically refrigerated coils. Agitation was started up and an equimolar quantity of ethyl chloroformate was then added slowly to the xanthate batch, at a rate that kept the temperature around 35°C. After all the ECF was added, it was stirred at the same temperature for at least one hour. The oil produced was then transferred to a wash tank where it was decanted from the brine. The brine was sent to distillation to recover the excess ethyl alcohol, and the oil was washed to remove the remaining brine.

The process is much more sensitive than would appear from the description, as impurities in the ethyl chloroformate can upset the reaction, producing unstable by-products which remain in the oil phase. The final product cannot be neutralised and cleaned up with soda ash or caustic, as it degrades.

In 1977, Tecnomin Ltd. started up a pilot plant in Chile to develop a better process for the production of xanthogen formates. The new process is covered by U.S. Patent 4,605,518, issued in August 1986. The improvement consisted in converting the xanthogen formate reaction into a semi-continuous process. This considerably improved product quality and allowed the by-product concentrations to be controlled, which permitted tailoring the reagent to improve recoveries on ores containing non-sulphide copper and clay fines.

In the patented process, the xanthogen formate reactor is operated with a high volume circulating loop, without agitation. At start-up after finishing a xanthate batch, the reactor is loaded with about 50% of the ECF required to fully react the xanthate batch, and the loop pump started up. Normally, the budgeted ECF amounts to about 95 to 97% of the theoretical. Important also is that the water content of the xanthate solution be such as to fully dissolve all the sodium chloride that forms in the xanthogen formate reaction. The reaction sequence then is based on the fact that the reaction is occurring in two phases: an aqueous xanthate phase, and an oil phase in which the ethyl chloroformate is soluble. The geometry of the piping and the reactor is designed so that flows of the raw material and the two phases can be controlled and dosed to the reaction zone, which is the interior of the main circulating pump. By adjusting the oil and brine phase ratios, one can drive the different by-product reactions and tailor the final composition of the collector. The details of the process are contained in the patent.

Table 17.- PROPERTIES OF ALKYL XANTHOGEN ETHYL FORMATES

Alkyl Group	Mol. Wt	Density g/ml 20°C	Refractive n_D 25°C	Viscosity cps 20°C	Minerec name
Ethyl	194.0	1.165	1.523	2.55	A
Isopropyl	208.2	-	1.511	-	2048
Butyl	222.2	1.123	1.515	3.04	B
Isobutyl	222.2	1.126	1.516	2.89	898
sec-butyl	222.2	1.113	1.511	2.61	201
2-isopentil	363.2	1.026	1.490	2.90	27

After the xanthate batch is fully reacted, the XF reactor is circulated, at temperature, for an hour or so, to complete the reaction. The mix is sent to a decantation/wash tank. The alcohol-laden brine is distilled to recover the excess alcohol, and the oil phase thoroughly washed with water, containing a small amount of soda ash.

Dialkyl thionocarbamates

Dow Chemical's Z-200 (ethyl isopropyl thionocarbamate) was the first selective collector for sulphide ores marketed. Dow's patented thionocarbamate production was promptly followed by a somewhat similar product made by the Minerec Corporation. The Minerec equivalent was produced by reacting a xanthogen formate with an alkyl amine, a reaction which produces a series of contaminants as co-products. In the early seventies, Minerec Corporation built a plant based on a catalytic process. The process was not as low cost as the Dow route, but did produce a higher purity product. This plant was then operated by the Essex Chemical Co., which purchased Minerec Corporation, but is now shut down and the plant dismantled due to ecological problems in Baltimore. More recently Shell, in Antofagasta (Chile), has been producing thionocarbamates by a similar process, apparently based on a different catalyst, as Minerec has a valid patent in that country. There is also production in Yugoslavia, Canada, Japan, Spain, and possibly Mexico and East Germany.

The Dow process is based on reacting a xanthic ester with an alkyl amine:

$$\text{Isopropyl-O-}\overset{\overset{\displaystyle S}{\|}}{C}\text{-S-}CH_3 + \text{Et-}NH_2 \longrightarrow \text{Isopropyl-O-}\overset{\overset{\displaystyle S}{\|}}{C}\text{-NH-Et} + CH_3SH$$

Due to better yields, the Dow process is lower cost than the direct xanthate plus amine reaction, but only if the methyl mercaptan by-product can be disposed of economically. Since Dow closed down its plant (and its process patent expired), the xanthic ester process has not been used because the mercaptan disposal problem has not been solved. The traditional laboratory synthesis is based on refluxing an alkyl isothiocyanate dissolved in a great excess of the corresponding alcohol, a process first reported by Hoffman in 1870, and probably the process employed by Douglass (1927), at Du Pont, in 1928. The reaction is:

$$\text{R-OH} + \text{R'-N=C=S} \longrightarrow \text{RO-}\overset{\overset{\displaystyle S}{\|}}{C}\text{-NHR'}$$

The first practical industrial process was developed by Harris (1954) in 1950. It consisted of reacting an alkyl xanthate with an alkyl halide, followed by reacting the ester formed by the first reaction with an alkyl amine to form the thionocarbamate:

$$RO-\overset{\overset{S}{\|}}{C}-SNa + R''X \longrightarrow RO-\overset{\overset{S}{\|}}{C}-SR' + NaX + R'NH_2 \longrightarrow RO-\overset{\overset{S}{\|}}{C}-NHR' + R''SH$$

A more recent process, in which thionocarbamates are produced catalytically from the direct reaction of a xanthate with an alkyl amine, was developed at Minerec by Crozier et al (1976). The catalyst employed is either nickel salts or palladium salts. The process is simple and has the virtue of producing a very pure thionocarbamate. It is difficult to carry out in vitro, as a slight excess of CS_2 is a key, and usually must be present as an impurity in the xanthate.

The alkyl thionocarbamates are colourless liquids, or solids, insoluble in water, but very soluble in organic solvents, so, when employed as collectors in flotation they can be mixed with frothers in any proportion to aid feeding to the flotation cells. Due to their impurity content, they have the characteristic smell of sulphur compounds. Alkyl thionocarbamates by the ester route normally have a purity of 90 to 86%, while the product of the catalytic process are 98 to 99% pure, with only excess alcohol as the contaminant, so a water wash is sufficient to ensure a good final product.

The Dow industrial process for the production of Z-200 was carried out in a single reactor. First the xanthate reaction was produced using a large excess of alcohol, and caustic: 4 moles isopropyl alcohol, and 1.135 moles of NaOH per mole of CS_2. The reaction temperature was maintained between 35° and 50°C by cooling. The reaction mix was then reacted with 1.01 moles of methyl chloride to esterify the xanthate. This reaction was carried out at a temperature kept between 44° and 55°C, by refluxing the reactor contents. On completion of this reaction, the amine was added, while keeping the temperature below 50°C. The final product was steam distilled to remove the mercaptan and the excess alcohol employed in the xanthate synthesis.

The Minerec catalytic process can be describe by the reaction:

$$RO-\overset{\overset{S}{\|}}{C}-SNa + (CH_3)_2CHNH_2 \xrightarrow{NiSO_4 \cdot 6H_2O} RO-\overset{\overset{S}{\|}}{C}-NHCH(CH_3)_2 + NaHS$$

To produce the isopropyl ethyl thionocarbamate, at Minerec, the xanthate was first prepared using a 20% excess alcohol, without alcohol recovery, and making the alcoholate using 75% caustic in a high intensity mixer-reactor. The xanthate was then dissolved in sufficient water to keep the sulphhydrate that will be produced by the next step in solution, and reacted at 80°C with a 30% excess of ethyl amine. The reaction is slow, and usually needs to be left overnight to obtain decent yields. The catalyst was filtered off (a somewhat messy job as the nickel salts produced are very fine). The collector oil was then decanted and washed. The resultant sulphhydrate brine is a serious disposal problem, so the Minerec, as well as the Dow plants, are no longer in operation. Yields were on the order of 88%, and purity over 98%. The methyl thionocarbamates, normally produced by Minerec, form an oil whose density is very close to that of the brine, so to effect the decantation, MIBC is added to dissolve the organic phase, and the mixture sold, after washing, as collectors.

Thionocarbamates are relatively stable compounds, but under acid conditions they hydrolyse:

$$RO\text{-}\overset{\overset{S}{\|}}{C}\text{-}NHR' + H_2O \longrightarrow R\text{-}OH + R'NH_2 + COS$$

They can be transformed into their S-alkylthiocarbamate equivalents by reacting with alkyl iodides:

$$CH_{32}CHO\text{-}\overset{\overset{S}{\|}}{C}\text{-}NHCH_2CH_3 \xrightarrow{(CH_3)2CH_2I} (CH_3)_2CHS\text{-}\overset{\overset{O}{\|}}{C}\text{-}NHCH_2CH_3$$

a structure that makes for interesting speculation on the mechanism of attachment of thionocarbamates to mineral surfaces.

Thionocarbamates are less corrosive than xanthogen formates, but should still be handled in steel equipment, avoiding copper and copper alloys. Plastics also must be tested prior to use as hoses or as pump seals. Thionocarbamates are flammable, and should be handled with chemical protection equipment because of possible skin rashes, etc.. They are toxic to fish.

Miscellaneous collector structures

Xanthic esters and long chain mercaptans are being recommended as specialty collectors capable of enhancing molybdenite and gold recoveries when these minerals are co-products of copper sulphide.

Xanthic allyl esters have been marketed by American Cyanamid for a number of years as specialty collectors for co-product molybdenum in copper ores (Re 3302). The use of these esters as supplementary collectors has been proven in a number of mines, and at this writing are the only structure that has consistently improved molybdenum recovery in a gamut of copper mines.

Mercaptans were laboratory tested extensively in the early thirties, but were rejected commercially because of odour problems. Pennwalt (U.S.A.), and Phillips Petroleum (U.S.A.), with strong mercaptan by-product positions, have been trying to introduce longer chain mercaptans as molybdenum collectors, with varying success.

For a number of years Cyanamid has been marketing various grades of *mercaptobenzothiazole* (Re 404, 407, 412) as supplementary for the flotation of tarnished and oxidised ores. They were first developed for the flotation of lead carbonate. A similar application is given to thiocarbanilide. These two structures are:

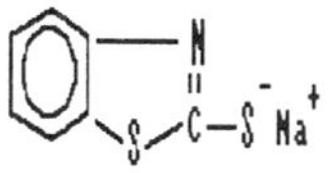

mercaptobenzothiazole

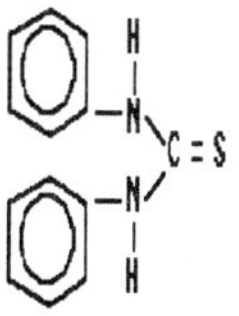

thiocarbanilide

Chapter 5

COMMERCIAL SULPHHYDRYL COLLECTORS

Only recently have collector and frother manufacturers and marketers started to identify the chemical composition of their reagents. A particularly good overview of reagent use as well as clear identification of their trade marked products is contained in the latest Cyanamid (1989) "Mining Chemicals Handbook". The tables in this chapter are based on Crozier (1977), but now include current commercial sources of collectors and frothers.

The application data have been updated using company literature, general references such as the "SME Mineral Processing Handbook" (1985), review articles, and mill visits. The tables should be a good starting point for a data base which can be continuously up-dated from trade literature and technical publications. Products no longer available commercially, such as the Dow xanthates and Z-200, and the Minerec xanthogen formates, have been included because many flotation laboratories have extensive collections of old sample bottles which can be used in screening tests when current reagents fail, and if a particular structure is spectacularly successful, custom manufacture can generally be made available. The Dow Z series nomenclature for xanthates, and the Cyanamid thiophosphate designations are also of interest because much of the old literature on flotation identifies reagents only by their trade names.

The trade designations and compositions of the collectors have been grouped under xanthates (Table 18), xanthic esters (Table 19), xanthogen formates, or Minerecs (Table 20), thionocarbamates (Table 21), dithiophosphates (Table 22), and miscellaneous collectors, such as mercaptobenzothiazole, etc. (Table 23). Tabulations for individual suppliers have been included, as those with extensive product lines do publish and distribute booklets. The old Dow products are easy to identify as the Z series are xanthates, with the exception of Z-200, which is a thionocarbamate. Cyanamid identifies its xanthates by the trade name Aero Xanthate and a number, its dithiophosphates by the designation Aerofloats and a number, and unusual compositions (such as xanthic esters) by the designation Aero promoters. A very brief summary of how and in what quantities to apply collectors to the recovery of particular minerals is listed, by collector type, in the adjunct tables designated 18a, etc.

The selection of a suite of reagents for effective processing of a particular mineral by flotation requires a considerable knowledge of surface properties and solution chemistry, and their effect on the differential rate of the valuable

Table 18.- STRUCTURES, AND SUPPLIERS OF ALKYL XANTHATES

R-O-(C=S)-S M Where M is sodium or potassium

Type	Manufacturer	Trade name
Potassium Ethyl X $CH_3CH_2O(C=S)S^-K^+$	Am. Cyanamid Dow Chemical	Aero X. 303 Z-3
Sodium Ethyl X $CH_3CH_2O(C=S)S^-Na^+$	Am. Cyanamid Dow Chemical UCB (Belgium) Kerley Chem.	Aero X. 325 Z-4
Potassium Isopropyl X $(CH_3)_2CHO(C=S)S^-K^+$	Am. Cyanamid Dow Chemical UCB (Belgium)	Aero X. 322 Z-9
Sodium isopropyl X $(CH_3)_2CHO(C=S)S^-Na^+$	Am. Cyanamid Dow Chemical UCB (Belgium) ZUPA Hoechst CANDINA Kerley Chem. Shellflot Sherritt-Gordon IQM RENASA	Aero X. 343 Z-11 PIPX Flex 415/P KI-80L SF-113
Sodium isobutyl X $(CH_3)_2CHCH_2O(C=S)S^-Na^+$	Am. Cyanamid Dow Chemical Hoechst CANDINA Shellflot Stanchem-CIL IQM ZUPA	Aero X. 317 Z-14 PIBX Flex 215/P SF-114
Sodium sec-butyl X $CH_3CH_2CH(CH_3)O(C=S)S^-Na^+$	Am. Cyanamid Dow Chemical	Aero X. 301 Z-12

X = xanthate

Table 18 cont'd.- STRUCTURES, AND SUPPLIERS OF ALKYL XANTHATES

R-O-(C=S)-S M Where M is sodium or potassium

Type	Manufacturer	Trade name
Potassium sec-amyl X Methylpropyl carbinol $CH_3(CH_2)_2CHOCH_3(C{=}S)S^-K^+$	Dow Chemical	Z-5
Potassium amyl X $CH_3(CH_2)_3CH_2O(C{=}S)S^-K^+$	Am. Cyanamid Dow Chemical Hoechst CANDINA Prospect Chem. Kerley Chem.	Aero X. 350 Z-6 PAX Flex 315/P KI-80L
Sodium isoamyl X $(CH_3)_2(CH_2)_2CH_2O(C{=}S)S^-Na^+$	Am. Cyanamid Stanchem-CIL Sherritt-Gordon Prospect Chem. Industrias Quimicas Mexicanas (IQM)	Aero X. 355
Potassium hexyl X $CH_3(CH_2)5O(C{=}S)S^-K^+$	Dow Chemical	Z-10

and the undesirable minerals in a particular ore. But the choice of specific equipment to employ - say selecting between the use of giant cells, flotation columns, and different schemes of "flash" flotation - is still an art, and the particular sequence of the steps used in a separation process directly affects how collectors, frothers, modifiers, and depressants interact, so it is impossible to give categoric recommendations on how or where to apply a given reagent. Collecting and collating case histories for flotation mills is a monumental task that has not been repeated since Professor Arthur Taggart put out the last edition of his " Handbook of Mineral Dressing" nearly fifty

Table 19.- STRUCTURES, AND SUPPLIERS OF XANTHIC ESTERS

R-O-(C=S)-S-R'

Type	Manufacturer	Trade name
Allyl Amyl	Am. Cyanamid Minerec Corp	Aero P. 3302 M-1750
Allyl hexyl	Am. Cyanamid Minerec Corp	Aero P. 3461 M-2023

years ago. His recommendations on how unusual ores have been dealt with is still probably the only reliable source for suggestions on the initial reagents to be used in an experimental programme to provide process design data. But it cannot cope with all the permutations of combined minerals that have contradictory flotation responses. Despite this, it certainly is the most prized possession of the mill manager of a boutique concentrator purchasing ores from small miners in a polymetallic district.

Table 20.- STRUCTURES, AND SUPPLIERS OF DIALKYL XANTHOGEN FORMATES

General structure	R-O-(C=S)-S-(C=S)-O-R'	
Type	Manufacturer	Trade name
Diethyl XF $CH_3CH_2O(C=S)S(C=O)OCH_2CH_3$	Minerec Corp Tecnomin Ltda Shellflot Prospect Chem. Dow Chemical	Minerec A T-2001 SF-203 DFC-404
Ethyl isopropyl XF $(CH_3)_2CHO(C=S)S(C=O)OCH_2CH_3$	Minerec Corp Dow Chemical	M-2048 DFC-411
Ethyl butyl XF $CH_3(CH_2)_3O(C=S)S(C=O)OCH_2CH_3$	Minerec Corp	Minerec B
Ethyl isobutyl XF $(CH_3)_2CH_2CH_2O(C=S)S(C=O)OCH_2CH_3$	Minerec Corp	M-898, 1995
Ethyl secbutyl XF $CH_3CH_2CHOCH_3(C=S)S(C=O)OCH_2CH_3$	Minerec Corp	M-201
Ethyl amyl XF $CH_3(CH_2)_3CH_2O(C=S)S(C=O)OCH_2CH_3$	Minerec Corp	M-1040
Ethyl methylamyl XF or Ethyl 2-heptanol XF $CH_3(CH_2)4CH_2OCH_3(C=S)S(C=O)OCH_2$	Minerec Corp	M-27

Table 21.- STRUCTURES, AND SUPPLIERS OF DIALKYL THIONOCARBAMATES

R-O-(C=S)-NH-R'

Type	Manufacturer	Trade name
Ethyl isopropyl TC $(CH_3)_2CHO(C=S)NHCH_2CH_3$	Dow Chemical Minerec Corp. Shellflot ZUPA Kerley Chem. Hoechst Am. Cyanamid	Z-200 M-1661, 2030 SF-323 KI-200 X-23 AP-3894
Methyl isopropyl TC $CH_3)_2CHO(C=S)NHCH_3$	Minerec Corp. Dow Chemical	M-1703 ZM-200
Methyl butyl TC $CH_3(CH_2)O(C=S)NHCH_3$	Minerec Corp. Am. Cyanamid	M-1331 Aero P. 6098
Methyl isobutyl TC $(CH_3)_2CHCH_2O(C=S)NHCH_3$	Minerec Corp.	M-1846
Ethyl isobutyl TC $(CH_3)_2CHCH_2O(C=S)NHCH_2CH_3$	Minerec Corp.	M-1669

Table 22.- STRUCTURES, AND SUPPLIERS OF THIOPHOSPHATE COLLECTORS

General structure - DIALKYL DITHIOPHOSPHATES

R-O S
P
R-O $S^- Na^+$

Type	Manufacturer	Trade name
Dialkyl dithiophosphates Na diethyl DTP $(CH_3CH_2O)_2(P=S)S^-Na^+$	 Am. Cyanamid Hoechst CANDINA Phillips 66	 Sod. Aerofloat HOSTAFLOT LET SPELD 3456 CO 500
Na diisopropyl DTP $[(CH_3)_2CHO]_2(P=S)S^-Na^+$	Am. Cyanamid Hoechst CANDINA Phillips 66	AF-211 AF-243 HOSTAFLOT LIP SPELD 3457 CO 510

Table 22 cont'd.- STRUCTURES, AND SUPPLIERS OF THIOPHOSPHATE COLLECTORS

Type	Manufacturer	Trade name
Na diisobutyl DTP	Am. Cyanamid	Aero P - 3477
$[(CH_3)_2CHCH_2O(P=S)S^-Na^+$		Aero P - 5430
	Allied Colloids	DPI-4560
	Hoechst	HOSTAFLOT LIB
	CANDINA	SPELD 4659
	Minerec	M-2044
	Phillips 66	CO 540
Na Di-sec-butyl DTP	Am. Cyanamid	AF-238
$[CH_3(CH_2)_2CHOCH_3]_2(P=S)S^-Na^+$	Allied Colliods	Procol CA825
	Hoechst	HOSTAFLOT LBS
	CANDINA	SPELD 3458
	Minerec	M-2176N
	Phillips 66	CO 530
	Lubrizol	Flotezol 100
Na Diisoamyl DTP	Am. Cyanamid	Aero P - 3501
$[CH_3(CH_2)_3CH_2O]_2(P=S)S^-Na^+$		Aero P - 5474
Na di-methylamyl DTP	Am. Cyanamid	AF-249
$[(CH_3)_2(CH_2)_3CH_2O]_2(P=S)S^-Na^+$		
DIARYL DITHIOPHOSPHATES These collectors are manufactured by reacting cresylic acid with P2S5. The Cyanamid number is the percentage of theoretical P2S5 used in the reaction.	CH_3– CH_3– P S S^-Na^+	
15% P2S5	Am. Cyanamid	AF-15
	CANDINA	SPELD 1335
	Hoechst	PHOSPHOKRESOL C
25% P2S5	Am. Cyanamid	AF-25
	Allied Colloids	Procol CA825
	CANDINA	SPELD 1334
	Hoechst	PHOSPHOKRESOL B
	Minerec	M-2074
Mixtures of Aryl DTP		
AF-25 plus 6% Thiocarbanilide	Am. Cyanamid	AF-31, AF-33
	Allied Colloids	Procol CA832
	CANDINA	SPELD 1333
	Hoechst	PHOSPHOKRESOL A
AF-25 neutralized with NH4OH	Am. Cyanamid	AF-241
	Allied Colloids	Procol CA833
	CANDINA	SPELD 2666
	Hoechst	PHOSPHOKRESOL E
AF-31 neutralized with NH4OH	Am. Cyanamid	AF-242

Table 22 cont'd.- STRUCTURES, AND SUPPLIERS OF THIOPHOSPHATE COLLECTORS

Type	Manufacturer	Trade name
Mixture of AF-238 plus sodium Aerofloat	Am. Cyanamid CANDINA Hoechst	AF-208 SPELD 5658 HOSTAFLOT LBS
MONOTHIOPHOSPHATES 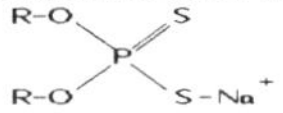 (R-O)(R-O)P(=S)S-Na+	Am. Cyanamid	Aero P. 3394
DITHIOPHOSPHINATES 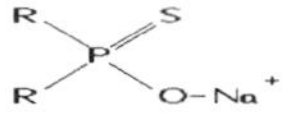 (R)(R)P(=S)O-Na+	Am. Cyanamid	Aerophine 3418A

Table 23.- STRUCTURES, AND SUPPLIERS OF MISCELLANEOUS COLLECTORS

Type	Manufacturer	Trade name
Mercaptans n dodecyl $C_{12}H_{25}SH$ t dodecyl or lauryl	Phillips 66	CO 100 CO 200
Mercaptobenzothiazole 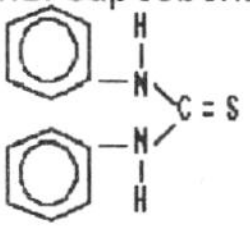	Am. Cyanamid Hoechst	Aero P. 400 Flotagen Captan
Mixtures of mercaptobenzithiazole with dithiophosphates	Am. Cyanamid	Aero P. 404 Aero P. 407 Aero P. 412 Aero P. 425
Thionocarbamate dithiophosphate mixtures	Am. Cyanamid	Aero P. 4037
Thiocarbanilide 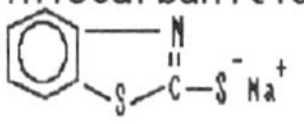	Am. Cyanamid	Aero T. 130

A thumb nail summary of the properties of the more common collectors is that the xanthates are work horse, relatively non-selective collectors. The shorter the alkyl group the more selectivity, which is frequently referred to as a "weaker" collector, while the longer chained xanthates, such as amyl or hexyl, are less selective, or stronger collectors.

The phosphorous pentasulphide based collectors, in general, have characteristics that parallel the xanthates, except that, for the same alkyl group, they tend to be more selective (weaker), and are easier to tailor by mixing with complementary collectors to enhance recovery of secondary values in an ore, such as gold, silver and molybdenum. The cresylic acid derivatives have frothing properties, which can be desirable or undesirable depending on the ore type, and/or the size distribution of the pulp particles, and the process water quality.

The thionocarbamates were first studied, in the late twenties, on a laboratory scale by Dr. William Douglass, of du Pont. The Dow Chemical Company first recognised the industrial value of ethyl isopropyl thionocarbamate (Z-200) in the early sixties. They popularised its use as a general purpose, highly selective collector for low grade porphyry copper minerals, and as a selective adjunct in cleaning more complex ores. They did not extend the alkyl groups available commercially. In the late sixties Dr. Fischer, at Minerec, recognised that methyl butyl and isobutyl thionocarbamates could replace Z-200 when the mill requirements combined good recovery and selectivity.

Xanthogen formates were first developed by Minerec, in the late twenties, to replace xanthates in acid circuits. They are unique in the copper-arsenopyrite and copper-zinc circuits, where moderately low pH flotation is mandated. They are also used when slimes and clays are a problem.

Mercaptans were mildly popular as collectors in the early thirties, when mill operators equated odour with collecting strength. They faded from use because of their smell, and, possibly more important, because the short chain mercaptans have frother properties. Recently the very long chain ones (lauryl mercaptan) have been touted as a replacement for fuel oil in the primary and by-product recovery of molybdenum. Trithiocarbonates (i.e., xanthates based on mercaptans) were also popular sixty years ago. Some are still used in South Africa.

A caution that cannot be repeated often enough when designing a test programme for an operating mill is looking out for reagent interactions. A common cause of anomalous results in a side by side rougher bank test is recycled scavenger concentrate, which will contain small quantities of the rest of the mill's reagents.

Table 18a.- STRUCTURES, AND MINERAL USAGE OF ALKYL XANTHATES OR ALKYL DITHIOCARBONATES

R-O-(C=S)-S M Where M is sodium or potassium

Type	Mineral	Dose g/t	Complementing	Dose g/t	Flotation Conditions
Potassium Ethyl X $CH_3CH_2O(C{=}S)S^-K^+$	a) Sphalerite	8 - 40	none	-	Natural pH 8 - 9 with 150 - 330 g/t $CuSO_4$
	b) Chalcopyrite	30	none	-	At pH 10.3 adjusted with lime
	c) Cinnabar	45	Creosote	60	Natural pH 8 with 300 g/t $CuSO_4$ activation
	d) other metal sulphides	8 - 80			
	e) Malachite	66	K amyl xanth.	66	pH 9.5 with lime at Kamoto, Zaire
Sodium Ethyl X $CH_3CH_2O(C{=}S)S^-Na^+$	a) Sphalerite	45 - 60	none	-	Natural pH 8 - 9.5 or adjusted with lime adding 330 - 500 g/t $CuSO_4$ for activation
			Coal tar	5	
			Fuel oil	15	
	b) Galena	33 - 450	none	-	Natural pH 7.5
			Coal tar	1	
		220	Butyl xanthate	80	For finely disseminated galena in ore with high carbon, natural pH 8.
	c) Bulk sulphides	125	Creosote	70	Sulphuric acid to pH 5
	d) other metal sulphides	8 - 80			
	e) Bornite	17	K amyl xanth.	20	pH 10.7 with Lime at Mufulira, Zambia
Potassium Isopropyl X $(CH_3)_2CHO(C{=}S)S^-K^+$	a) Chalcopyrite	65	Barrett oil	40	At pH 10.2 with lime
	b) Chalcopyrite, bornite	12	Z-200	12	At pH 10.2 with lime
	c) Free gold, auriferous sulphides	60	Aerofloat 25	1.3	At natural pH 7.8

Table 18a cont'd.- STRUCTURES, AND MINERAL USAGE OF ALKYL XANTHATES OR ALKYL DITHIOCARBONATES

Type	Mineral	Dose g/t	Complementing	Dose g/t	Flotation Conditions
Sodium isopropyl X $(CH_3)_2CHO(C{=}S)S^-Na^+$	a) Sphalerite	30 - 100	none	-	At pH 10.0 - 10.8 adjusted with lime and 250 - 370 g/t $CuSO_4$ activation
	b) Marmatite,	55	none	-	At pH 10.8 adjusted with lime and 205 g/t $CuSO_4$ activation
	c) Marmatite, Sphalerite	50	none	-	At pH 11.0 adjusted with lime and 110 g/t $CuSO_4$ activation
	d) Chalcopyrite	45	none	-	At pH 8.5 adjusted with lime
	e) Chalcocite	16	none	-	At pH 11.4 adjusted with lime
	f) Chalcocite, native Cu	100	Aerofloat 249 Fuel oil	12 165	At pH 9.5
	g) Galena	8 - 60	none	-	At pH 7.0 - 10.5, either natural or adjusted with lime
	h) Stibnite	16	none	-	At pH 7.7 with lime, using 2.7 kg/t lead acetate activation
	i) Pyrite with cobalt values	4 - 40 60	none Aerofloat 25	- 8	At pH 5 - 8.5 with sulphuric acid At pH 9 with sulphuric acid and 80 g/t $CuSO_4$ activation
	j) Pyrrhotite	40	none	-	At pH 10.0 with lime
Sodium isobutyl X $(CH_3)_2CHCH_2O(C{=}S)S^-Na^+$	a) Pyrite	10 - 20	none		At natural pH, powerful collector
	b) Chalcopyrite	12 - 15	Aerofloat 238	12	At ph 9 with lime

Table 18a cont'd.- STRUCTURES, AND MINERAL USAGE OF ALKYL XANTHATES OR ALKYL DITHIOCARBONATES

Type	Manufacturer	Mineral	Dose g/t	Complementing	Dose g/t	Flotation Conditions
Sodium sec-butyl	Am. Cyanamid	a) Chalcopyrite	60	none	-	At pH 10.0 with lime
$CH_3CH_2CH(CH_3)O(C{=}S)S^-Na^+$	Dow Chemical	b) Chalcopyrite, chalcocite	70	none	-	At pH 12.0 with lime and 300 g/t NaSH for activation
		c) Chalcopyrite, galena with Au and Ag values	16 - 50	Thiocarbanilide	16	At pH 11.5 with lime
				Aero Promoter 404	2	At natural pH 6.0 - 8.0
		d) Galena	20	none	-	At natural pH 8.0
		e) Sphalerite	8 - 16	none	-	At pH 11.0 - 11.5 with lime
				Burner oil	4	and 80 - 250 g/t $CuSO_4$ activation
		f) Cinnabar	125	none	-	At pH 8.0 with soda ash and 200 g/t $CuSO_4$ for activation
		g) Free Au, auriferous pyrite	100	none	-	At pH 8.6 with lime, plus 4 g/t fatty acid as frother aid
Potassium sec-amyl X	Dow Chemical	a) Sphalerite	32	none	-	At pH 10.4 with lime, and 185 g/t $CuSO_4$
Methylpropyl carbinol		b) Chalcopyrite	40	Creosote	16	At pH 9.6 adjusted with lime
$CH_3(CH_2)_2CHOCH_3(C{=}S)S^-K^+$		c) Stibnite	125	none	-	At natural pH 6.5, using 1.2 kg/t lead acetate activation
		d) Anglesite	330	none	-	At pH 10.2 using 3.3 kg/t NaSH for sulphidisation
		e) Cobaltite	95	none	-	At pH 8.8 with soda ash, adding 500 g/t $CuSO_4$ and 450/g/t NaSH to activate
		f) Pyrrhotite	82	Aerofloat 25	55	At natural pH 8.4, adding 20 g/t $CuSO_4$

Table 18 cont'd.- STRUCTURES, AND MINERAL USAGE OF ALKYL XANTHATES OR ALKYL DITHIOCARBONATES

R-O-(C=S)-S M Where M is sodium or potassium

Type	Mineral	Dose g/t	Complementing	Dose g/t	Flotation Conditions
Potassium amyl X $CH_3(CH_2)_3CH_2O(C=S)S^-K^+$	a) Chalcopyrite	8 - 40	none	-	At pH 10.0 - 10.5 with lime
			Z-200	8	At pH 10.0 - 10.5 with lime
		8 - 30	Aero Prom. 404	1 - 12	At pH 11.5 - 11.8 with lime
		21	Z-11 or Z-9	6	At pH 9.5 with soda ash
	b) Chalcopyrite, galena	45	none	-	At pH 8.4 with lime
	c) Sphalerite	220	none	-	At pH 12.1 with lime, and 825 g/t $CuSO_4$ pulp aerated 15 min to enhance selectivity
	d) Bulk bismuthite, native bismuth, molybdenite	20	Aerofloat 208 Aeroflotat 25	9 3	At natural pH 6.9
Sodium isoamyl X $(CH_3)_2(CH_2)_2CH_2O(C=S)S^-Na^+$	e) Stibnite	165	none	-	At pH 8.0 with soda ash, adding 740 g/t lead acetate for activation
	f) free gold	80	none	-	At pH 8.6 with soda ash
	g) Cerussite	80	none	-	At pH 10.0 with soda ash, adding 1 kg/t NaSH for sulphidisation
	h) Malachite	50	Palm oil	16	At pH 9.0 using 900 g/t NaSH to sulphidise
	i) Cuprite	1,650	none	-	At pH 9.0 with lime, adding 500 g/t NaSH to sulphidise
Potassium hexyl X $CH_3(CH_2)5O(C=S)S^-K^+$	a) Chalcopyrite	8.3	Aerofloat 238	8	At pH 11 with lime

Table 19a.- STRUCTURES, AND MINERAL USAGE OF XANTHIC ESTERS

R-O-(C=S)-S-R'

Type	Mineral	Dose g/t	Complementing	Dose g/t	Flotation Conditions
Allyl Amyl	a) Chalcopyrite, chalcocite, molybdenite	7.5	Z-11, AX-343	4.5	At pH 10.5 with lime
	b) Cu, zinc, moly sulphides; sulphidised Cu carbonate	2 - 20	Fuel oil	80	
Allyl hexyl	a) Similar to AP-3302	2 - 20			

Table 20a.- STRUCTURES, AND MINERAL USAGE OF DIALKYL XANTHOGEN FORMATES

R-O-(C=S)-S-(C=S)-O-R'

Type	Mineral	Dose g/t	Complementing	Dose g/t	Flotation Conditions
Diethyl XF $CH_3CH_2O(C=S)S(C=O)OCH_2CH_3$	a) Chalcopyrite	8 - 16	Z-11, AX-343	4	At pH 10.5 with lime
		80	none	-	At pH 5 with sulphuric acid
		16	Amyl xanth.	20	At pH 7.5 with sulphuric acid in cleaning magnetite concentrate
	b) Cement copper	40	none	-	At pH 4.5 with lime in an LPF circuit, also in sea water pulp.
	c) Galena	40	none	-	At pH 9.5 - 10.0 with soda ash
Ethyl isopropyl XF $(CH_3)_2CHO(C=S)S(C=O)OCH_2CH_3$					
Ethyl butyl XF $CH_3(CH_2)_3O(C=S)S(C=O)OCH_2CH_3$	a) Cu and Pb sulphides	8 - 80			Alkaline circuits
Ethyl isobutyl XF $(CH_3)_2CH_2CH_2O(C=S)S(C=O)OCH_2CH_3$	a) Chalcocite	12 - 16	none	-	At pH 10.0 - 10.5 with lime
	b) Chalcopyrite	100	none	-	At pH 10 with lime
Ethyl secbutyl XF $CH_3CH_2CHOCH_3(C=S)S(C=O)OCH_2CH_3$	a) Sphalerite	8 - 30	none	-	At pH 9 - 9.5 (lime) adding 400 - 800 g/t $CuSO_4$ for activation
Ethyl amyl XF $CH_3(CH_2)_3CH_2O(C=S)S(C=O)OCH_2CH_3$					
Ethyl methylamyl XF or Ethyl 2-heptanol XF $CH_3(CH_2)4CH_2OCH_3(C=S)S(C=O)OCH_2$	a) Chalcopyrite	20	none	-	At pH 9 with lime
	b) Cu sulphides in Cu-Zn ores	8 - 80			

Table 21a.- STRUCTURES, AND MINERAL USAGE OF DIALKYL THIONOCARBAMATES

R-O-(C=S)-NH-R'

Type	Mineral	Dose g/t	Complementing	Dose g/t	Flotation Conditions
Ethyl isopropyl TC $(CH_3)_2CHO(C=S)NHCH_2CH_3$	a) Chalcopyrite	14	Aero P 404	1	At pH 10.5 with lime
	b) Chalcocite	5	Isoprop. Xanth.	9	At pH 11 with lime
	c) Sphalerite	33	none	-	At pH 9.3 (lime) plus 370 g/t $CuSO_4$ for activation
Methyl isopropyl TC $CH_3)_2CHO(C=S)NHCH_3$					
Methyl butyl TC $CH_3(CH_2)O(C=S)NHCH_3$	a) Chalcopyrite	31	Ethyl xanth.	9	At pH 8.5 with lime
	b) Chalcocite	8 - 20	none	-	At pH above 12
	c) Cement Cu	295	none	-	At natural pH 5 of dump leach iron launders
	d) Sphalerite	12	Ethyl xanth.	12	At natural pH 8 with 300 g/t $CuSO_4$
Methyl isobutyl TC $(CH_3)_2CHCH_2O(C=S)NHCH_3$					
Ethyl isobutyl TC $(CH_3)_2CHCH_2O(C=S)NHCH_2CH_3$	a) Chalocite	8			

Table 22a.- STRUCTURES, AND MINERAL USAGE OF THIOPHOSPHATE COLLECTORS

General structure - DIALKYL DITHIOPHOSPHATES

R-O, R-O, P, S, $S^- Na^+$

Type	Mineral	Dose g/t	Complementing	Dose g/t	Flotation Conditions
Dialkyl dithiophosphates Na diethyl DTP $(CH_3CH_2O)_2(P{=}S)S^-Na^+$	a) Chalcocite	10	Z-200	3	At pH 11.8 with lime
	b) Miscell. Cu Sulphides	18	Z-14	4	At pH 11 with lime
	c) Sphalerite	12 - 30	None	-	At pH 9 - 9.5 (lime) adding 160 - 205 g/t $CuSO_4$ to activate
		40	Z-200	20	At pH 9.3 (lime) adding 270 g/t of $CuSO_4$ to activate
Na diisopropyl DTP $[(CH_3)_2CHO]_2(P{=}S)S^-Na^+$	a) Sphalerite	12 - 50	None	-	At natural pH 8 - 8.5 adding 410-600 g/t $CuSO_4$ to activate. At pH 10.9 (lime) and 250 g/t $CuSO_4$ to activate
		33	Z-6	12	At pH 11.6 (lime) plus 205 g/t $CuSO_4$ to activate the zinc
Na diisobutyl DTP $[(CH_3)_2CHCH_2O(P{=}S)S^-Na^+$	a) Cu, Zn, Ag ores	8 - 40			

Table 22a cont'd.- STRUCTURES, AND MINERAL USAGE OF THIOPHOSPHATE COLLECTORS

Type	Mineral	Dose g/t	Complementing	Dose g/t	Flotation Conditions
Na Di-sec-butyl DTP $[CH_3(CH_2)_2CHOCH_3]_2(P{=}S)S^-Na^+$	a) Chalcopyrite, chalcocite	20	Z-11	4	At pH 10.5 with lime
	b) Cu, Zn and Pb sulphides	8 - 40			Alkaline circuits
Na Diisoamyl DTP $[CH_3(CH_2)_3CH_2O]_2(P{=}S)S^-Na^+$	a) Cu, Zn and Pb sulphides	8 - 40			
Na di-methylamyl DTP $[(CH_3)_2(CH_2)_3CH_2O]_2(P{=}S)S^-Na^+$	a) Cu sulphides	8 - 40			
DIARYL DITHIOPHOSPHATES These collectors are manufactured by reacting cresylic acid with P2S5. The Cyanamid number is the percentage of theoretical P2S5 used in the reaction.					
15% P2S5	a) Cu, Zn, Pb and Ag sulphides	2 - 80			
25% P2S5	a) Free Au, auriferous pyrite	80	Z-4	80	At natural pH 7.2
	b) Galena	66	None	-	At natural pH 8.5
	c) Argantite	40	Z-6	40	At natural pH 8.8
	d) Descloizite	80	Z-6	40	At ph 11 (lime) plus 1.6 kg/t Pb nitrate to activate
	e) Metallic Fe	100	Z-6	40	At pH 4 (HCl)

Table 22a cont'd.- STRUCTURES, AND MINERAL USAGE OF THIOPHOSPHATE COLLECTORS

Type	Mineral	Dose g/t	Complementing	Dose g/t	Flotation Conditions
Mixtures of Aryl DTP	a) Galena	37 - 50	None	-	At natural pH 7.5 - 8.5
AF-25 plus 6% Thiocarbanilide	b) Cu, Pb and Ag sulphides; oxidized gold ores	20 - 80			
AF-25 neutralized with NH4OH					
AF-31 neutralized with NH4OH	a) Chalcopyrite	20 - 29	Z-6	4 - 33	At pH 9 - 9.5 with soda ash
	b) " , Galena	50	Z-6	10	At pH 9.8 with soda ash
	c) Galena	16	None	-	At natural pH 8
		140	None	-	At pH 9.5 with NaOH
	c) Argentite	4	None	-	At natural pH 8
Mixture of AF-238 plus sodium Aerofloat	a) Chalcopyrite, Au and Ag values	29	Z-6	4	At pH 9.5 - 10.5 with lime
	b) Sphalerite	23	None	-	At natural pH 8.5 adding 330 g/t $CuSO_4$
	c) Native Au, Cu and Ag	4 - 40			

MONOTHIOPHOSPHATES

R–O, R–O, P, S, $S-Na^+$

DITHIOPHOSPHINATES

R, R, P, S, $O-Na^+$

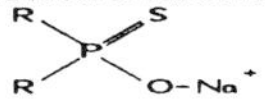

Table 23a.- STRUCTURES, AND MINERAL USAGE OF MISCELLANEOUS COLLECTORS

Type	Mineral	Dose g/t	Complementing	Dose g/t	Flotation Conditions
Mercaptobenzothiazole 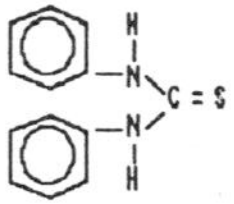					
Mixtures of mercaptobenzithiazole with dithiophosphates	a) Chalcopyrite	57	Z-11	6	At pH 9.8 with lime
Thionocarbamate dithiophosphate mixtures					Substitute for Z-200
Thiocarbanilide 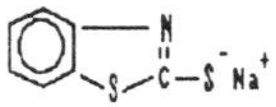	a) Ag sulphides	33	AF-238	12	At natural pH 8
Thiophosphoryl Chloride 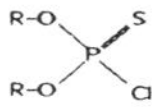	a) Bulk Sulphides				Used in acid circuits

Chapter 6

PROPERTIES OF FLOTATION FROTHS

Flotation frothers and froth hydrodynamics

As we have seen, the choice of a frother, or a frother combination, for a particular flotation application depends on the frothing mechanism involved. For sulphide flotation it is not unusual that two or more frothers will be blended, as one must complement the collector used to form interaction complexes, and at the same time, flotation efficiency depends on a mechanically satisfactory froth. In a majority of the cases a frother containing at least a portion that is an alcohol, such as MIBC, is the best choice to match the collector, and a higher molecular weight frother, such as the pine oils or the polypropylene glycols, can be added to modify the physical properties of the froth and/or the bubble size when coarser particles are to be floated.

Dynamic froth stability

Froths break down through rupture of bubbles due to film thinning, as shown by Ewers and Sutherland (1952), who state that the key factor is movement of inter-bubble liquid away from the point of potential rupture: "The surface moves from a region of low surface tension (high surface pressure) to a region of high surface tension. When the surface tension is highest at the centre of the disturbance, the film will be stable; when the surface tension is lowest at this point the surface film and hence the substrate will move away from this point, and the film will rupture". Thus, low molecular weight frothers, such as alcohols, which tend to diffuse rapidly, will even out surface tension differences along the bubble film, and produce less stable froths. This concept of film flow hindrance explains the better stability of three-phase froths with inter-bubble mineral loads. Here the mineral particles act as pegs locking the bubbles in position, as do slimes, colloids, and other compounds affecting inter-bubble liquid viscosity.

The stabilising effect of the mineral is the reason that flotation in a

Hallimond tube, in the absence of a frother, is possible. Because of this, great care should be taken in extrapolating results based on Hallimond tube data, because conclusions of the relative merits of a collector with a particular ore are questionable in the absence of a frother.

Surface tension differences between frothers is probably of trivial importance in selection of frothers, as in commercial flotation the normal frother concentrations are such that the lowering of surface tension in a mill pulp is hardly measurable. In fact, a significant lowering of the pulp surface tension is probably deleterious, as many have found when a detergent got into their flotation circuit by accident. The observed lack of load carrying capacity of a froth, or foam, generated by a detergent, may well be due to non-interaction of the collector with the detergent on the mineral surface; i.e. the detergent coats the bubble surface to form a stable foam, but does not link to the collector on the mineral surface.

The effect of the collector on froth properties

Frother Power: Wrobel (1953) defined this term as the volume of froth generated in a standard machine under standard operating conditions. Fig. 16 shows froth volume data for systems without a mineral pulp, reported by Leja and Schulman (1954) for straight, branched chain and cyclic alcohols having a varying number of carbon atoms, used alone or in conjunction with straight chain xanthates. In all cases, reagent concentration was one thousandth molar (M/1000).

As can be seen from Fig. 16, some general conclusions can be reached on the effect of carbon atoms in the alkyl group, or chain length, on the froth properties generated by alcohols. Froth volume, which Wrobel equated to frother power, increases with the number of carbon atoms up to 6 or 7, and then drops drastically when the alcohol has more than 8 carbon atoms. A xanthate, with less than 6 carbon atoms in the alkyl group, alone does not produce frothing, while lauryl xanthate will produce a froth on its own, but will kill the froth if mixed with an alcohol, possibly explaining why long chain xanthates are depressants. With strong frothers, such as cresylic acid or pine oil, lauryl xanthate produces a non-mineralisable foam.

The curve for froth volume versus number of carbons in the alcohol shows a steady increase in froth volume, up to 7 carbon atoms. Mixing the alcohol with longer chained xanthates, such as potassium amyl xanthate (KAX), increases froth volume for the short chained alcohols and strongly increases frother power, when combined with alcoholic frothers with up to 7 or 8 carbon atoms, while a shorter chain xanthate such as potassium ethyl xanthate (KEX) decreases frother power for alcohols with 5 or less carbon atoms, and only moderately increases the froth volume for the longer chain alcohols. Lekki and Laskowski (1975) repeated some of the above frother power measurements, in the presence of minerals, and, in general, confirmed Leja and Schulman's pure frother/collector data.

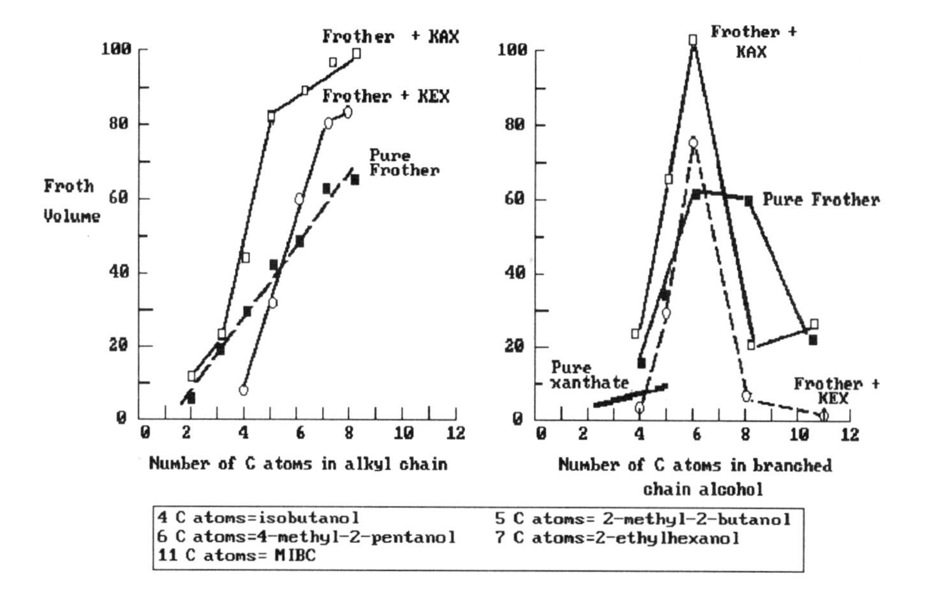

Fig. 16.- Frother volume versus alcohol chain length

These data support the observation in operating mills that control over the froth hydrodynamics is easier through changes in the collector than by changing dosage or type of frother employed. It also explains the fact that longer chain xanthates are "stronger" collectors, because their effect on froth texture is to make it less selective due to retention of particles, which automatically results in increased overall recovery coupled with reduced concentrate grade. At least in sulphide flotation, the correlation between contact angle and recovery could well be fortuitous, as under normal operating dosages the contact angle for the different xanthates, if present, does not vary a measurable amount.

As noted previously, in the presence of a mineral, froth stability is enhanced and the froth character changes. For example, amyl xanthate combined with alpha-terpineol (the main constituent of pine oil) is a strong frothing combination, but in the presence of a sulphide mineral these reagents will tend to produce little or no froth and only a thick dry film of mineral will float. Salts (electrolytes) in solution in the pulp will generally significantly reduce froth stability during sulphide flotation, even if present only in very low concentrations. On the other hand, in coal flotation, sea water can be a good self frothing media. These also are phenomena that can be explained by frother-collector interaction on the mineral.

The influence of the frother on the rate of flotation

From two different papers by Lekki and Laskowski (1971, 1974) it is possible to compare the effect of varying collector and frother dosages on flotation rate, when the flotation is carried out in a Hallimond tube (Fig. 17 and 18) or in a laboratory flotation machine (Fig. 19). Note that to obtain flotation rate curves with a Hallimond tube, the xanthate dose must be an order of magnitude greater than in a froth flotation cell, and that at the high collector concentration required in the Hallimond tube the phenomenon of "over-oiling" does not appear, although the interaction between the frother and collector on the recovery at constant flotation time is obvious. But the same experiment carried out in a laboratory machine produces iso-recovery curves that very clearly reflect the practical experience in mills: i.e., that over feeding either the collector or the frother results in a rise and fall in the overall metal recovery. Because of this hill-like recovery response in flotation, mill men intuitively look for a reagent system where the peak of the hill is flat, to avoid

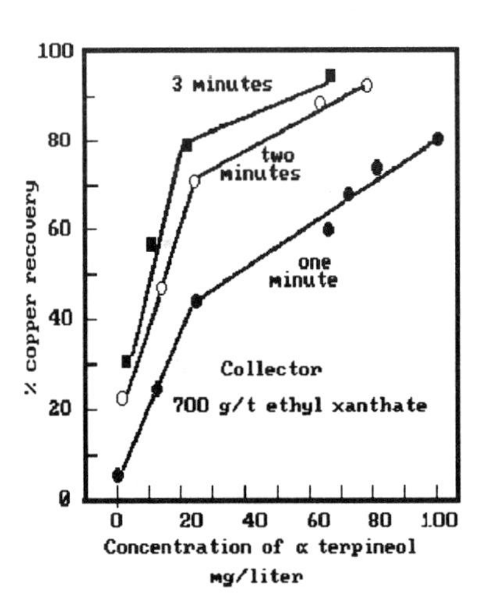

Fig. 17.- Effect of frother concentration on chalcocite recovery in the presence of 700 g/t ethyl xanthate

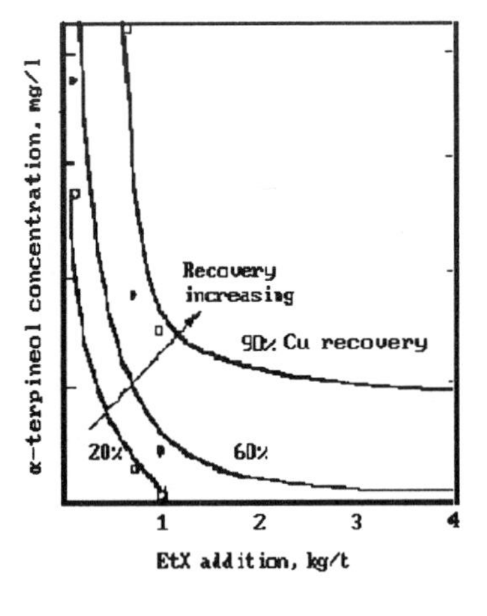

Fig. 18.- Isorecovery curves varying both frother and collector conc. after 1 min float in a Hallimond tube

too sensitive a control requirement on reagent dosage as a function of changes in the mineral fed to the mill. An interesting complement to the alcohol carbon chain versus froth volume data, shown in Fig. 20, is some recent work by Klimpel and Hansen (1987), where they have shown that the reaction product of a six carbon alcohol with a specified amount of propylene oxide gives an unusually strong frother with an enhanced ability to float coarse particles. A typical example on coal is illustrated in Fig. 21, and on copper in Fig. 20. There is no question that some of the effect is due to inherently enhanced collecting properties of the frother itself, in the case of coal, and better frother-collector compatibility, in the case of minerals.

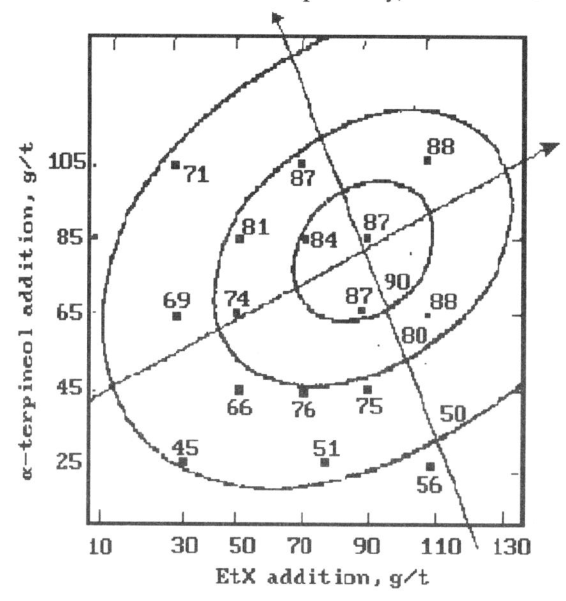

Fig. 19.- Iso-recovery curves for chalcocite as in Fig. 18 but using a batch flotation machine.

Besides the increase in coarse particle recovery, demonstrated by the alcohol-propylene oxide adducts over the same alcohol alone, they also noted that increasing the branching of the alcohol backbone gradually decreases the effective particle size range floated by the six-carbon propylene oxide adduct. Fig. 21 also shows the extreme of this logic: a highly branched triol (glycerol), reacted with propylene oxide, actually is a more effective fine particle frother than any of the alcohols alone. It is also important to note in the work of Klimpel and Hansen (1987) that each specific frother chemistry has a unique particle size range; the benefits of blending frothers for improved overall valuable recovery is obvious.

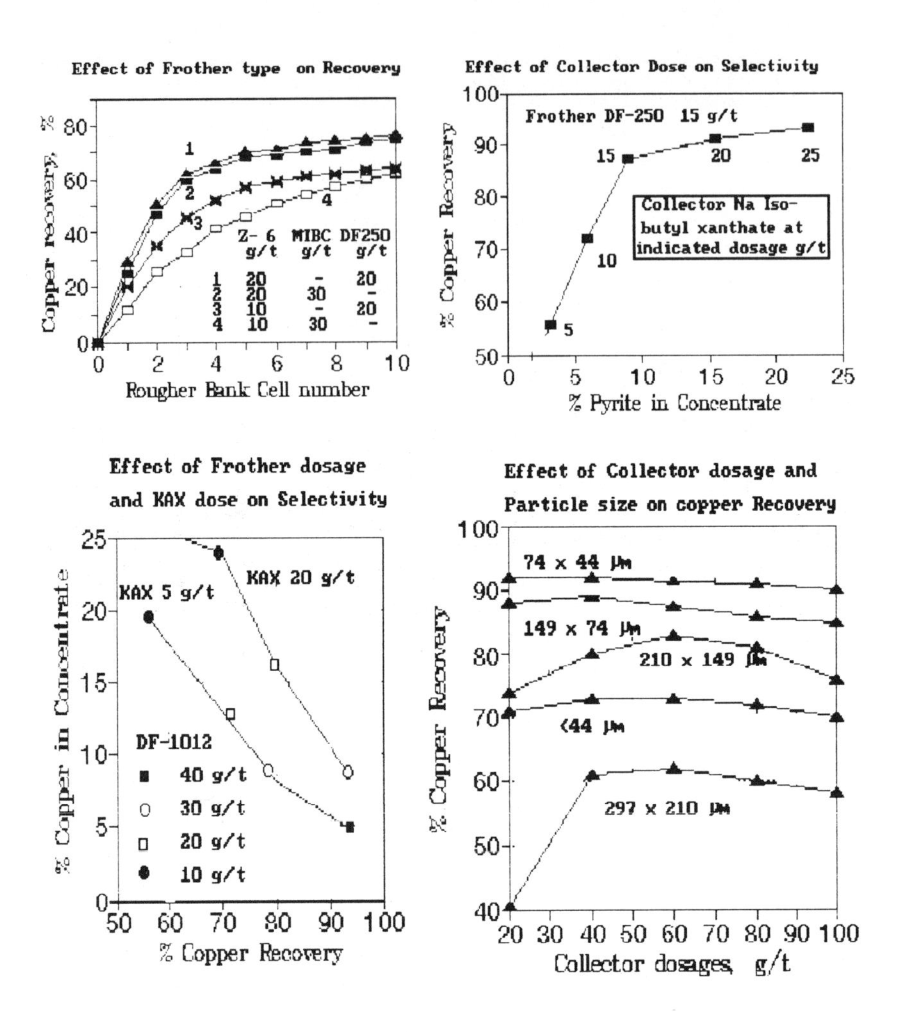

Fig. 20.- Rate of flotation of copper ores

Probably the most extensive combined laboratory and plant scale testing of the influence of frother chemical structure and frother dosage on rate of recovery is also the work of Klimpel and Hansen. As might be expected, the rate of industrial mineral flotation was found to be significantly slower than that associated with non-mineral systems, regardless of the frother type used. They emphasised the interaction of frother structure and dose with particle size, and a number of industrially important trends were observed. The first is the rate influence of increasing the molecular weight of polypropylene glycol frothers (such as the Dowfroth frothers). A detailed study was

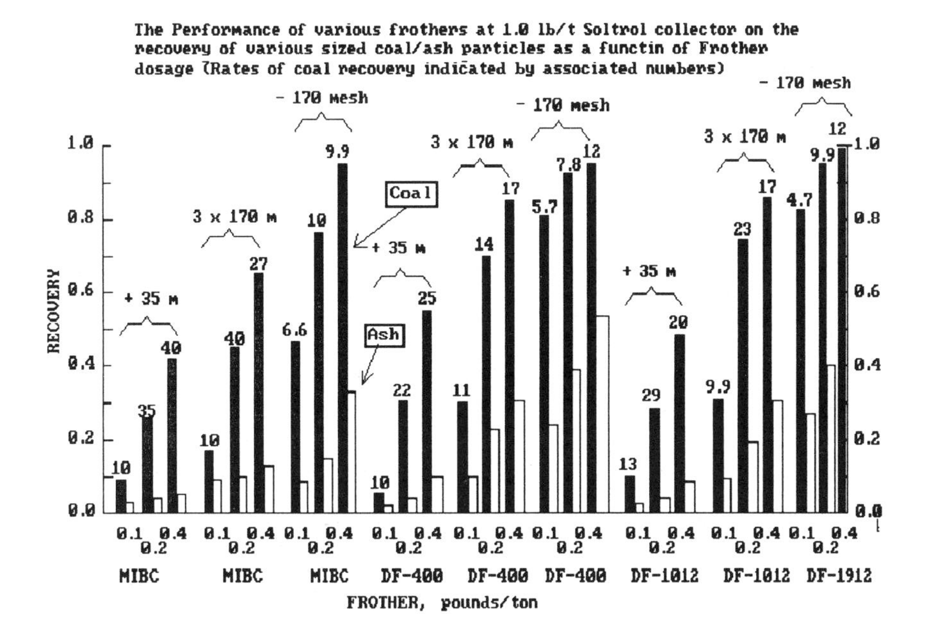

Fig. 21.- Performance of frothers on recovery of coal/ash particles as a function of frother dosage

conducted on compounds having the underlying structure

$$CH_3-(O-C_3H_6)_n-OH$$

where 'n' was varied gradually so as to give molecular weights from 200 to 400. Increasing the value of 'n' clearly increased the rate of recovery of minerals,coals and potash. However, almost always, this increase in rate of recovery was accompanied by a decrease in selectivity. In mineral systems, the alcohol frothers were almost always slower in rate than the polypropylene glycol frothers, over all mineral particle sizes, especially the coarser sizes. The alcohol frothers were often more selective, especially with the finer particles. In coal flotation, the alcohol frothers were found to have excellent rate character, but almost always the rate associated with the alcohols was much more sensitive to dosage than the rate associated with propylene glycol frothers. A second important trend observed is the strong influence of frother dosage with particle size, as illustrated in Fig. 21. This is an illustration on coal of the R/K trade-off that is frother dosage driven, rather than collector dosage driven. The three frothers shown are among the three most common coal frothers in commercial use today: MIBC, Dowfroth 400, and Dowfroth 1012. In the figure, the recoveries, as a function of particle size intervals, are shown in the appropriate columns labelled with the rate, K, for each recovery indicated near that column. The poor coarse particle recovery of MIBC is shown along with its superior fine particle selectivity. The polypropylene glycol frothers are clearly superior coarse particle frothers. The influence of the R/K trade-off with collector dosage is also evident, with the rate slowing or even decreasing at higher dosages. As this example implies, from the wide variation of rate associated with changing frother dosage, the use of frother dosage change is a key control variable in industrial flotation operations. These data also show why it is common in plant practice to first use frother dosage to "fix" short-term plant flotation problems, followed by changing collector dosage only if the frother adjustment is not sufficient.

Over oiling

Seventy years ago, when frothers with collector properties, such as cresylic acid, were commonly used as the sole flotation reagent, it was known that as the dosage of the flotation "oil" was increased, mineral recovery first increased and then decreased. This depression with excessive collector or frother was known as "over oiling". As soon as a separate collector was introduced, it was noted that the same behaviour was exhibited independently by the collector and the frother; i.e., increasing the dosage of either the collector or the frother results in a rise and then a drop in metal recovery. Curiously, there is very little published on this effect, and it seems to be practically unknown among the researchers on flotation.

Klimpel (1987) has studied this over oiling extensively in industrial scale sulphide and coal flotation processes, and has called the effect "the R/K trade-off", and attributes the changes in recovery to changes in the rate of flotation. This terminology is taken from the observation that as collector dosage is increased from starvation dosages, both the equilibrium recovery, R, and the rate, K, at which this recovery occurs, increase. However, as collector dosage is further increased, the value of R continues to increase, but the rate of flotation, K, passes through a maximum. Excess frother dosage can also lead to the R/K trade-off, but the effect in plant environments is most often due to excessive collector. If the rate is slowed sufficiently, the associated plant capacity may be insufficient to correct for the time delay in levitation, and the observed plant recovery actually falls at "excessive" collector dosage.

Fig. 22 is one of the many examples which illustrates the R/K trade-off on a copper mineral circuit. The K values indicated are the result of mathematically fitting laboratory time-recovery profiles. Klimpel has also shown that the effect is very prevalent in many industrial plants, and that industrial circuit performance can be mathematically predicted with good accuracy by simply using appropriate batch laboratory rate data and realistic residence time distributions for the industrial cell.

The depressive effect of excess collector has been used industrially. In one US mill in which a copper/moly separation was operated, it was not uncommon to have problems depressing the copper with certain changes in the feed ores. When this occurred, rather than increasing the dosage of the depressant, additional copper collector was added to the moly circuit, and the copper was successfully depressed at a lower cost.

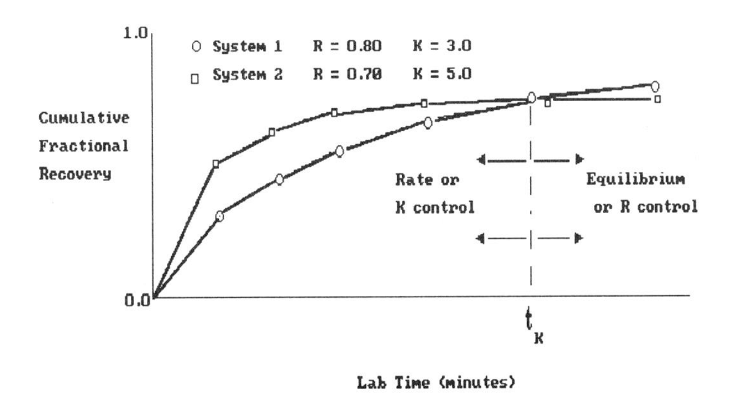

Fig. 22.- Illustration of two laboratory time-recovery profiles run under different conditions

Chapter 7

CHEMICAL PROPERTIES OF FROTHERS

Reagents commonly employed as frothing agents can be divided into slightly soluble in water, and those that are completely miscible, such as polyethers and polyglycol ethers, which correspond to the more modern frothers. Table 3 (page 8) summarises the frother types and their chemical structures, and Table 55 (page 177) the use in the US flotation market.

Partially soluble frothers

Aliphatic alcohols: a series of different mixtures of C_6 to C_8 alcohols were marketed, at one time, by du Pont and still are by Cyanamid, as tailored frothers for different types of ores. The largest volume alcoholic frother used today is methylisobutyl carbinol (MIBC or 4-Methyl-2-pentanol).

Some properties of alcohols used as frothers are shown in Table 24. Note the generally low solubility, which resulted in the assumption that all effective frothers should be nearly insoluble. The synthetic glycol polymers reversed this premise.

Table 24.- PROPERTIES OF ALCOHOLS USED AS FROTHERS

Alcohol	Formula	Frz. Pt. °C	B.Pt. °C	Density g/ml	Solub. g/l
n-Pentanol	$CH_3(CH_2)_3CH_2OH$	-78.5	137.3	0.8144	23.0
Isoamyl Alc.	$(CH_3)_2CHCH_2CH_2OH$	-117.0	132	0.813	25.0
Hexanol	$CH_3(CH_2)_4CH_2OH$	-52.0	156.5	0.819	6.0
Heptanol	$CH_3(CH_2)_5CH_2OH$	-34.0	176	0.822	1.8
MIBC	$(CH_3)_2CHCH_2CHOHCH_3$	-90.0	132	0.808	17.0
Caprylic Alc.	$CH_3(CH_2)_5CHOHCH_3$	-38.6	179	0.822	12.0
4-Heptanol	$CH_3(CH_2)_2CHOH(CH^2)^2CH^3$	-41.2	161	0.8183	4.5

The alcoholic frother mixtures include:

Mixed C_6 to C_9 alcohols: sp. gr. 0.856, viscosity 5 cps, solubility in water, low; flash pt. 552°C. A more selective frother than MIBC.

Mixed C_4 to C_7 alcohols + hydrocarbon oil: amber liquid; sp.gr. 0.82; viscosity 4.5 cps; solubility in water 5 g/l; flash pt. 44°C; used mainly on copper/moly ores. Designed for copper-molybdenite ores, also recommended for flotation of minerals such as talc, graphite, sulphur and coal. Mixed C_4 to C_7 alcohols produce a livelier froth than MIBC.

Mixed C_5 to C_8 alcohols: sp.gr. 0.81-0.83; viscosity 6.9 cps; solubility in water 10 g/l; boiling range 135-190°C; flash pt. 55°C; produces a heavier froth than AF 73 and AF 76, but less persistent than polypropylene glycols, pine oils and cresylic acids.

Higher alcohols (still bottoms): strong froth, less persistent than polypropylene glycols, used in coal flotation.

Natural oils: in the early days of flotation, eucalyptus oil was a popular frother cum collector (because the first industrial scale flotation mills were in Australia). Later, pine oil became the dominant natural oil frother, because of its greater availability in the western world. The quality standard was set by the Yarmour trade-named oils distilled by the Hercules Corp. (U.S.A.). Essentially all the pine growing countries (Canada, Finland, China, etc.) have supplied different grades of pine oils at one time or another, but from the flotation point of view the problems that mill operators have faced is the difficulty in specifying a particular grade and obtaining the same frother performance from different suppliers. This problem, and the fact that the natural oils all have collector properties, have made them less popular with modern mill operators who now favour MIBC and synthetic non-collecting soluble frothers. Those using pine oils remark on its small-bubble, closely knit, froths that do not allow excessive particle fallout, but break down readily in the launder. The close-knit pine oil froth favours recovery but lowers concentrate grade. Increasing the amount of pine oil fed to the cells tends to flatten the froth, decrease its volume, and cause surface spatter. An emergency supply of pine oil can be very useful at mills with ores which tend to have periods of over- frothing. It can also be used in blends with MIBC to regulate the characteristics of the froth and compensate for changes in ores which may affect selectivity and rougher recovery.

A typical assay of a commercial pine oil is:

Alpha Terpineol	60-70%
Tertiary alcohols	10
Borneol and fenchyl alcohol	10-15
Camphor	10-15

α - Terpineol   Fenchyl Alcohol

Typical properties are: sp.gr. 0.927 - 0.940, distillation range 200 - 235°C, flash point 172°C (C.C.).

The eucalyptus oils have greater collector action because they are richer in ketones than the pine oils, which contain primarily camphor, while eucalyptus oil ketones are cineoles and piperitones rather than camphor, and borneols and piperoteols, as well as terpenes. Typical eucalyptus oils available today have a specific gravity of 0.921 - 0.923 (25°C), distillation range 174 -177°C.

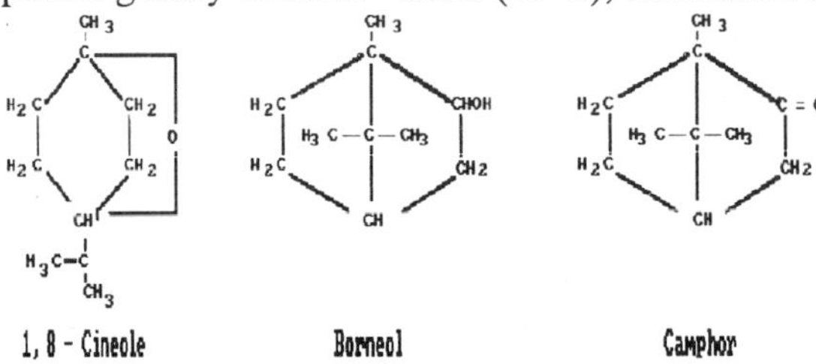

Table 25.- PROPERTIES OF TERPENES AND KETONES USED AS FROTHERS

	Boiling Point °C 760 mm	Boiling Point °C 100 mm	Freeze Pt. °C	Refract. Ind n_D^{20}	Density g/ml 20°C
Alcohols					
α-Terpineol	219	149	36	1.4831	0.9336
Borneol	212	-	210	-	1.01
Fenchyl alc.	201	133	39	1.4734	0.935
Ketones					
Piperitone	233/5	-	-	1.4845	0.9324
Camphor	209 sb	-	179	-	1.000
Fenchone	193	122	6	1.41619	0.9452
Ethers					
1,8-Cineole	174	108	1	1.4574	0.9245

Typical specifications for pine oil to be used as a frother are: a clear, yellow oil, with a typical smell, that boils between 170 and 228°C, with density of 0.915 to 0.935 g/ml. At least 44% by weight of the composition should be alcohols (expressed as terpineols). At least 78% must distils between 170 and 220°C. Its average solubility in water should be approximately 2.5 g/l. Eucalyptus oil used in flotation has similar specifications, as far as alcohol content is concerned. Its main odorous component is the ether 1,8-cineol, rather than the ketone, camphor. Its solubility in water is typically 1.4 g/l.

Cresylic acids: the same quality control problems as for pine oils apply to the cresylic acids, as these products are also distilled from impure starting materials (coal tars), and in addition have strong collecting properties. Their popularity declined when acid circuit mills went out of fashion, though they are still popular in coal flotation because of their availability from coking ovens. The froths produced by cresylic acids are generally similar in structure to those of pine oil, but of a somewhat larger bubble size. An increase in frother dosage decreases froth volume and produces a tendency to effervesce. The lower boiling fractions produce a more fragile froth and are somewhat more selective.

Typical specifications for flotation grade cresylic acids were: distillation range of 190 - 235°C, with no more than 2% below 190°C, no less than 75% distilling below 235°; density, 1.01 to 1.04 g/ml and solubility in water, 1.7 g/l. According to the Condensed Chemical Dictionary, today's commercial grade has not less than 50% distilling over 204°C. If it boils below 204° it is called cresol. A typical commercial cut boils between 220 - 250°C, and has as typical compositions: meta, para-cresols 0-1%; 2,4- and 2,5-xylenols 0-3%; 2,3- and 3,5-xylenols 10-20%; 3,4-xylenols 20-30%, and C_9 phenols 50-60%.

Some properties of the individual isomers are:

	Sp. grav.	melt pt °C	boil pt °C	
Meta cresol	1.034	12	203	liquid
Ortho cresol	1.047	30.9	191	solid
Para cresol	1.039	35.3	202	solid

Froth characteristics and collector properties of cresylic acids are very dependent on the meta cresol and phenol content, so specifications of cresylic acid purchases should be rigorous if froth control problems are to be avoided in the concentrator. The purer cresylic acids are reputed to give a more fragile froth and to require a stronger collector for good flotation. Impure cresylic acids contain pyridines and sulphur impurities, which were much sought after 50 to 70 years ago in flotation mills when the frother was also employed as a collector, but in conjunction with selective collectors these cresylic acids can give process control problems.

Alkoxy paraffins: these structures were developed in South Africa by Dr. Powell (1951). The best known products are TEB (triethoxybutane) and the Powell accelerator, so named because it has a strong effect on the flotation rate. They never have had extensive application outside South Africa, except in Chile (El Teniente). The froths these reagents produce are similar to those of pine oil, except that over-feeding does not affect the froth character as much. The structure of TEB is:

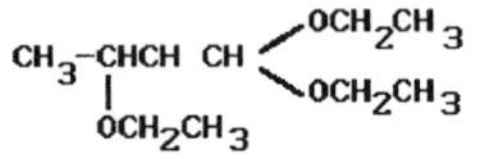

$$OCH_2CH_3$$
$$CH_3-CHCH\ CH$$
$$OCH_2CH_3$$
$$OCH_2CH_3$$

The Powell Accelerator has the following structure:

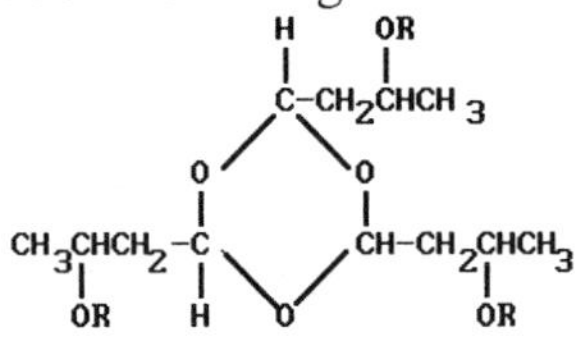

H OR

C-CH_2CHCH $_3$

O O

CH_3CHCH_2-C CH-CH_2CHCH_3

OR H O OR

This is a more powerful frother than TEB, but otherwise has similar characteristics: low water solubility, good stability, and a strong effect on flotation rate.

Non-ionic polypropylene glycol ethers: The Dow Chemical company offers Dowfroth 4082, and Union Carbide R 17 and R 48, which are partially soluble. Their structure is:

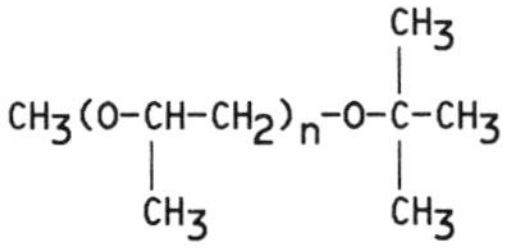

$$CH_3$$
$$CH_3(O-CH-CH_2)_n-O-C-CH_3$$
$$CH_3 \qquad CH_3$$

Completely water miscible frothers

Polyglycol ethers: these frothers, which are completely miscible in water, were first developed by Tveter (1952), at Dow Chemical Co., and Booth (1954) at the American Cyanamid laboratories. Until they came on the market, mill metallurgists assumed that good frothers could only be very slightly soluble in water. These glycol ethers are produced as by-products of synthetic brake fluids. They are marketed by the Dow Chemical Company under the name Dowfroths, by Union Carbide as Ucon frothers, by Cyanamid as Aerofroths and ICI as Teefroths. With MIBC they account for nearly 90% of all frother use in metallic ore flotation. The polypropylene glycols produce compact, lasting froth structures that break down readily in the launders. Unlike the froths from slightly soluble frothers, these frothers produce more tightly knit, more selective froths, that do not spatter on over-feeding the reagent. The Dowfroths are designated by numbers that are proportional to the polymer's molecular weight. Dow informs its customers that the higher molecular weight products are more powerful frothers; the strongest are Dowfroth 400 and 1400. Because of the high activity of the polyglycol ethers, all can cause persistent froths which can cause operating problems if over-dosed.

Table 26.- PROPERTIES OF WATER SOLUBLE FROTHERS

	Dowfroth					Union Carbide	
	200	250	1263	1012	1400	Ucon 200	UconR55
Mol. Wt.	206	250	348	400	400	400	-
Viscosity cps 25°C	7	12	19	27	33	75	-
Density g/ml, 25°C	0.97	0.98	0.978	0.988	1.007	1.009	0.9735
B.P. Start °C 760 mm	243	252	>250	293	-	-	-
Freeze Pt °C <	-50	-50	-41	-50	-45	-50	-50
pH	7.2	7.2	-	7.0	-	-	-
Solubility	total	total	total	partial	total	total	total
Chemical formula n=	3	4	5	6	6		

	Cyanamid				
	AF 65	AF 67	AF 68	AF 70	AF 73, 76, 77HP
Mol. Wt.	400	similar	similar	see	mixed C_4 to C_8
Viscosity		to	to	MIBC	alcohol
cps 25°C	71	DF-250	DF-1012		frothers
Density g/ml, 25°C	1.007				
B.P. Start °C 760 mm	180				
Freeze Pt °C <	-50				
pH	-				
Solubility	total				

The Dowfroths are methoxy polypropylene glycols (polypropylene glycol methyl ethers), with the general formula:

$$CH_3-(O-C_3H_6)_n-OH$$

with Dowfroth 200, 250 and 1012 having molecular weights of approximately 206, 250 and 400. Dowfroth 1400 is a diol of 400 molecular weight, with the formula:

$$H-(O-C_3H_6)_n-OH$$

The American Cyanamid products (Booth, 1954) are condensations of propylene oxide with propylene glycol. The formula for Aerofroth 65 is approximately:

```
         CH3  CH3        CH3       CH3
         |    |          |         |
HO-CH2CH-O-CH-CH2-O-CH-CH2-O-CH-CH2-OH
```

Union Carbide supplies similar products based on polypropylene glycols and polyethylene glycols, under the trade names PPG and Ucon.

Polyglycol glycerol ethers: One of the two new families now being produced by the Dow Chemical Company is of the general formula:

```
R-O-(CH-CH2-OH)n
     |
     CH3
```

where R is from 4 to 6 carbon atoms. The characteristics of such frothers were described earlier in this section. An example of the second family is the reaction product of glycerol with propylene oxide to a 450 molecular weight. The products available in industrial quantities have the following properties:

Type	XK 35004.00L	XK 35004.01L	XK 35004.02L
Avg. Molecular Wt.	450	250	700
Viscosity, 25°C, cks	510	600	220
Sp. gr. at 25°C	1.05	1.09	1.03
Flash pt. (PMCC) °C	+205	199	260
Solubility in water	miscible	misc.	misc.

XK 35004.01L and .02L are weak frothers that can be blended with other polyglycolethers to fine tune froth properties. They are stable in the range pH 3.5 to 12.3, and usually employed in alkaline flotation.

Chapter 8

FLOTATION MODIFIERS

The classical treatment of this subject is given by Sutherland and Wark (1955) in their book "Principles of Flotation". By very elegant experiments based on contact angle measurements, they were able to show that flotation could occur only below a critical pH for a given mineral and reagent combination and dosage. Based on this concept, a rational explanation for the mechanism of activation and depression was developed which allows a systematic approach to the subject. Their data on critical flotation pH for potassium ethyl xanthate is shown at three temperatures in Table 27. In Table 28 similar data is shown for ethyl dithiophosphate and xanthate, and for iso-amyl xanthate for galena, pyrite and chalcopyrite.

Table 27.- CRITICAL pH FOR FLOTATION

Mineral	10°	Room Temp	35°
Pyrrhotite	-	6.0	-
Arsenopyrite	-	8.4	-
Galena	10.8	10.4	9.7
Pyrite	10.2	10.5	10.0
Marcasite	-	11.0	-
Chalcopyrite	13.0	11.8	10.8
Covellite	-	13.2	-
Activated Sphalerite	-	13.3	-
Bornite	-	13.8	-
Chalcocite	-	>14.0	-

If these type of data are to be used, the actual curves published by Sutherland and Wark should be consulted, as the critical pH increases quite sharply with concentration for most collectors at the values quoted above.

The effect of a depressant, such as NaCN, is to significantly reduce the critical pH for flotation. Data reported by Sutherland and Wark (1955) for the addition of 20 mg/l of NaCN to a pulp containing 25 mg/l of potassium ethyl xanthate are shown in Table 29.

Table 28.- EFFECT OF COLLECTOR ON CRITICAL pH FOR FLOTATION

Collector	Galena	Pyrite	Chalcopyrite
Na diethyl dithiophosphate, 32.5 mg/l	6.2	8.5	9.4
Potassium ethyl xanthate, 25 mg/l	10.4	10.5	11.8
Potassium iso-amyl xanthate, 31.6 mg/l	12.1	12.3	>13

Sutherland and Wark (1955) provide a variety of graphs on the effect of the cyanide ion on the critical pH, and also show that the effect of the cyanide ion can be calculated from the ionisation constant for the different cyanides as a 45° straight line on log-log paper.

Table 29.- EFFECT OF SODIUM CYANIDE ON CRITICAL pH

Temperature	10°C	35°C
Chalcocite	12.8	10.7
with 20 mg/l NaCN	8.4	7.3
Pyrite	10.6	10.0
with 20 mg/l NaCN	7.1	6.9
Galena	10.9	9.6
with 20 mg/l	10.9	9.6

Activating agents

These are chemicals that are used to permit flotation of a mineral that is difficult or impossible to float with only the use of a collector and a frother.

Copper sulphate is the best example of an activator. It was first introduced by Bradford in 1913 in Australia in zinc blende flotation, and later was applied to the activation of gold bearing pyrite in that country. It is universally used in the flotation of sphalerite, which is otherwise impossible to float. It is also used to reactivate minerals which have been depressed with cyanide, such as chalcopyrite, pyrite, pyrrhotite and arsenopyrite. Thus, alternate use of copper sulphate and cyanides can be used very effectively in differential flotation. In mill use it should be remembered that the salt is quite corrosive and is best handled in plastic equipment.

Lead nitrate or acetate are used to activate stibnite and to reactivate copper sulphides depressed with cyanide. They are sometimes employed to improve recovery of tarnished gold. They are also activators for silicates and carbonates in soap flotation, and for sodium chloride in fatty acid flotation to separate potash from halides.

Sodium sulphide can act as an activator at low concentration for oxidised and tarnished ores. Morenci has found that the ammonium salt is an even more effective activator for their tarnished copper ore. Controlling the effective concentration of the sulphhydrate ion is extremely tricky, as sulphides are so easy to oxidise. Because of this, the few industrial applications of sulphidisation have employed a circuit in which the floatable sulphide minerals are taken off first, prior to the addition of sodium sulphide or NaSH, to float the remaining oxide minerals. The choice of sulphide or NaSH depends on the pH required in flotation, as the sulphide contributes twice as much alkalinity as NaSH.

Hydrogen sulphide has been used in Butte to precipitate copper in solution and allow recovery by flotation. It will depress gold and silver, and Cu-Fe minerals in moly separations.

pH modifiers such as lime, soda ash, caustic soda, sulphuric acid, etc., can also be considered activators because for all mineral/reagent combinations there is a critical pH above which flotation will occur. Because the flotation process is a surface phenomena that is extremely sensitive to the ion content of the flotation water, the use of soda ash (caustic) versus lime as a pH controller depends on whether the calcium cation will form insoluble precipitates with naturally occurring anions in the pulp.

Generally, because of cost and availability, lime is the chosen pH modifier except in the flotation of pyrite, particularly gold bearing pyrite, and to a lesser extent in galena flotation, where soda ash is the first choice.

Inorganic depressants

These assist in the separation of minerals when the floatability of two or more minerals are too similar for a particular collector to effect a separation. Besides lime, the other most used depressant is the cyanide ion.

Sodium cyanide is a strong depressant for iron sulphides: pyrite, pyrrhotite, marcasite and arsenopyrite, and also for sphalerite. It will act as a less effective depressant for chalcopyrite, enargite, tennantite, bornite and most other sulphide minerals, with the possible exception of galena.

Lime is used for depressing pyrite and other iron sulphides, galena, marmatitic zinc and some copper minerals, as well as cobalt sulphides. It will depress gold in xanthate flotations, and interferes with sulphidisation processes. For inorganic minerals, it will retard recovery of silicates when cationic collectors are involved.

Zinc sulphate is used in conjunction with cyanide, or alone, for the depression of sphalerite while floating lead and copper minerals.

Chromic acid and dichromates will depress galena in copper-lead-zinc separations.

Permanganates are used for the selective depression of pyrrhotite and arsenopyrite in the presence of pyrite. It will depress sphalerite, and is also used in copper/moly separation.

Sodium silicate is a complicated reagent because of its variable composition. It is used as a depressant of quartz and other silicate minerals. Also used in coagulation of slimes and as a modifiers in fine particle flotation. Generally helpful if maintaining concentrate grade is a problem.

Sodium hydroxide is used for the depression of stibnite. It is helpful in gold flotation, and will peptise slimes. It is a saponifier for fatty acids in inorganic flotation, and is used in scrubbing ores and in scheelite flotation.

Sodium and calcium sulphites and hyposulphites and sodium hydrosulphite are used in the flotation of lead and copper minerals while depressing pyrite and sphalerite.

Ferrocyanide is used in the depression of copper sulphides in copper/moly separation and in the separation of some copper sulphides from sphalerite.

Ferro and ferricyanides will help separate cobalt and nickel sulphides from copper sulphides.

Sulphuric acid is used for quartz depression. It will reactivate pyrite depressed with lime, and is used in gold flotation to clean iron salts that interfere with gold recovery.

Hydrofluoric acid is used in the depression of quartz in the flotation of feldspar with cationic collectors.

Sulphur dioxide, usually in conjunction with causticised starch, is used to depress galena from copper sulphides.

Organic depressants

Organic depressants are large molecules usually with a molecular weight above 10,000. The mechanism of their depressant action is not clear, though Lovell (1982) thought that these are depressants because they contain a large number of hydrated polar groups. The natural products are generally polysaccharides, while new synthetic depressants include polyglycol ethers and polyphenols. A curious extreme depressant is a xanthate prepared by reacting sugar with caustic and carbon disulphide; it has not found a practical use as it depresses everything.

The natural depressants include the following:

Quebracho and tannic acid are polyphenols used for the depression of calcite and dolomite in fatty acid flotation of fluorite or scheelite and also pyrite. The formula for polyphenols can be represented:

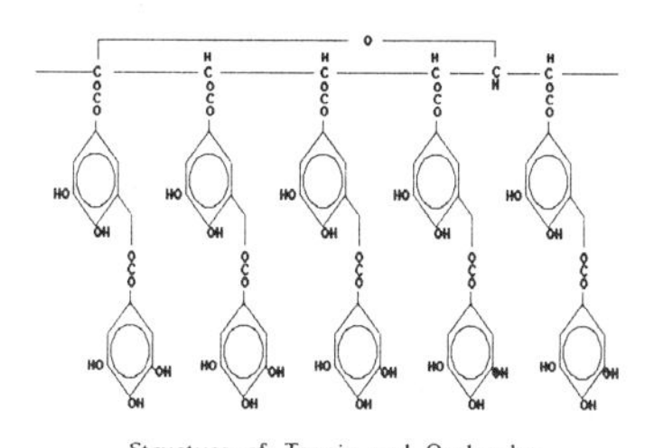

Structure of Tannin and Quebracho

Starch and glues, either unmodified, as extracted from maize potatoes, or partially hydrolysed to produce the more soluble dextrins, are used for the depression of mica, talc and sulphur in sulphide flotation. They also act as flocculants, particularly with haematite, and as clay dispersants. Starch is a highly polymeric carbohydrate built up of glucose units with molecular weights ranging from 50,000 to several million daltons. Corn starch consists of two types of polymers: a linear amylose composed of 200 to 1,000 pyranose rings bonded together through oxygen atoms in positions 1 and 4, and a branched amylopectin with 1,500 or more pyranose rings with periodic branching at positions 1 and 6.

Glucose units

Corn starch is more strongly adsorbed on haematite and geothite than on quartz; its adsorption decreases with increasing pH. Adsorption on quartz is minimal in the absence of calcium ions. In highly alkaline solutions, calcium ion is adsorbed on quartz as a hydroxy complex, $CaOH^+$, which is thought to be responsible for the effectiveness of calcium ions in anionic silica flotation.

Natural gums, such as gum arabic, gum traganth, and guar, have strong flocculent properties, but can be treated to be selective depressants for talc and siliceous materials. Their chemical structure is:

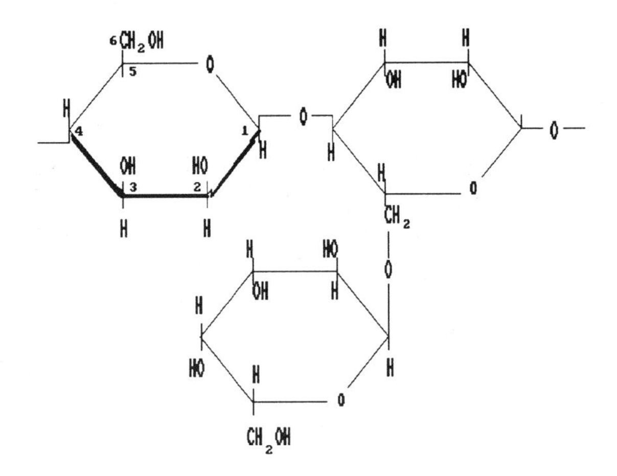

Chapter 9

SULPHIDES - DEPRESSION AND ACTIVATION

Sodium sulphide

The sulphhydrate ion has been known as an activator of oxidised minerals since the sulphidising process was patented [A.Schwartz (1905), US Patent No. 807501], practically at the same time as the birth of flotation, and has frustrated metallurgists ever since. Non-predictable performance is no surprise if one considers that small concentration of HS^- will depress the flotation of metallic sulphides, and not activate carbonates or oxides; a medium amount will erratically activate the flotation of non-sulphide copper minerals such as malachite [$CuCO_3.Cu(OH)_2$], azurite [$(CuCO_3)_2.Cu(OH)_2$], and chrysocolla [$Cu(Si_4O_{10})(OH)_{12}.8H_2O$]; more reliably it has been used to activate lead oxides, such as cerrusite [$PbCO_3$], and anglesite [$PbSO_4$], but with generally high dose rates; a large amount will desorb collectors and depress activated minerals. The surface adsorption mechanism that provides these contradictory qualities to sodium sulphide is still not completely clear. In solution, it is very easily destroyed by dissolved oxygen, its activity is strongly pH dependent, and it significantly affects the system alkalinity, more so if sodium sulphide is employed as the flotation modifier than if the reagent is sodium sulphhydrate.

Historically, it was successfully employed in xanthate or thiophosphate flotation of cerrusite using as much as 20 kg of sodium sulphide per tonne of ore. As cerrusite characteristically has clay-like properties, and is a major constituent of slimes, sulphidisation normally has been combined with the use of sodium silicate as a peptising agent. Its main use as a depressant has been to remove copper from molybdenite concentrates during the cleaning step. When used as a rougher depressant for sulphides, it appears to operate primarily due to its oxygen scavenging properties.

Chemical properties of sulphides

Sodium sulphide: Na_2S; formula weight 78.04; ambient stable crystal

$Na_2S.9H_2O$, formula weight 240.18; commercially available form, flakes which contain 60% by weight Na_2S with an approximate density of 1.75.

Data available on the solubility of sodium sulphide is conflictive, due primarily to the ease with which it is oxidised in solution in contact with air. The best data based on a literature review is that reported by Seidell-Linke (1965). Their interpolations and phase data are reproduced in Table 30.

Table 30.- SODIUM SULPHIDE SATURATION SOLUBILITY IN WATER AND PHASES IN EQUILIBRIUM

t°C	Density gm/ml	Solubility gm/100g sat.soln	Solid phase
0	-	11.0	$Na_2S.9H_2O$
5	-	12.3	"
10	-	13.4	"
15	-	14.5	"
20	1.1682	16.5	"
25	-	17.5	"
30	1.1972	19.8	"
40	1.2287	23.4	$Na_2S.9H_2O$
48.5		transition	to $Na_2S.6H_2O$
50	-	27.6	$Na_2S.6H_2O$
60	-	29.0	"
70	-	30.8	"
80	-	33.4	"
90	-	36.8	"
91.5		transition to	$Na_2S.5\frac{1}{2}H_2O$
	Volumetric solubility, gm Na_2S/lt soln		
20		192.7	
30		237.0	
40		287.5	

Ref: Seidell-Linke (1965), page 1113

Ionisation as a function of pH: On dissolving Na_2S or NaSH in water, ionisation proceeds in various stages. First

$$Na_2S + 2H_2O \rightarrow 2Na^+ + H_2S + 2OH^- \quad \text{or}$$

$$NaSH + H_2O \rightarrow Na^+ + H_2S + OH^- \quad \text{followed by}$$

$$H_2S \rightarrow H^+ + HS^-$$

where the equilibrium constant is

$$K_1 = \frac{(H^+)(HS^-)}{(H_2S)} = 9.1 \times 10^{-8} \quad ; \quad pK = 7.04$$

while the third stage is

$$HS^- \rightarrow H^+ + S^=$$

and the equilibrium constant is

$$K_2 = \frac{(H^+)(S^=)}{(HS^-)} = 1.1 \times 10^{-12}; \qquad pK = 11.96$$

These equilibrium constants indicate that, on a molar basis, below pH 5 there are no sulphide ions present, and at pH 5 only 10% of the H_2S has ionised to HS^-. At pH 7, half the H_2S has ionised to HS^-, while between pH 8 and 11, over 90% of the sulphide ions are HS^-. The next ionisation node is at pH 12 when the moles of HS^- and $S^=$ are equal, while by pH 13, 90% of the ionisation is to $S^=$. For rough estimation of the ionic concentration, in grams per litre, the curves in Fig. 23 can be multiplied by the Na_2S concentration.

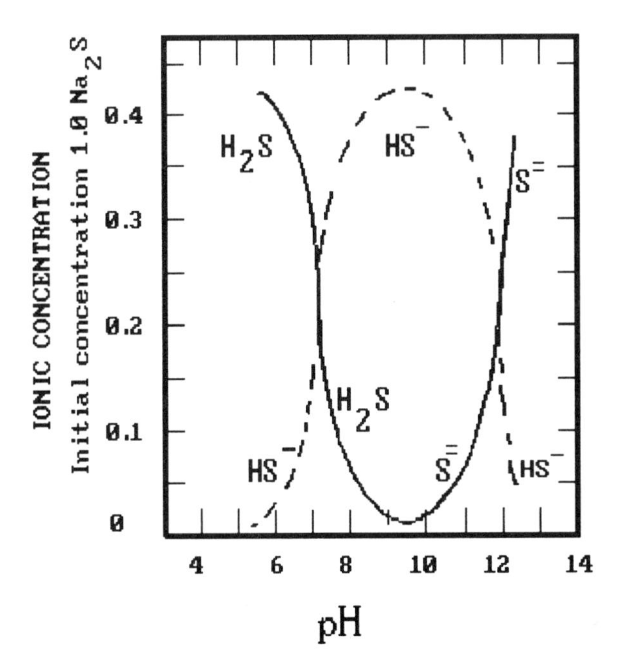

Fig. 23. Ionisation of sodium sulphide solutions

Sodium hydrosulphide: NaSH, formula weight 56.06; ambient hydrate, $NaSH.2H_2$, formula weight 92.09; fused hydrate, $NaSH.3H_2O$, formula weight 110.1; melting point 22°C.

Surprisingly, there is very little information available in the literature on the solubility of NaSH in water. Seidell-Linke (1965) indicate that at 20°C, in water containing 2% NaCl, solubility of NaSH is 35.8 g/100 g of saturated solution. NaSH is available commercially as a solution containing 16% sulphide; a hot solution containing 40 to 44% active NaSH; crystals (tri hydrate) which are 50.9% active, and which on melting (at 22°C) dissolve in their own water of hydration; and flakes containing 70 - 72%, which melt at 55°C.

When dissolving flake NaSH in water, caution is necessary, as the more impure products can decompose with evolution of H_2S, a gas that is as toxic as hydrogen cyanide (TLV for both is 10 ppm), and more dangerous because it paralyses the sense of smell, the odour threshold being close to the fatal concentration. It forms explosive mixtures in air in the range 4.3 to 46% H_2S.

The pH as a function of concentration is shown in Table 31. Fig. 24 shows the effect on solution pH of the addition of equal weights of NaSH and Na_2S. As can be seen, at any given dose, the pH for NaSH is about a quarter unit lower than for Na_2S. The effect of temperature is shown in Table 32.

Table 31.- EFFECT OF PH ON THE EQUILIBRIUM VALUES OF SULPHIDE ION, HYDROSULPHIDE ION AND HYDROGEN SULPHIDE

pH	For a g/l of Na_2S the conc. of each ion is:		
	H_2S, g/l	HS^-, g/l	$S^=$, g/l
4.0	0.4363	3.8538×10^4	4.1085×10^{12}
4.5	0.4354	1.2163×10^3	4.1005×10^{11}
5.0	0.4328	3.8226×10^3	4.0000×10^{10}
5.5	0.4245	1.1857×10^2	3.9972×10^9
6.0	0.4003	3.5356×10^2	3.7692×10^8
6.5	0.3391	9.4722×10^2	3.1933×10^7
7.0	0.2286	0.2019×10^2	2.1530×10^6
7.5	0.1126	0.3146×10^2	1.0605×10^5
8.0	0.0432	0.3819×10^2	4.0711×10^5
8.5	0.0147	0.4095×10^2	1.3806×10^4
9.0	0.0047	0.4188×10^2	4.4649×10^4
9.5	0.0015	0.4209×10^2	1.4191×10^3
10.0	4.7415×10^4	0.4188×10^2	4.4649×10^3
10.5	1.4660×10^4	0.4095×10^2	1.3805×10^2
11.0	4.3229×10^5	0.3818×10^2	0.0407×10^2
11.5	1.1259×10^5	0.3145×10^2	0.1060×10^2
12.0	2.2852×10^6	0.2018×10^2	0.2152×10^2
12.5	3.3885×10^7	0.0946×10^2	0.3191×10^2

Ammonium hydrosulphide: NH_4SH, formula weight 51.11; melting point 118°C; solubility at 0°C is 128.1 g/100 g H_2O. The commercially available products are ammonium hydrosulphide crystals, and a 40 - 44% solution. Ammonium sulphide is used more as a flotation activator, because the ammonia ion appears to clean the mineral surface and enhance the sulphidisation reaction.

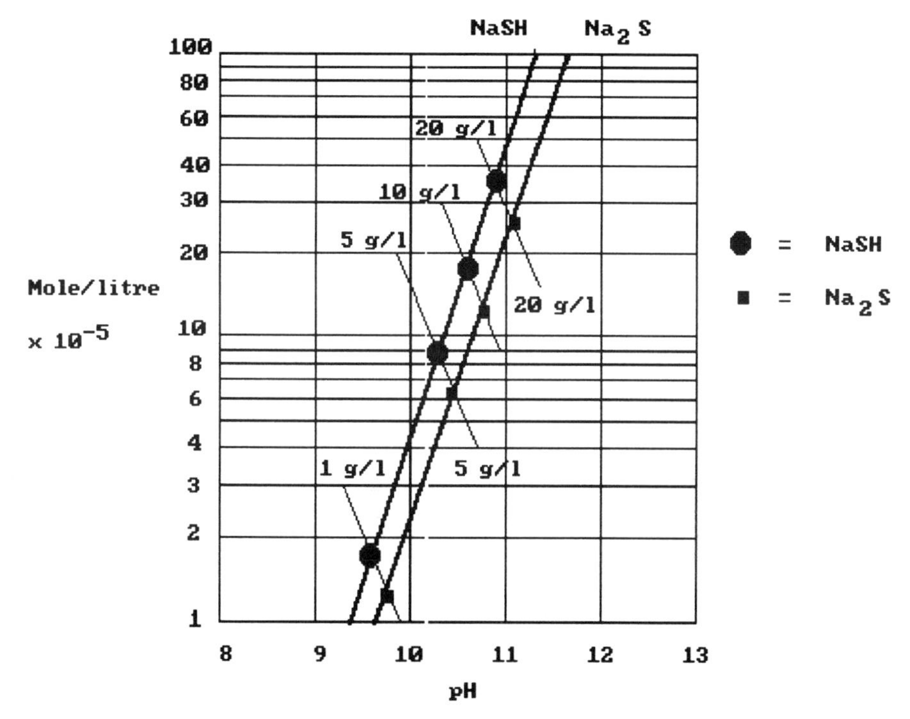

Fig. 24.- Relative effect of the addition of NaSH and Na_2S on solution pH

Table 32.- EFFECT OF TEMPERATURE ON THE PH OF SATURATED SOLUTIONS OF NASH AND NA_2S

Temperature	solution pH 10°C	15°C	20°C
pure water	7.27	7.17	7.08
5 g/l NaSH	10.48	10.29	10.12
5 g/l Na_2S	10.64	10.45	10.27

Molybdenum concentrate cleaning

The most important application of HS^- as a depressant is in the up-grading of molybdenite concentrates obtained as a by-product of copper sulphide flotation. It is the key reagent in the depression of chalcopyrite during the

Table 33.- REAGENT PRACTICE IN MOLYBDENUM RECOVERY FROM PORPHYRY ORES, COPPER CIRCUIT

Plant & company	Heads % Cu	Heads % Mo	%-200 mesh	pH	Main Mineral	PRIMARY FLOTATION CIRCUIT Collectors	PRIMARY FLOTATION CIRCUIT Frothers	concentrate % Cu	concentrate % Mo	% Solids Thickn	% Solids Flt
United States											
Bingham, Utah, Kennecott	0.7	0.025	60	8.5	Cpy/Cte	Reco. F.O.	Cres.Ac.,MIBC	28	0.9	45	35
Ray, Arizona, Kennecott	0.9	0.018	60	11.5	Cpy	Reco. F.O.	MIBC(Shell 10)	18	0.2	60	15
Chino, New Mexico, Kennecott	0.9	0.008	55	11.0	Cpy	Reco. F.O.	Shell 10,P.O.	20	0.2	50	15
Mc Gill, Nevada, Kennecott	0.9	0.016	52	10.8	Cpy	Re404,Z-200	MIBC, P.O.	20	0.2	60	45
Sierrita, Arizona, Duval	0.3	0.03	47	11.0	Cpy	Z-6,3302,F.O.	MIBC	25	2.5	50	30
Mineral Park, AZ, Duval	0.4	0.04	60	11.5	Cpy	Re3302,Z200	MIBC	18	1.3	50	20
Esperanza, Arizona, Duval	0.4	0.03	60	11.5	Cpy	Re3302,Z-6	MIBC	22	1.5	50	20
San Manuel, Arizona, Magma	0.7	0.018	60	10.5	Cpy/Cte	Minerec,F.O.	MIBC(Shell 10)	28	0.5	55	30
Mission, Arizona, ASARCO	0.7	0.018	56	11.5	Cpy	Re238,Z-6	MIBC,P.O.	25	0.4	60	35
Silver Bells, AZ,ASARCO	0.7	0.008	48	11.0	Cpy/Cte	Re238.Z-10	MIBC,Cr.Ac.	30	0.45	50	20
Pima, Arizona, Cyprus	0.5	0.013	50	11.5	Cpy	Z-6,F.O.	MIBC(Shell 10)	26	0.25	50	35
Bagdad, Arizona, Cyprus	0.5	0.015	50	11.5	Cpy/Cte	Z-4,Z-6	MIBC,P.O.	28	1.5	50	20
Morenci, AZ, Phelps Dodge	0.8	0.007	55	10.5	Cpy	Z-200,Minerec	Cres.Ac.	22	0.14	45	35
Twin Buttes, Arizona, Anamax	0.6	0.03	53	11.0	Cpy	Z-6	MIBC	28	1.0	50	35
Inspiration, AZ, Inspiration	0.7	0.007	55	10.5	Cpy	Minerec 898	MIBC,AF-65	35	0.3	60	35
Pinto Valley, Cities Service	0.5	0.011	35	10.5	Cpy/Cte	Z-14,NaAerofl	Dow250	26	0.2	30	15
Canada											
Brenda, B.C., Noranda	0.2	0.05	40	8.0	Cpy	Z-6	MIBC	22	3.0	65	18
Gaspé, Quebec, Noranda	0.6	0.015	67	10.0	Cpy/Cte	Re3302,Z-6	BHB	27	0.3	50	25
Island Copper, B.C.,Utah Int.	0.5	0.017	50	11.5	Cpy	Z-6,Re238	MIBC	24	0.6	65	35
Lornex, B.C., Rio Algom	0.5	0.016	50	9.5	Cte/Cpy	Z-6,Z-11	Dow250,P.O.	33	0.3	50	35
Gibralter, B.C., Placer	0.4	0.01	70	10.5	Cte/Cpy	IPX-343	MIBC,P.O.	30	0.5	55	30
Chile											
Chuquicamata, CODELCO	2.0	0.06	46	11.0	Cte	RE238,Z-11	Dow250,MIBC,PO	40	0.9	60	35
El Teniente, CODELCO	1.5	0.04	67	4.2	Cte/Cpy	Minerec A	Diw1012	42	0.5	-	-
El Salvador, CODELCO	1.2	0.024	50	11.0	Cte	Z-200,Z-11,F.O	MIBC,P.O.	42	0.8	60	45
Andina, CODELCO	1.8	0.015	70	9.0	Cpy	Minerec2030	MIBC,Dow250	28	0.25	55	40
Peru											
Toquepala, Southern Peru	1.2	0.018	60	11.5	Cpy/Cte	Z-11	Pine Oil	31	0.3	60	30
COMECOM											
Balkhash, Kazakhstan, USSR	0.4	0.01	62	12.0	Cpy/Cte	Z-11,Z-14,Kero	Pine Oil	16	0.15	45	45
Almalyk, Uzbekstan, USSR	0.7	0.01	46	11.5	Cpy	Z-7, Kero	Pine Oil	17	0.2	60	24
Kadzharan, Armenia, USSR	1.2	0.05	60	9.0	Cpy	Z-7. Kero	Pine Oil	16	1.5	60	25
Medet, Bulgaria	0.3	0.008	60	8.5	Cpy	Z-14, Kero	Pine Oil	13	0.2	50	20

Table 34.- REAGENT PRACTICE IN MOLYBDENUM RECOVERY FROM PORPHYRY ORES, MOLY CIRCUIT

Plant & company	Moly/copper separation section practice					Final Cu concentrate		Final Moly concentrate	Overall Recovery	
	Feed pH	Heat Treatment	Reagents used: Depressor	Reagents used: Other	Number Clners	%Cu	%Mo	%Cu	%Cu	%Mo
United States										
Bingham, Utah, Kennecott	11.0	Roast	Utah process	NaCN, Nokes	6	30	0.1290	1.0	90	56
Ray, Arizona, Kennecott	11.5	-	$Na_4Fe(CN)_6$	Nokes, $Na_2Zn(CN)_4$	9	20	0.0882	1.8	82	31
Chino, New Mexico, Kennecott	11.0	Steam, roast	NaSH	Nokes,	14	20	0.1 82	1.25	79	52
Mc Gill, Nevada, Kennecott	9.0	Steam	$Na_4Fe(CN)_6$	NaCN	9	20	0.1560	1.5	78	15
Sierrita, Arizona, Duval	9.0	Steam	Nokes	-	8	26	0.2080	3.0	90	76
Mineral Park, AZ, Duval	7.0	Steam	$Na_4Fe(CN)_6$	-	8	20	0.1090	0.3	76	62
Esperanza, Arizona, Duval	7.0	-	$Na_4Fe(CN)_6$	Roast to MoO_3	8	25	0.1283	1.5	87	74
San Manuel, Arizona, Magma	11.5	Pres. Steam	$Na_4Fe(CN)_6$	$Na_2Zn(CN)_4$, H_2O_2	6	30	0.0685	1.0	92	70
Mission, Arizona, ASARCO	11.5	-	$Na_4Fe(CN)_6$	NaCN	10	28	0.0885	1.0	89	67
Silver Bells, AZ, ASARCO	11.0	-	Na_2S, NaSH	-	8	30	0.1885	0.8	84	30
Pima, Arizona, Cyprus	11.0	Roast	NaSH, $(NH_4)_2S$	Roast for talc	6	28	0.1042	1.8	83	33
Bagdad, Arizona, Cyprus	8.5	-	Nokes	NaCN	12	30	0.1590	1.0	88	64
Morenci, AZ, Phelps Dodge	10.5	cook	$Na_4Fe(CN)_6$	$Na_2Zn(CN)_4$, H_2O_2	8	26	0.0685	1.0	83	-
Twin Buttes, Arizona, Anamax	10.5	-	NaSH	-	8	29	0.2573	1.1	76	35
Inspiration, AZ, Inspiration	10.5	autoclave	Nokes	-	7	38	0.1792	0.5	76	40
Pinto Valley, Cities Service	11.0	-	Nokes	-	7	27	0.1188	1.12	87	36
Canada										
Brenda, B.C., Noranda	8.0	-	NaSH	NaCN	12	28	0.2092	0.1	88	81
Gaspe, Quebec, Noranda	10.0	Steam	NaSH	$Na_2(CrO_4)$	20	30	0.0688	0.5	90	56
Island Copper, B.C., Utah Int	10.5	-	NaSH	$Na_2Zn(CN)_4$, NaCN	10	24	0.0675	1.0	88	67
Lornex, B.C., Rio Algom	9.5	-	Nokes	-	7	34	0.0683	1.0	88	64
Gibralter, B.C., Placer	10.5	-	-	-	-	30	0.1580	1.0	85	45
Chile										
Chuquicamata, CODELCO	9.6	-	Anamol D	NaCN	5	42	0.2090	0.1	90	56
El Teniente, CODELCO	-	-	Nokes	NaCN	25	42	0.1595	0.5	83	42
El Salvador, CODELCO	11.0	-	Anamol D	NaCN	7	46	0.1097	1.0	81	60
Andina, CODELCO	9.0	-	NaSH	-	8	28	0.0388	1.0	90	50
Peru										
Toquepala, Southern Peru	8.0	-	$Na_4Fe(CN)_6$	$Na(ClO_4)$	8	32	0.1287	1.2	88	37
COMECOM										
Balkhash, Kazakhstan, USSR	11.0	Steam	Na_2S	-	7	23	0.0560	na	85	41
Almalyk, Uzbekstan, USSR	11.0	-	Na_2S	-	6	17	0.0450	na	83	40
Kadzharan, Armenia, USSR	10.5	-	Na_2S	-	6	16	0.2079	0.7	80	65
Medet, Bulgaria	8.0	Steam	Na_2S	-	8	20	0.1041	1.8	82	26

Index to abbreviations for Tables 33 and 34

Z-4	=	Na ethyl xanthate
Z-6	=	K amyl xanthate
Z-7	=	K butyl xanthate
Z-10	=	Na hexyl xanthate
Z-11	=	Na isopropyl xanthate
Z-14	=	Na isobutyl xanthate
Kero	=	kerosenes
F.O.	=	fuel oil
Z-200	=	ethyl isopropyl thionocarbamate
Miner 1661	=	" " "
Miner 2030	=	" " "
Miner 898	=	ethyl isobutyl xanthogen formate
Minerec A	=	ethyl ethyl xanthogen formate
Reco	=	Na dicresyl dithiophosphate
Re 404	=	mercaptobenzothiazole
Re 3302	=	allyl amyl xanthic ester
Cres. ac.	=	cresylic acid
P.O.	=	pine oil
MIBC	=	methyl isobutyl carbinol
AF 65	=	polyglycol ether
Dow 250	=	polypropylene glycol ether

molybdenite cleaning process, supplied from sodium sulphhydrate in the western world and from sodium sulphide in Russia. Table 33 lists the reagents employed in the seventies in moly/copper separations.

A very important factor in choosing the primary collector/frother system for ores that contain molybdenite is the effect of the reagents on the subsequent moly separation from the copper concentrate, as the maximum copper allowed (0.5%), and molybdenum grade (50%) specified in the final moly concentrate, is stringent. The flotation circuit used in moly recovery nearly always is based on depressing the copper and other minerals and floating off the molybdenite into a concentrate with less than 1% copper; if this is not attained, the concentrate must be leached (usually with ferric and ferrous chloride) to remove surplus copper. The copper depression processes employed can be divided into chemical (as mentioned), and thermal, which include roasting and steaming. Roasting will normally decompose the collector on the mineral surface and vaporise the frother; it may also oxidise the copper mineral surface to some extent and provide natural depression. The concentrate, after thermal treatment, is then floated in up to 25 stages of cleaning, using fuel oil and MIBC as the flotation reagents.

Note that there are only six mills that do not use some form of sulphhydrate ion in the copper/molybdenite separation. Flotation recovery of molybdenite from porphyry copper ores is governed primarily by copper metallurgy, but is adjusted to accelerate and optimise molybdenum recovery as well, particularly where molybdenum is an important by-product or co-product, such as at Sierrita, Brenda or Chuquicamata. As can be seen from the above tables, there is no consistent pattern in the relative efficiency of the recovery of copper or molybdenum, except for the generally better recovery from ores with a lower copper content. This improved recovery is probably due to the greater attention paid to optimising molybdenum recovery and the lesser crowding out effect of other sulphide minerals. The regional average mine head assays, and corresponding overall recoveries, tend to confirm this surmise, viz:

Table 35.- REGIONAL MINE PERFORMANCE

Region	number of mines	Heads		Recoveries	
		% Cu	% Mo	% Cu	% Mo
United States	16	0.66	0.02	83.9	50.3
Canada	5	0.42	0.021	87.8	62.6
Chile-Peru	5	1.62	0.031	86.4	49.8
COMECOM	4	0.67	0.02	84.8	43.0
Total average	30	0.78	0.022	84.8	50.8

(Note: these data are deliberately pre-1980 because they provide a more comprehensive technology comparison)

The tables show that in 67% of the plants producing molybdenite, some form of a xanthate is used as the collector. Half of them used it as a single collector, with or without the addition of fuel oil as a moly promoter. Eight plants used a dithiophosphate coupled with a xanthate and fuel oil; none used only a thiophosphate. Five employed a thionocarbamate; only two of them did not add a second collector. Two mines - El Teniente and Inspiration - used xanthogen formates. Finally, five operations used a xanthic ester (Cyanamid RE 3302), usually combined with a xanthate. Xanthic esters are the only family of collectors that show a statistically significant correlation with molybdenite recovery.

Reviewing frother practice, we find that 18 of the 30 operations employed MIBC as the primary frother, 12 in combination with still bottoms (Shell 10), which is a heavier alcoholic fraction. The next most popular frother was pine oil, which is reputed to improve molybdenite (unconfirmed statistically), and which was used by 12 operations. In four cases it was combined with MIBC, in 3 cases with a soluble frother (Dowfroth 250), and never with cresylic acid. Cresylic acid was used in 3 mines, in two cases combined with MIBC. Soluble frothers were used in 6 cases, of which only in one case - that of El Teniente (Dowfroth 1012) - it was used as the sole frother.

An analysis of all the data available for reagent practice in copper sulphide concentrators (Crozier, 1978) showed no statistically significant correlation between molybdenite recovery and either collector or frother selection, with the exception of a weak one with xanthic esters, despite the amount of published opinions on improved moly recovery with fuel oil additions to the collector suite, or the use of pine oil as a frother. The probable reason for this contradiction is that in mills for which the data was analysed, the main collector-frother combination was selected to enhance the recovery of the most valuable mineral component - usually copper. The exceptions are Sierrita in the U.S.A., with copper heads of only 0.3%, and Brenda in Canada, with 0.2 Cu in the concentrator feed. In both cases the molybdenum recovery is much higher than the norm (76% and 81%, respectively), indicating that the circuit design optimised moly and copper recovery. It is significant that in both mines a xanthic ester (Re 3302) is used as the collector, and MIBC as the frother. Two mines in Canada and two in the United States, which have low copper heads (0.4%) also have abnormally high molybdenum recoveries (61%); again Re 3302 was used with a xanthate as the collector.

Copper depression in the molybdenite separation

In the recovery of molybdenite from a copper concentrate, it is the norm to depress the copper and pyrite gangue, and float off the molybdenite, although in very unusual cases the opposite has been opted for by the metallurgist. The more important copper depressants employed in moly separation have been:

1.- Sodium ferrocyanide and ferricyanide
2.- Sodium zinc cyanide
3.- Sodium cyanide
4.- Nokes reagent (P_2O_5 dissolved in NaOH)
5.- Arsenic Nokes, or Anamol D (AS_2O_3 dissolved in Na_2S)
6.- Sodium sulphide and sulphhydrate
7.- Ammonium sulphide
8.- Sodium thioglycolate

In addition, according to Table 34, in the U.S.A., 9 out of 16 mills employ a thermal process to strip the rougher flotation reagents from the copper concentrate prior to moly recovery. This luxury was not indulged in the rest of the world except for the COMECOM countries, where 2 out of the 4 mills listed strip the primary reagents. The stripping processes employed were:

1.- Steaming

2.-Pressure steaming
3.-Skin roasting (a light roast)
4.-Cooking (indirect heating of the pulp)
5.- Conditioning with acid and hydrogen peroxide
6.-Conditioning with sodium hypochlorite

Needless to say, the skyrocketing oil prices of the early seventies discouraged heat treatment of copper concentrates, and encouraged the development of more sophisticated copper depression schemes, as also the use of hydrometallurgy to effect the final copper removal from the moly concentrate.

Post reagent stripping: the practice used in the removal of molybdenite from the copper concentrate was to rougher float a molybdenite concentrate, and then submit this concentrate to multiple cleaning steps (from 5 to 20). This operation has been an obvious candidate for the substitution of multiple flotation banks by flotation columns; but this substitution is not easy because the specifications on the minimum copper, pyrite and insoluble content of commercial molybdenite concentrates is stringent, and the value of the co-product to the mill does not allow the luxury of Mo losses to the copper tails.

Sodium sulphide as a depressant

Because hydrogen sulphide is a weak acid, prediction of mineral depression by its ions requires a knowledge of the system pH. The classic work elucidating this subject was published by Sutherland and Wark (1955). They point out, based on contact angle data, that, just as in the case of the cyanide ion, "there is a critical hydrosulphide ion concentration below which adsorption of xanthate is possible, above which adsorption is impossible" (see Fig. 25). They add that the critical hydrosulphide ion concentration for 25 mg/l of potassium ethyl xanthate is: 0.01 gm HS^-/l for galena, 0.30 mg/l for chalcopyrite, 1.3 mg/l for bornite, 1.7 mg/l for covellite, 2.5 mg/l for pyrite, and 6.4 mg/l for chalcocite. Similar curves can be measured for other collectors and at different concentrations.

The mechanism of depression of sulphide minerals by the HS^- ion is controversial. Leja (1982) suggests that the presence of colloidal sulphur on the mineral surface could be a depressive factor, that thio anions and sulphanes may be formed on the mineral surface, or that the sulphhydrate ion could destroy the collector coating on the bulk floated mineral, if depression is during the cleaning process. Klassen and Mokrousov (1963), writing in the late fifties, quote Russian data showing that the depressive effect of sulphides is considerably more effective, the shorter the alkyl group

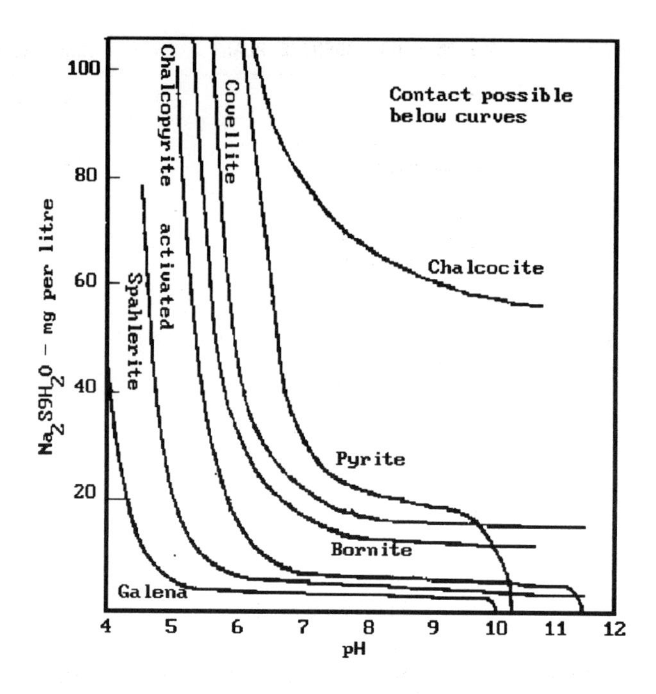

Fig. 25.- Relationship between pH and the concentration of sodium sulphide necessary to prevent contact at the surface of various sulphide minerals, in the presence of 25 mg/l of KEtX

is on a xanthate. They also emphasise the decomposition of the sulphhydrate ion in solution by oxygen, suggesting that in many cases the depression of a sulphide mineral may well be due to oxygen scavenging in the pulp, pointing out that "sulphide minerals depressed by sodium sulphide regain floatability if sufficient time is allowed for the dissolved oxygen to react with the sulphide". Also of interest is their observation of the catalytic effect of mineral surfaces on the decomposition of the sulphides, saying:

> "...variations in the activating or depressing action of sodium sulphide with time are due to the fast oxidation of hydrosulphide and sulphide ions in the pulp. Studying the rate of oxidation of these ions, Shvedov and Shorsher found that, on agitation of an aqueous solution of sodium sulphide, the bivalent sulphur ions disappear from solution in 35 minutes. When finely ground pyrite is introduced into the same solution (10%), the sulphide ions disappear from the solution even faster: they cannot be detected after 20 minutes. In both cases, parallel with the disappearance of

bivalent sulphur ions, a quantity of $SO_4^{=}$ ions appear." ....

"Mitrofanov considers that the sulphide ion oxidation proceeds catalytically on the surface of particles. The average time of oxidation of monolayers of $S^{=}$ and HS^{-} ions on the surface of such minerals as galena is on the order of a fraction of a second."

Sulphides in the copper/moly process: the first use, in the western world, of sodium sulphide in the depression of copper in a molybdenum by-product plant appears to have been in 1934 at Cananea (Mexico). The process also was patented in Russia in 1936. The Nokes reagents, a sulphhydrate variant, were patented in 1949 by Kennecott for use in El Teniente.

Table 36.- MOLY PLANTS USING THE SULPHIDE PROCESS - Shirley (1979)

Anamax	NaSH only depressant used - hot water in the rougher.
Andina	NaSH only depressant.
Mission	NaSH used only in the rougher and first two stages of cleaning prior to roasting.
Pima	Preconditioning with ammonium sulphide prior to rougher flotation. A mixture of NaSH (85%) and $(NH_4)_2S$ (15%) used as primary depressant. Sodium cyanide and Zn cyanide complex used in various of the cleaner stages.
Brenda	NaSH used a primary depressant. NaCN used in final cleaners.
Island Copper	NaSH used a primary depressant. NaCN used in final cleaners.
Lornex	NaSH used a primary depressant. NaCN used in final cleaners.
Gibralter	Preconditioning with $(NH_4)_2S$, with NaSH then used as the primary depressant.
Gaspé	NaSH used a primary depressant. NaCN used in final cleaners.
Russian Plants	Na_2S used as a primary depressant. Plant feeds generally steamed prior to the sulphide addition. Hot pulp floated in the rougher.
Inspiration	Ammonium sulphide preconditioner prior to rougher and phosphate Nokes as primary depressant
Pinto Valley	Ammonium sulphide preconditioner prior to rougher and phosphate Nokes as primary depressant

Air (oxygen) and water are the most important variables with respect to sulphide consumption in the depression process. The effective amount of sulphhydrate ion is based on its ionic concentration, therefore pulp dilution is a major factor in consumption. Shirley (1979) reports (Table 36) that at a Canadian mine treating chalcopyrite, where the molybdenite rougher is operated at 25 to 30% solids, NaSH consumption is about 10 kg/t, while other mines running at 35 to 40% solids have consumption rates just above 10 kg/t. A further reduction in sulphhydrate consumption can theoretically be obtained by replacing the air in the flotation machines with nitrogen sourced from an oxygen plant feeding a flash smelter. Apparently the laboratory reductions in reagent consumption are not always duplicated consistently in the full-scale operations.

In many applications the sulphide fed into the copper/molybdenite roughers is supplemented with sodium cyanide or the ferro or ferricyanides in the cleaners to try and reduce the number of stages required to meet concentrate specifications.

The chemistry of the Nokes reagents

The Nokes patent (US Patent No. 2,492,936) specified a series of formulations which produced HS^- for the depression of copper in a moly separation. Only two have been used commercially: LR-744, developed for El Teniente in 1949; and arsenic Nokes, better known as Anamol D, as it was developed by Anaconda for Chuquicamata, but fell under the claims of the Nokes patent.

LR-744, or Phosphate Nokes is easily prepared, but involves a very dangerous operation as H_2S is evolved, and can be evolved copiously if the reaction runs away. As has been noted above, hydrogen sulphide is as toxic as hydrogen cyanide, as well as having a very broad explosive limit in air, so precautions must be taken even in laboratory preparations. At an industrial scale it is prepared in a jacketed stirred tank with adequate cooling water availability. The apparatus and the area must be explosion proof, using gasoline refinery standards for the electrical installations. The recipe for the preparation is to dissolve caustic soda in water (highly exothermic), cool the solution to 40-45°C, then slowly add phosphorous pentasulphide, keeping the tank stirred vigorously and at a temperature lower than 75°C. The ratio of NaOH to P_2P_5 should be around 3:2 (El Teniente uses 13:10), and amounts should be such that the resultant solution will be 25% by weight.

The reaction has been studied by Castro and Pavez (1977), who postulate the following main reaction

$$P_2S_5 + 10NaOH \longrightarrow Na_3PO_3S + 2Na_2S + 5H_2O$$

while Gutierrez and Sanhueza (1977) indicate that they have identified the following hydrolysis products

$$Na_3PS_4 + H_2O \longrightarrow Na_3PS_3O + H_2S$$

$$Na_3PS_3O + H_2O \longrightarrow Na_3PS_2O_2 + H_2S$$

$$Na_3PS_2O_2 + H_2O \longrightarrow Na_3PSO_3 + H_2S$$

$$Na_3PSO_3 + H_2O \longrightarrow Na_3PO_4 + H_2S$$

Castro and Pavez (1977) determined that at a 13:10 ratio, the solution assayed an equivalent 27% Na_2S, and at 1:1, 30%. Using this reagent, El Teniente is one of the very few mines that does not need to supplement the depressors to meet concentrate copper specifications, nor do they heat the concentrate of the rougher pulp. The improved effectiveness is attributed to unidentified phosphorous compounds.

Anamol D or Arsenic Nokes is covered by the Nokes patent, but in use its addition rate is adjusted to maintain a pulp emf above -200 mV, which is covered by an Anaconda patent issued to J.F. Delaney (US Pat No 3 655 044, 1972). It was for many years the standard depressant in Anaconda's Chuquicamata mine. Its use was discontinued by CODELCO because of concern over handling arsenic trioxide on site. Anamol D is prepared by dry mixing arsenic trioxide and sodium sulphide, or can be prepared by adding arsenic trioxide to a sodium sulphide or sulphhydrate solution. Ease of dissolution depends on the purity of the arsenic trioxide. The ratio of sodium sulphide to arsenic trioxide employed varies between 3 and 4:1. In a mixture containing 4 parts sulphide, only 27% of the Na_2S reacts. Castro and Pavez (1977) suggest that the following reactions occur:

$$As_2O_3 + 3Na_2S + 2H_2O \longrightarrow Na_3AsO_2S_2 + Na_3AsO_3 + H^+$$

$$As_2O_3 + 3Na_2S + 2H_2O \longrightarrow Na_3AsO_4 + Na_3AsOS_3 + H^+$$

These reactions indicate that the depressant species are HS^-, sodium arsenate and a mixture of mono-, di-, and tri-thioarsenate. With Anamol D, the HS^- ion is more important as a depressant than the arsenic compounds, which probably function as oxidation inhibitors to protect the HS^-; while in the LR-744 there is only a small residual HS^- concentration, suggesting that the phosphorous compounds are probably the more important depressants for copper.

Laboratory testing to develop, or improve, a moly/copper separation system requires considerable experience with the process; but today it is obvious that column flotation will provide a much lower capital hardware alternative, and should change the depression steps to a simpler scheme. Because of the aging effects on wet concentrate, it is prudent to plan the experimental work on-site, even though all the analytical and mineralogical work can be done at a better equipped research facility.

Sulphidisation of refractory ores

In studying the processing of oxide ores, it is important to pay attention to the composition of the gangue. There are two main varieties of gangue: siliceous and carbonaceous. The first is mainly silica and the latter calcium carbonate. In addition, pyrites will weather to limonite, and if alumina is present there will be colloidal clays. The main oxidised copper minerals are: the two basic carbonates, azurite and malachite; the oxide, cuprite; and various silicates, of which chrysocolla is the most important. The principal lead oxides are: the carbonate, cerussite; and the sulphate, anglesite. The main zinc oxide minerals are: the carbonate, smithsonite; and the silicate, calamine.

The first use of sulphidisation was with Australian lead-zinc-silver ores, the pioneer flotation industry. The excessive consumption of sodium sulphide and difficulty in controlling the flotation process has kept the Australian CSIRO active in research on methods of improving the process. Very early on, the stage addition of the reagents was instituted, and in the past decade the use of ion specific electrodes to control the residuals of SH^- and collector molecules has been successfully applied. Jones and Woodcock (1979) studied the recovery of values from oxidised lead minerals from old dump material, which they treated by flotation of a sulphide concentrate (galena and marmatite) followed by sulphidisation and flotation of an anglesite concentrate. Sulphidisation by slug addition of Na_2S gave relatively uncontrolled sulphidising conditions, but potential controlled sulphidisation, using a sulphide ion-selective electrode (ISE), resulted in good control and improved metallurgy. The optimum E_s value for three stage flotation was found to be -600 mV (the absolute value may not be reproducible because of differences in electrode construction). They noted that sulphidisation time, at -600 mV, over the of range 1 - 5 minutes, had little effect on oxide recovery. Flotation time was important and three 10 minute stages (total 30 minutes) was needed for good lead recovery. A decant wash prior to sulphidising improved metallurgy and decreased sulphide consumption. Excess $S^=$ in solution displaced xanthate from the mineral surface, the amount displaced varying dependent on E_s. At E_s more negative than -600 mV, maximum displacement occurred.

The laboratory procedures followed are of general application for sulphidisation studies, so will be reviewed in detail. Flotation tests were made in a Denver D1 lab flotation machine, using a 500 ml glass cell. Impeller speed was fixed at 1800 rpm. The E_s values were measured as the potential between an ISE and a saturated calomel electrode (SCE). pH measurements were made with a glass electrode/SCE pair. E_{Pt} values are the potential of a platinum electrode versus SCE. Oxygen concentrations were expressed as per cent saturation as measured with a Beckman 777 oxygen analyser.

Flotation tests were conducted on 500 gm portions riffled from a bulk sample of residue dump material supplied by The Zinc Corporation, Broken Hill, N.S.W., Australia. The sample assayed 3.69% total lead, 2.74% total zinc, and 76 g/t silver. About 3/4 of the lead was present as anglesite. The zinc was mainly present as marmatite. Soluble salts, including zinc and manganese, were present. Sizing of the material was 2.5% plus 0.417 mm, 26% minus 0.074 mm, with 80% passing 0.208 mm. Commercial reagents were employed in testing. Flake sodium sulphide (62% Na_2S) was used as water solution containing 5, 10, or 20% sodium sulphide, depending on the E_s values required. Potassium amyl xanthate (KAX) from American Cyanamide and MIBC from Shell (Australia) were used as 1% water solutions. Sodium metasilicate (BDH technical reagent) was used as a 10% $Na_2SiO_3.5H_2O$ solution, and copper sulphate was used as a 5% $CuSO_4.5H_2O$ solution.

The flotation procedure used was based on a scheme provided by The Zinc Corporation. A 500 g sample was wet ground in an iron ball mill for 20 min., giving a product sizing 5% plus 0.1457 mm, 58% minus 0.074 mm, and 80% passing 0.104 mm. After transfer to the cell and conditioning for 1 min. with 225 g/t copper sulphate, the sulphide minerals (galena and marmatite) were floated with 135 g/t KAX and 18 g/t MIBC, using 5 min. roughing and 10 min. scavenging. Conditions for this sulphide float were kept constant. The rougher and scavenger concentrates from each test were assayed separately, but the results for a combined sulphide concentrate are reported.

After sulphide flotation, oxide flotation was conducted either immediately, or, as in plant practice, after decantation of mineralised liquor. With no decantation, flotation was accomplished by conditioning the pulp with 1800 g/t sodium silicate for 1 min., sulphidising with sodium sulphide as described later, and then floating with KAX and MIBC. However, in some tests the pulp was allowed to settle, and a constant volume (1200 ml) of slightly turbid supernatent liquor was decanted to remove dissolved salts. This decanted liquor was added to the final tailings for filtration. The settled pulp was repulped with 1200 ml of tap water, and, after conditioning with sodium silicate, sulphidised and floated.

Two different sulphidising-flotation techniques were investigated.

1.- A single slug addition of Na_2S was made. After 1 min, 225 g/t $CuSO_4.5H_2O$ was added, and after a further 1 min, 90 g/t KAX and 4.5 g/t MIBC were added. Rougher flotation of an "oxide" concentrate (i.e., sulphidised anglesite) was conducted. After addition of a further small amount of KAX and MIBC, a 10 minute scavenging was conducted.

2.- Three sulphidising periods of 1, 3 or 5 min at controlled E_s levels were used with the air inlet on the cell closed. After each sulphidising period, KAX and MIBC were added and a 5 or 10 min flotation period was used. No Na_2S was added during the flotation periods, and the concentrates were kept separate. In most tests the E_s value was kept at the same level in each of the three stages. In some tests, however, a different value of E_s was used in each stage.

All solid products from each test were dried, weighed, and assayed. After fusion with Na_2O_2, cooling and dissolution of the melt with 1:1 HCl, and appropriate dilution, total lead and zinc were determined by atomic absorption spectrophotometry (AAS). Some products were assayed for silver by AAS or fire assay.

Leach Precipitation Float (LPF): the process, where copper from partially oxidised ores was recovered by a sulphuric acid leach followed by metallic copper precipitation with iron particles in the pulp and flotation, is now obsolete, though the term is used to describe certain sulphidisation schemes. Currently, oxide copper ores are heap leached with sulphuric acid, the dissolved copper separated from iron in solution by liquid-liquid extraction, and metallic copper recovered by direct electrowinning, a very low capital and cost process; it rescued the financially crippled Arizona copper mining industry in the early eighties by providing instant cash flow.

The classic LPF process was still used in the USSR into the sixties, but in the post World War II western economy LPF had been displaced by vat leaching and cementation with scrap iron. When the vast amount of cheap scrap iron from the battle fields was exhausted, the LIX copper extraction process was introduced. This process produces copper in an acid solution pure enough to be directly electrowon. A precocious example of direct electrowinning was the Anaconda Company operation in Chuquicamata, Chile (1913) , where the oxide copper ore was low in iron, and therefore could be sent directly to an electrolytic refinery, without elaborate pre-purification.

Extreme sulphidisation is now frequently referred to as an LPF process. The closest to the original LPF process was the procedure used in the Butte concentrator of the Montana Mining Division cf the Anaconda company. This plant was built in 1963; its design included a sand/slime separation to allow efficient recovery of the oxide copper in the slime circuit. Palagi and Stillar (1976) describe the process as follows: the slime flotation plant treats mainly -200 mesh material sourced from the primary cyclones (5.6% +65 mesh, 18.2% +200 mesh, and 70% -400 mesh). In sweet ore there was relatively little soluble copper in the slime, so it was floated normally. In sour ore, the slime's grade ran up to 0.1% Cu higher than the sand grade, with up to 40% of the acid soluble copper in the plant feed reporting to the slime fraction. With sour ore, to recover the copper in the slime, this was acidified to pH 4.0 and piped to the special slime treatment plant where the pulp was stored for 20 minutes in an agitated leach and precipitation tank, where the pH was lowered to 2.0. After the acid soluble copper was in solution, the slime was reacted with hydrogen sulphide gas in an in-line static mixer, where artificial covellite was precipitated. The amount of H_2S fed was automatically controlled by an EMF probe held at 0 to +150 mV. The pulp was then floated with diethyl xanthogen formate and Cyanamid 3302 as the collectors, and the concentrate upgraded in the sand float cleaner, as it was essentially free of colloidal fines. Separate treatment improved slime copper recovery by 15 to 20 points.

According to Bolles (1985), Morenci originally treated a monzonite porphyry containing primarily chalcocite and pyrite. Later, considerably greater amounts of "oxide" copper was encountered: mainly cuprite, azurite, malachite, and brochantite. The bulk of the oxidised copper was/is formed subsequent to the breaking of the ore as imperceptible coatings on the chalcocite. Oxidation by weathering begins when the ore is broken, and continues until it is delivered to the crushing plants. Typically, the proportion of oxidised copper increases by two to three times between drilling for blasting and delivery to the mill. For 1972, he quotes an average assay of: total copper, 0.83%; oxide copper, 0.16%; pyrite, 5%; quartz, 43%; feldspars, 21%; micas, 4.3%; molybdenite, 0.015%; gold, 0.0014 oz/t; silver, 0.0454 oz/t. Bolles further says: "..the Morenci LPF process converts approximately 40% of oxide copper minerals to a sulphide which can be recovered by conventional flotation methods in an alkaline circuit", and describes it as follows:

"Acid soluble copper content of the Morenci ores often

> process was devised to recover part of this copper bearing mineral. The process involves leaching the acid soluble copper minerals with dilute sulphuric acid in a revolving drum. Precipitation of the dissolved copper by a unique sulphide precipitant was done simultaneously in the same vessel. The treated slurry was made alkaline and subsequent grinding and flotation operations were performed on the alkaline pulp. Flotation of the precipitated cupric sulphide occurred together with the naturally occurring sulphide copper."

The chemical reaction took place within a slowly rotating drum 6 ft in diameter and 26 ft long. The drum was lined with acid-resisting brick and was equipped with rubber-covered lifters for discharge of treated ore slurry. Contact within the drum averaged 5.5 min.

Dry ore was fed into the drums by conveyor belts; water was then added to dilute to approximately 78% solids within the drum, followed by sulphuric acid to lower the pH to 1.6. Finely ground calcium sulphide in a water slurry was drawn from a circulating loop pipeline and injected into the drum. Control of this reagent is vital to the success of the process. This was accomplished with an oxidation-reduction meter whose electrodes were immersed in a sample stream of the discharge solution. A negative millivolt reading was maintained, which resulted in not-quite-complete precipitation of dissolved copper.

Normal operation results in minor evolution of hydrogen sulphide gas, which increases with excess use of precipitant. Therefore, fume control is essential. This was accomplished by directing the flow of air that sweeps the drum mixer into the ball mill, where lime additions completely react with the H_2S in the gases during passage through the ball mill.

The calcium sulphide precipitant was an impure product produced locally from pyrite and burnt lime. These components, mixed in proper proportions, were finely ground (dry), then moistened and pelletised. After hardening by gentle heating and screening to remove fines, the pellets were fed with fine coal into a reducing kiln. Reduced pellets, containing both calcium sulphide and metallic iron, were magnetically cleaned, then stored for wet grinding before use.

Chapter 10

MILL TESTS - CASE HISTORIES

San Manuel - a non-sulphide moly separation process

At San Manuel, the final copper concentrate, which contains between 1.0 and 1.2% molybdenite, is thickened to 50% solids and sent to two conditioners in series. In the first conditioner, a sodium zinc cyanide complex is added to the pulp, which is kept at a pH of 6.5 to 7.0, using sulphuric acid. The second conditioner is fed oxide. Conditioning takes between 22 and 25 minutes. From the conditioner the pulp is diluted to 20% solids with fresh water, and fuel oil is added as the moly collector while it is sent to the rougher flotation stage, where the pH is controlled with sulphuric acid. The feed to the first cleaner combines the concentrate from the rougher and the tails from the second and third cleaners. In the first cleaners and the scavengers, sodium ferrocyanide is added as the copper depressant, and MIBC as the frother. In the next two cleaning stages the reagents added are potassium ferricyanide and sodium hypochlorite plus MIBC as the frother, following which potassium ferricyanide is added at each stage. The froth in the final stages is kept in check with Exfoam (Dearborn Chemical Co). The amount of each reagent fed in the different flotation stages is controlled by measuring the pulp's Redox potential. The San Manuel flowsheet thus consists of two distinct stages: (1) collector removal in the conditioners and the rougher flotation using hydrogen peroxide and sodium zinc cyanide, and (2) ferro and ferricyanide copper depression in the cleaner and scavenger circuits. The moly concentrate from the final cleaners may or may not be leached to meet the copper specs. San Manuel's copper and moly concentrate assays are shown in Table 37.

Because Farlow Davis's paper on the San Manuel moly plant was only circulated as a mimeographed handout at the local section meeting of the AIME in Arizona, and therefore difficult to obtain, the following pages are reproduced verbatim, as they provide his explanation of the mechanisms involved in the copper depression process at San Manuel.

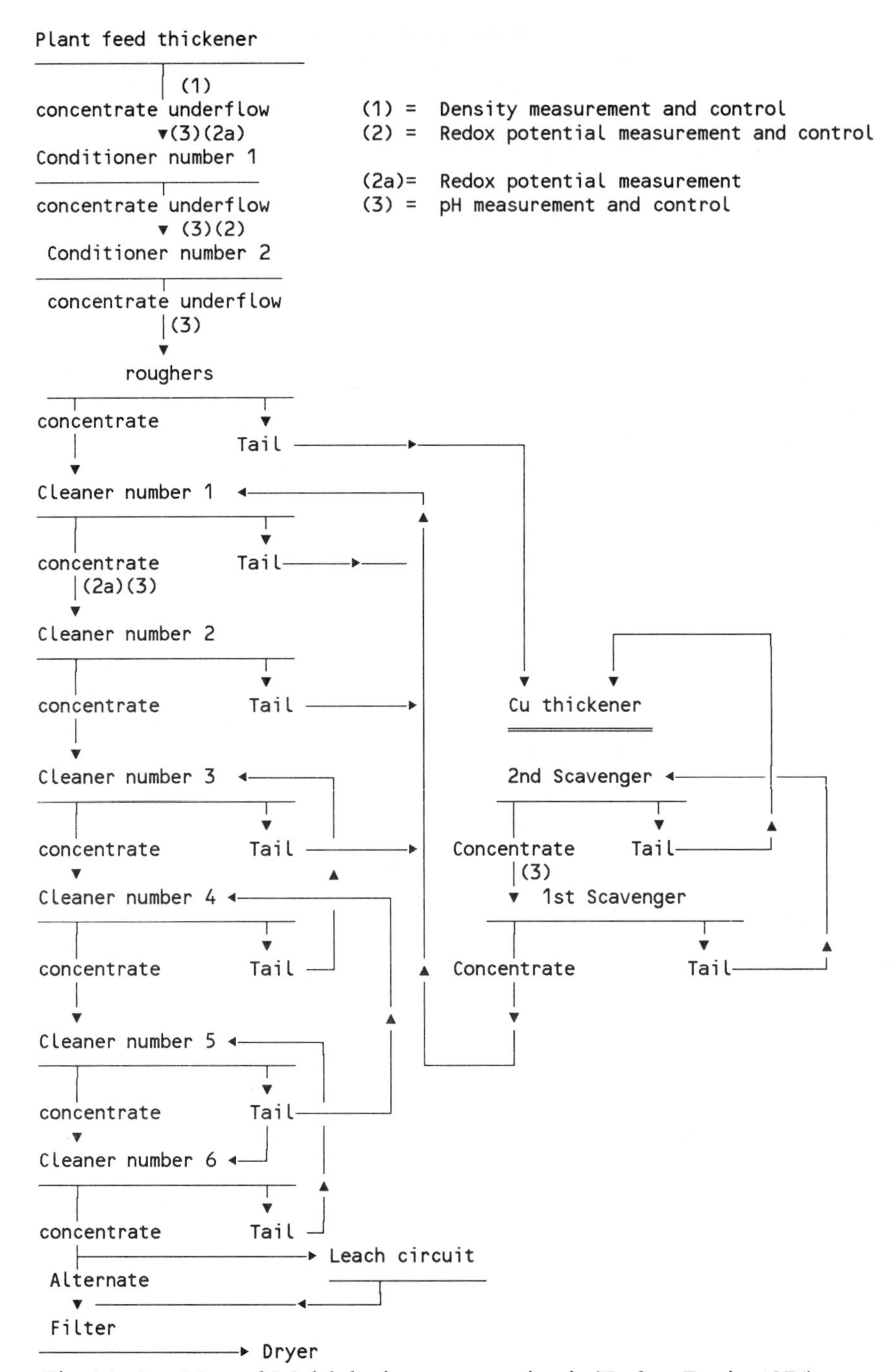

Fig. 26.- San Manuel Molybdenite recovery circuit (Farlow Davis, 1976)

Table 37.- SAN MANUEL COPPER AND MOLY ASSAYS

	Copper Concentrate Raw	Copper Concentrate Floated	Molybdenum Concentrate
Cu	27.6%	27.9%	0.5%
MoS_2	1.12%	0.2%	96.2%
Re	0.0017%	-	0.08%
Au oz/t	-	-	0.38

Instrumentation: The molybdenite plant thickener is equipped with a density measuring device and vari-drive pump operating in conjunction with a recorder-controller to maintain a constant density and stable flow to the conditioning tanks. Manual control of the underflow can produce surges in the conditioners and rougher, detrimentally affecting the copper-molybdenite separation.

The pH of the conditioners, rougher and first scavenger are automatically controlled by the addition of sulphuric acid with positive displacement pumps.

Oxidation-reduction potential (ORP) measurements are taken from the #1 conditioner, #2 conditioner and second cleaner. .... Using a platinum electrode and a silver-silver chloride reference, the ORP measurement depends on the ratio of the reducing or oxidising activities or concentrations and not upon the absolute amounts present. Thus, in an operational sense, ORP can be understood about as well as pH.

Laboratory work has shown a strong response of ORP to the reagents used in the conditioning steps of the molybdenite circuit. This strong response of the ORP to reagent conditions controlling the oxidation-reduction reactions in the pulp suggested the use of this function in regulating rates of reagent addition.

Sodium zinc cyanide: This is the reactant product of sodium cyanide and zinc oxide. It serves two purposes: (1) to provide a desired level of HCN in the conditioning step with hydrogen peroxide; and (2) to remove, either wholly or partially, the collector coating on the mineral surface. The addition rate is critical for an effective conditioning step with hydrogen peroxide. As will be explained later, too little or too much are equally unsatisfactory. It is added to the #1 conditioner to prepare the collector on the chalcopyrite for nullification by the hydrogen peroxide in the second conditioning step.

The process of cyanide attacking the chalcopyrite surface would be driven by two factors: (1) the property of cyanide as a reducing agent; and (2) the ready complexation of Cu^{+} and Fe^{++} ions at the mineral surface with a slow solution of both. On slightly altered chalcopyrite surfaces, cyanide would reduce cupric sites to cuprous, at the same time losing an equivalent free cyanide as cyanate. A monolayer of adsorbed cupro and ferro cyanogen complexes would form which, given time, would solubilise leaving vacant sites for further cyanidation and solution. The process is one of mobile equilibrium.

Since at least some surface oxidation seems to be required for effective mineral flotation by soluble xanthates or oily collectors, the reducing action of cyanide plus the presence of a cyanide monolayer would represent two strong deterrents to mineral surface-collector interaction.

While the foregoing may apply to clean or slightly altered chalcopyrite surfaces, the chemistry is complicated by chemisorbed and physically absorbed reagents responsible for the hydrophobic condition of the mineral surface. These would include the primary and secondary collectors in addition to penetrations of the collector hydrocarbon members by fuel oils, alcohols and perhaps other products of collector oxidation and decomposition.

Admittedly, the action on the mineral surface under these conditions is not clear. If the cyanide does penetrate the absorbed collector, then some complexing and solution would be expected. It is possible that metal collector bond cleavage might take place leaving the collector ions in proper condition for reaction with the hydrogen peroxide; or, at least, leaving the surfaces clean for reaction with the ferro-ferri cyanide complex.

Hydrogen peroxide (fed as a 50% solution) serves two purposes. First, the pulp is conditioned with hydrogen peroxide to transform all the water soluble sulphhydryl type collectors present, such as the normally used isopropyl xanthate, to insoluble materials. Secondly and possibly most important, the residual peroxide in the pulp leaving the conditioning step oxidises a portion of the sodium ferrocyanide fed to the rougher flotation to form ferricyanide.

At the inception of using hydrogen peroxide in the molybdenite plant, S-3302 started to be used in the copper plant as the primary collector with satisfactory results in the molybdenite plant. Since then, Minerec SM-8 has replaced the S-3302. Neither of these oily collectors seem to be acted upon by peroxide in the conditioning step. Several other oily collectors have been tested at San Manuel since hydrogen peroxide was put in use and in every case there was no noticeable effect in the conditioning step. Since these oily collectors are the primary contributors to the hydrophobic state of the mineral surface, it follows that any mechanism accounting for the depression of chalcopyrite must also account for the removal from, or neutralisation of, these collectors from the mineral surface. Straightforward oxidation of these molecules is not a satisfactory explanation. If, as has been suggested, the cyanide and the pH create an environment in the first conditioner whereby the collector ions are removed from the mineral surface, then oxidation of the collector may not be necessary if the mineral surface could be altered to prevent collector ion reattachment.

The hydrogen peroxide dimerises the isopropyl xanthate into what is commonly called dixanthogen. Laboratory tests have indicated that hydrogen peroxide will react in a similar manner with all the commonly used sulphhydryl type collectors. The dimerisation of these collector is a relatively slow one requiring minutes to take place even under the most ideal conditions. The main factors directly affecting the conditioning time are the pH of the pulp, the efficiency of the conditioner and the concentration of hydrogen peroxide in the pulp. The size of the conditioners was chosen on empirically derived data from the pilot plant tests and on the maximum volume of pulp that could be expected to be treated. It was soon realised that the large plant conditioners were not as efficient as the smaller pilot plant ones, and their capacity had to be increased.

To insure the dimerisation reaction going to completion within the time available, the pH of the conditioner wherein the hydrogen peroxide is added normally should be maintained between 6.5 and 7.0. Up to a point, the lower the pH, the faster the reaction takes place. Although the plant can be operated below 6.0 pH, it is more difficult because of the heavy metal ions put into solution, which in turn accelerate the decomposition of the hydrogen peroxide. In addition, the corrosion and maintenance costs would increase at lower pHs. With the conditioning time limited by the size of the conditioners, the reaction time required can be adjusted to some extent by increasing or decreasing the pH or the amount of hydrogen peroxide.

Other minor factors affecting the reaction time are temperature and the type and amount of reagent being used. The process has operated satisfactorily with pulp temperatures varying from 60° F to 90° F throughout the year. Different xanthates have been used with no necessary adjustments from one to the other. Oily collectors, as has been stated, apparently do not react with hydrogen peroxide in the conditioning step. It has been found that over-feeding the primary collector (SM-8) in the copper plant will adversely affect the copper-molybdenite separation. Chemical changes in the molybdenite plant will not remedy the effect. An increase in the secondary collector

(Z 11) in the copper plant will not result in the same situation, since apparently a change in the hydrogen peroxide feed rate will result in a satisfactory separation.

Another factor which indirectly affects the reaction time and directly affects the amount of peroxide that must be added is the concentration of the hydrocyanic acid in the feed to the second conditioner. Hydrocyanic acid is needed to stabilise the hydrogen peroxide in the rather hostile atmosphere of a copper concentrate pulp. Besides the normal catalytic decomposition of peroxide in such an environment, hydrogen peroxide reacts with chalcopyrite, San Manuel's primary copper mineral. This reaction will take place in preference to the slow dimerisation reaction. Hydrocyanic acid can be used to inactivate catalysts causing decomposition of the peroxide. It was also pointed out that if there were not enough hydrocyanic acid present to "poison" all the catalysts present, the catalytic decomposition would take place at a faster rate. In other words, enough hydrocyanic acid has to be present and anything less than enough is unsatisfactory. A sodium zinc cyanide-sodium cyanide mixture is added to the pulp in the first conditioner. It was found that the cyanide feed rate could be set at a fixed value which would provide a satisfactory amount of hydrocyanic acid during normal fluctuations in the amount of feed to the plant. Of course, drastic changes in the amount of feed would call for an adjustment to the cyanide feed rate. Also, it can be seen that lowering the pH would also call for an increase in the cyanide feed rate.

The function of sodium ferrocyanide and potassium ferricyanide is to produce a hydrophyllic surface on the copper and the iron minerals which, in effect, depress these minerals while allowing the molybdenite to float. Even though ferrocyanide has been effectively used to depress chalcocite in several plants, its ability to depress chalcopyrite in most cases has been proven to be unsatisfactory. The use of ferricyanide had not been seriously considered as a commercial depressant because of its relatively high cost and the general opinion among experimenters that there was little or no difference between using it or ferrocyanide. In the case of copper sulphide minerals such as chalcocite and covellite, there is probably little difference. However, in the case of chalcopyrite where there are both copper and iron positions on the surface, the problem becomes more complex. In all probability the ferrocyanide and the ferricyanide absorb equally well on the copper positions: however, absorption at only the copper positions would result in only partial depression of the mineral. Partial depression refers to a state in which the rate of flotation of a mineral is slowed down, but that the mineral will float if given enough time.

It is believed that in order to obtain complete depression, ferrocyanide or ferricyanide ions must also be absorbed at the iron positions as well as the copper positions. If the iron atoms on the mineral surface are in the reduced form, ferricyanide will be absorbed effectively; and if the iron atoms on the surface are in the oxidised state, ferrocyanide will effectively absorb on them. The iron in chalcopyrite is generally considered to be in the reduced state; therefore, ferricyanide would be an effective depressing ion. So, in the treatment of a commercial copper concentrate pulp containing chalcopyrite, two approaches can be taken. These approaches are: (1) the addition of ferrocyanide with a small amount of ferricyanide to take care of the reduced iron positions, or (2) the addition of ferricyanide with a small amount of ferrocyanide to take care of the oxidised iron positions.

In San Manuel's process, the first method of approach is utilised in the rougher and the first cleaner. However, in both cases, it is unnecessary to add ferricyanide as such. In the rougher feed there is residual hydrogen peroxide which reacts with part of the ferrocyanide addition to form ferricyanide. This is a reversible reaction. Residual sodium hypochlorite existing in the second cleaner tailing oxidises part of the first cleaner ferrocyanide to ferricyanide.

Although pyrite has not been mentioned in this discussion, it appears that the proposed theory applies equally as well to its depression. The iron positions in pyrite are also generally considered to be in the reduced state; and, therefore, ferricyanide would be the effective depressant if the surfaces of the mineral were fresh and clean. However, if the surfaces were subjected to an oxidising atmosphere, ferrocyanide would be the effective depressant.

There are, of course, factors other than the ratio of ferricyanide to ferrocyanide that have to be taken into consideration in obtaining an effective depression. The pH of the pulp should be between 7.0 and 8.0 for the best results. Also, the addition of too much fuel oil will have a tendency to interfere with the depression.

In the five final cleaners, potassium ferricyanide is added. In the pilot plant test work, the results indicated that, if ferricyanide was added in the final cleaners, a satisfactory concentrate grade could be produced without the need of adding sodium hypochlorite; whereas, if sodium ferrocyanide were used in the final cleaners, sodium hypochlorite would have to be added to produce an acceptable grade concentrate. As it turned out, hypochlorite had to be added in both cases.

At times, ferrocyanide has been substituted for ferricyanide with excellent results. However, when it is used, the addition rate of hypochlorite has to be increased. The increased amount of hypochlorite required is probably due to a portion of the hypochlorite reacting with ferrocyanide producing ferricyanide.

The function of sodium hypochlorite addition to the second cleaner is not fully understood. It is believed that the sodium hypochlorite performs several functions at once: namely, it disperses the slime faction and oils, dissolves a great quantity of slimes, dissolves molybdenite that has been smeared on copper mineral particles, and adjusts the pH of the pulp. In addition, it tarnishes the surface of some iron and copper mineral particles, and it oxidises some of the ferrocyanide added to the first cleaners to ferricyanide.

In the pilot plant test work it was not necessary to use sodium hypochlorite in the circuit. However, its use in the molybdenite plant is essential for the continuous production of acceptable molybdenite concentrate. Various substitutes have been tested to replace hypochlorite but, to date, none have been successful.

Exfoam 636: The reagent is a formulation sold by Dearborn Chemical Company. It has been found to contain approximately 60% kerosene and some glycols. Its use in the final three cleaning stages is essential in the production of an acceptable grade of molybdenite. It is not used to kill the froth but simply to loosen the froth somewhat to provide bubble drainage.

Fuel oil: a 50-50 mixture of #2 diesel and kerosene is added to promote molybdenite in the rougher circuit. The addition of fuel oil should be held to a minimum because an excess seems to interfere with the depression of iron and copper minerals. Also, fuel oil tends to flocculate the slime particles, creating a difficult separation in the cleaners. An emulsifying agent, Dowfax 9N9, is added to the fuel oil used in the flotation circuits.

Sulphuric acid is added automatically at several points to control pH.

Methyl isobutyl carbinol is used as a frother; it must be added at various points to produce a satisfactory froth.

Water: The proper use of water is extremely important to the successful separation of molybdenite from copper sulphide concentrate. At San Manuel fresh (mine) water is used for dilution in the rougher, first and second cleaners, and scavenger floats. The water does not vary in pH and soluble iron content and, for these reasons, it is not suitable for use in the final four cleaners. Domestic water is used there for dilution. In a molybdenite

Table 38.- MOLYBDENITE PLANT REAGENT CONSUMPTION

Reagent	gm/metric ton of Copper Conc. Treated
Sodium cyanide	100 - 225
Zinc oxide	20 - 35
Sulphuric acid	750 - 975
Hydrogen peroxide	750 - 1,125
Sodium ferrocyanide	700 - 1,250
Fuel oil	225 - 275
Caustic	350 - 525
Chlorine	300 - 550
Exfoam 636	150 - 250
Potassium ferricyanide	200 - 225
Methyl isobutyl carbinol	30 - 50

cleaner circuit, it is generally thought that the more dilution the better. This has proven to be the case at San Manuel; however, the amount of dilution is limited by the necessity of maintaining a satisfactory froth condition throughout the cleaner circuit. The addition of frother to the cleaner stages to obtain a froth is not completely satisfactory because of its tendency to build up in the final cleaners. Therefore, the solids density maintained in the various cleaners is important to proper cleaning. This does not mean to imply that a final value can be set for each cleaner since optimum values will vary from day to day, but general limits can be set.

Table 39.- MICROSCOPIC ANALYSIS OF THE MOLY PLANT STREAMS

All figures are weight %	Feed	Rougher Tail	2nd Scavenger Tail	Final Moly Concentrate
Chalcopyrite	80.91	80.37	88.97	1.80
Covellite	1.45	1.18	0.66	0.01
Chalcocite	0.39	0.30	-	Trace
Bornite	0.51	0.29	0.12	Trace
Native Copper	Trace	Trace	-	-
Molybdenite	2.60	0.25	5.87	97.72
Gangue	6.56	8.38	1.90	0.31
Pyrite	7.58	9.23	2.48	0.05
Assays				
Cu	27.61	27.92	31.48	0.50
MoS_2	1.12	0.2	3.72	96.20
Re	0.0017			0.080
Au oz/ton				0.38

CODELCO Andina - slime sulphidisation in a sand/slime separation

In 1977, Andina pushed up mill feed from 13,000 to 14,000 t/d, with a consequent drop in what was already an unsatisfactory recovery, so corrective action had to be taken.

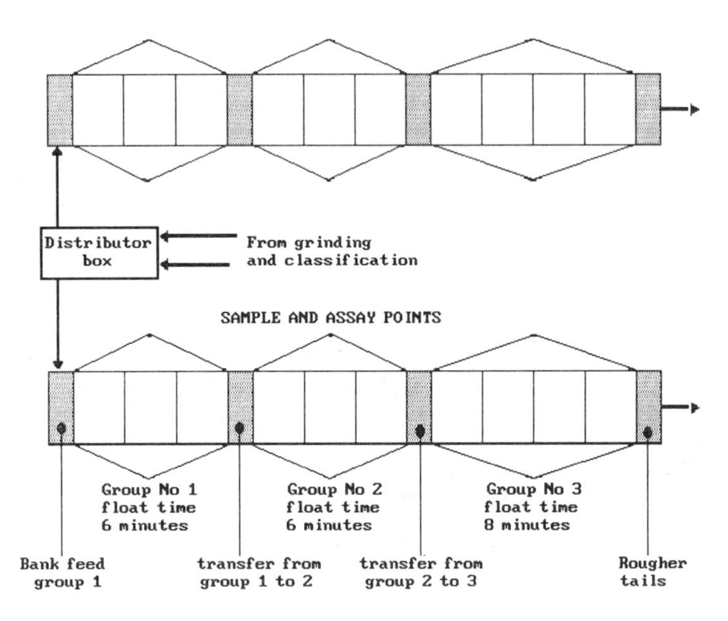

Fig. 27.- Andina rougher circuit in 1976, showing sampling points.

The starting point for the experimental programme was from the 1976 circuit, shown in Fig. 27. It consisted of two banks of 10 each Agitair No 120 flotation cells having a unit volume of 240 ft^3, and a total rougher residence time, per bank, of 20 minutes. The cells were in three groups of 3, 3 and 4 cells. The effect of increasing the tonnage was a reduction in residence time in the flotation bank, and a coarser grind with increased middlings. The metallurgical department was asked to study seven different schemes to improve mill performance:

1.- Increase rougher flotation time.
2.- Install a sand/slime separation of the rougher feed with separate flotation in two banks.
3.- A 6 minute float of all the feed, then a sand/slime separation, and individual flotation of the slimes and sands in the remaining cells.
4.- A 12 minute float of all the feed, then a sand/slime separation, and

individual flotation of the slimes and sands in the remaining cells.

5.- NaSH addition to the head of flotation.

6.- NaSH addition after 6 minutes of flotation.

7.- NaSH addition after 12 minutes of flotation.

Laboratory flotations were made to check the effect of a 3 minute increase in flotation time. To do this, samples were taken of the heads, concentrate and rougher tails, every day for a period of two months. The pulp sample of the tails of the third group were then floated in the laboratory for 3 additional minutes without adding collector or frother. Average recovery increased from 85.59% to 87.82% for the period. So two additional cells were added to each bank, the maximum number that could fit in the underground cavern. The new circuit is shown in Fig. 28.

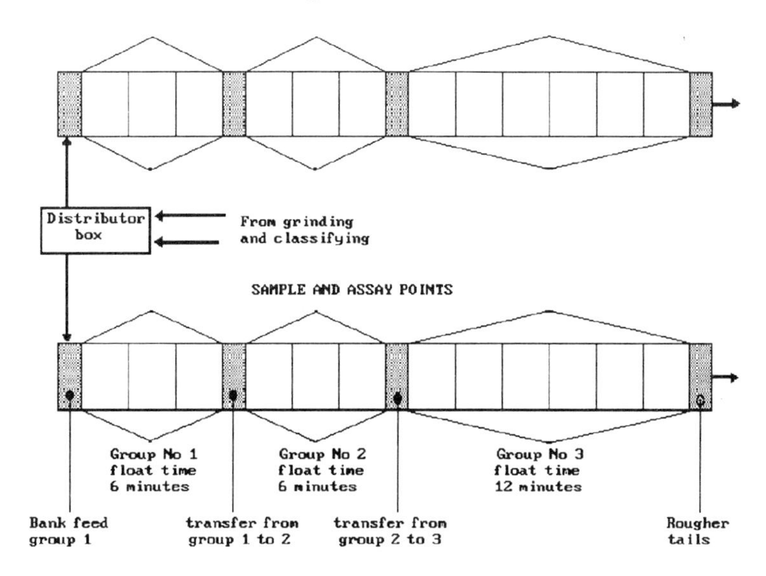

Fig. 28.- Andina rougher circuit in 1977, showing sampling points

To evaluate the effect of a sand/slime separation, ten tests were performed. Samples were taken of the rougher feed, concentrate and tails, under normal plant operating conditions, to obtain a standard to test as a comparison. The feed sample was screened at -200 mesh, to simulate a cyclone cut. To approximate the cyclone, the coarse sample had about 15% of the -200 mesh fraction added to it. The +200 mesh sample was repulped to 40% solids, with the addition of 15 g/t of amyl xanthate as a kicker for the middlings, and floated in the laboratory for 8 minutes. The -200 mesh fraction was floated for 8 minutes, at normal solids, and without added reagents.

Table 40.- LABORATORY RESULTS OF SAND/SLIME TESTING WITH SPLIT AT THE BALL MILL AND WITHOUT ADDED REAGENTS

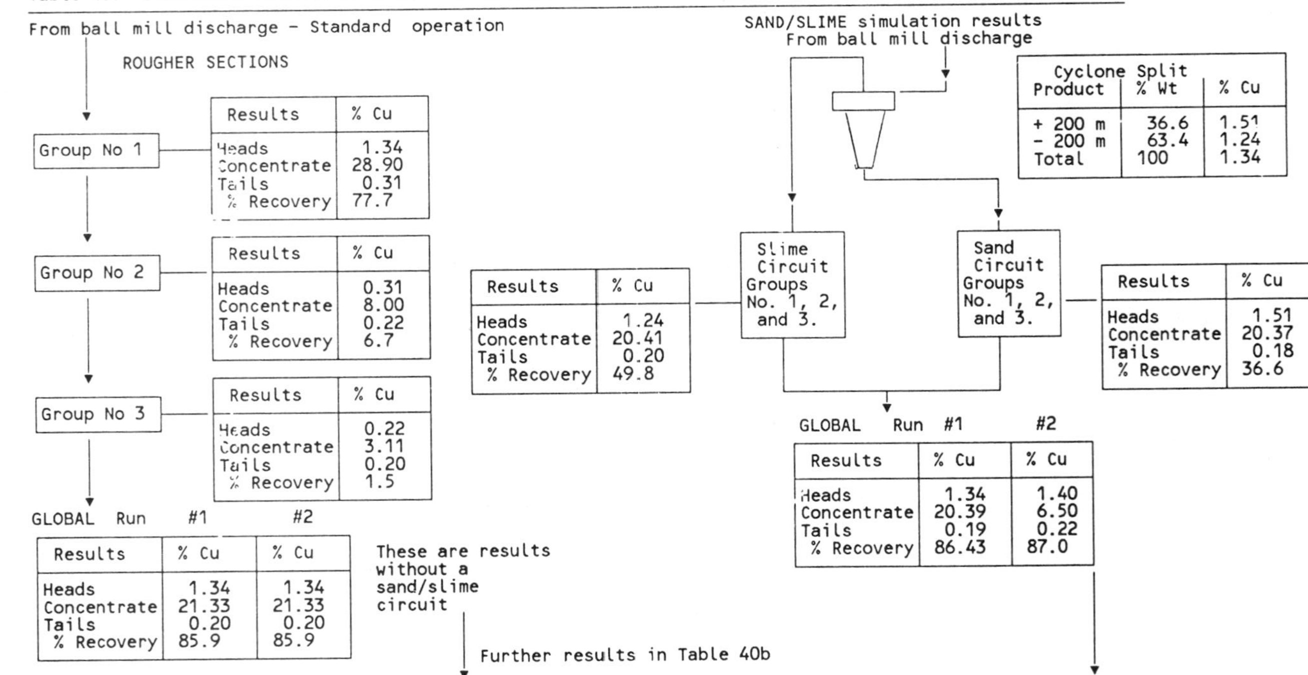

From ball mill discharge - Standard operation — ROUGHER SECTIONS

Group No 1

Results	% Cu
Heads	1.34
Concentrate	28.90
Tails	0.31
% Recovery	77.7

Group No 2

Results	% Cu
Heads	0.31
Concentrate	8.00
Tails	0.22
% Recovery	6.7

Group No 3

Results	% Cu
Heads	0.22
Concentrate	3.11
Tails	0.20
% Recovery	1.5

GLOBAL Run #1 #2

Results	% Cu (#1)	% Cu (#2)
Heads	1.34	1.34
Concentrate	21.33	21.33
Tails	0.20	0.20
% Recovery	85.9	85.9

These are results without a sand/slime circuit

Further results in Table 40b

SAND/SLIME simulation results — From ball mill discharge

Cyclone Product	Split % Wt	% Cu
+ 200 m	36.6	1.51
- 200 m	63.4	1.24
Total	100	1.34

Slime Circuit Groups No. 1, 2, and 3.

Results	% Cu
Heads	1.24
Concentrate	20.41
Tails	0.20
% Recovery	49.8

Sand Circuit Groups No. 1, 2, and 3.

Results	% Cu
Heads	1.51
Concentrate	20.37
Tails	0.18
% Recovery	36.6

GLOBAL Run #1 #2

Results	% Cu (#1)	% Cu (#2)
Heads	1.34	1.40
Concentrate	20.39	6.50
Tails	0.19	0.22
% Recovery	86.43	87.0

Table 40b.- LABORATORY RESULTS OF SAND/SLIME TESTING WITH SPLIT AT THE BALL MILL AND WITHOUT ADDED REAGENTS.

Rougher results without sand/slime - 20 minute flotation time

GLOBAL Run	#3	#4	#5	#6	#7	#8	#9	#10
Results	% Cu	% Cu	% Cu	% Cu	% Cu	% Cu	% Cu	% Cu
Heads	1.45	1.54	1.62	1.59	1.55	1.54	1..6	1.41
Concentrate	11.89	9.91	12.41	9.80	16.50	13.74	5.67	12.26
Tails	0.27	0.22	0.22	0.32	0.28	0.28	0.25	0.21
% Recovery	83.3	87.7	88.0	82.6	83.4	83.5	86.7	86.6

Rougher results without sand/slime - 24 minute flotation time

GLOBAL Run	#3	#4	#5	#6	#7	#8	#9	#10
Results	% Cu	% Cu	% Cu	% Cu	% Cu	% Cu	% Cu	% Cu
Heads	1.45	1.54	-	-	-	1.54	1.46	1.41
Concentrate	7.34	8.18	-	-	-	8.79	5.12	8.43
Tails	0.20	0.21	-	-	-	0.26	0.24	0.19
% Recovery	88.6	88.6	-	-	-	85.7	87.7	88.5

Sand/slime results

GLOBAL Run	#3	#4	#5	#6	#7	#8	#9	#10
Results	% Cu	% Cu	% Cu	% Cu	% Cu	% Cu	% Cu	% Cu
Heads	1.45	1.54	1.62	1..59	1.55	1.54	1.46	1.41
Concentrate	9.62	12.68	11.87	10.13	15.57	13.07	6.95	13.14
Tails	0.27	0.21	0.20	0.30	0.27	0.27	0.24	0.20
% Recovery	83.3	88.0	89.4	83.3	83.7	83.7	86.3	87.1

The results for the first two tests, where the overall flotation time was unchanged, are shown in Table 40 and 40b; the detailed results are included, with a summary for the extended flotation times at the bottom of the table.

The third series involved a bulk float for the first 6 minutes (the first group of rougher cells), followed by a sand/slime separation, and sand and slime circuits. Here another ten experiment series was run. Samples were taken of the plant feed, the concentrate and tails of the first group; a common concentrate for the second and third group, and the overall rougher tail. With these data, standard recoveries for the first group and the combined group 2 and 3 were calculated. As in the first series of runs, the tails from rougher group 1 was sampled and screened to simulate a cyclone split. The +200 mesh fraction was repulped to 40% solids, 15 g/t of amyl xanthate was added and a 6 minute float was run. The -200 mesh fraction was also floated for 6 minutes without further addition of reagents. An example of the data obtained is shown in Table 41.

Table 41.- LABORATORY RESULTS OF SAND/SLIME TESTING WITH SPLIT AFTER 6 MINUTE ROUGHER

From ball mill discharge

Group No 1

Results	% Cu
Heads	1.46
Concentrate	28.10
Tails	0.55
% Recovery	63.6

Group No 2

Results	% Cu
Heads	0.55
Concentrate	5.60
Tails	0.34
% Recovery	14.8

Group No 3

Results	% Cu
Heads	0.34
Concentrate	0.80
Tails	0.25
% Recovery	8.32

GLOBAL Run #9 #10

Results	% Cu (#9)	% Cu (#10)
Heads	1.46	1.41
Concentrate	5.67	12.26
Tails	0.25	0.21
% Recovery	86.7	86.6

Results without a sand/slime circuit

Tails from Group 1

Cyclone Product	Split % Wt	% Cu
+ 200 m	39.7	0.61
- 200 m	60.32	0.51
Total	100	0.55

Slime Circuit Groups No.2 and 3.

Results	% Cu
Heads	0.51
Concentrate	4.93
Tails	0.26
% Recovery	13.6

Sand Circuit Groups No. 2, and 3.

Results	% Cu
Heads	0.61
Concentrate	4.55
Tails	0.23
% Recovery	10.2

GLOBAL Run #9 #10

Results	% Cu (#9)	% Cu (#10)
Heads	1.46	1.41
Concentrate	5.59	12.78
Tails	0.24	0.20
% Recovery	87.3	87.2

The fourth series of ten experiments consisted of normal flotation for the 12 minutes corresponding to the first two banks of the unexpanded roughers, followed by a sand/slime split and separate flotations in the remaining cells of the roughers. Samples were taken of the plant feed, the combined concentrate of the first and second groups, the tails from the second group, concentrate from the third group, and the overall rougher tails. With these data, standard recoveries for the combined groups 1 and 2, and for group 3, were calculated. As in the first series of runs, the tails from rougher group 2 were sampled and screened to simulate a cyclone split. The +200 mesh fraction was repulped to 40% solids, 15 g/t of amyl xanthate was added and a 4 minute float was run. The -200 mesh fraction was also floated for 4 minutes without further addition of reagents. The 4 minute float is intended to simulate the third group of cells. One example of the data obtained is shown in Table 42.

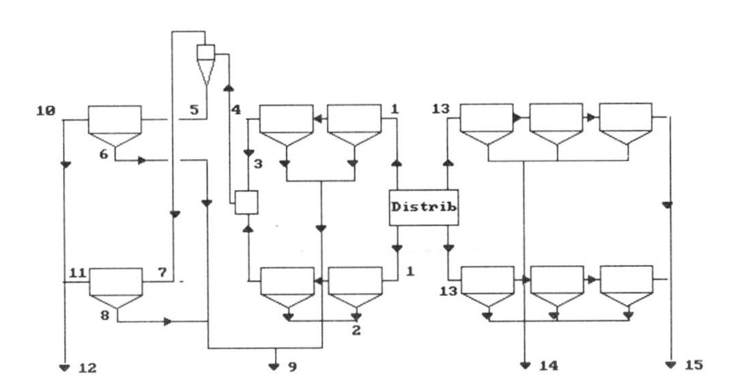

Sample point No.	Stream
1	Section A feed
2	Combined concentrate groups 1 & 2 section A
3	Tail group 2 section A
4	Cyclone discharge
5	Sand circuit feed
6	Sand circuit concentrate
7	Cyclone overflow
8	Concentrate slime circuit
9	Concentrate section A
10	Feed section B
11	Tail slime circuit
12	Tail section B
13	Feed section B
14	Concentrate section B
15	Tail section B

Fig. 29.- Rougher circuit for Andina's sand/slime mill test

Table 42.- LABORATORY RESULTS OF SAND/SLIME TESTING WITH SPLIT AFTER 12 MINUTE ROUGHER

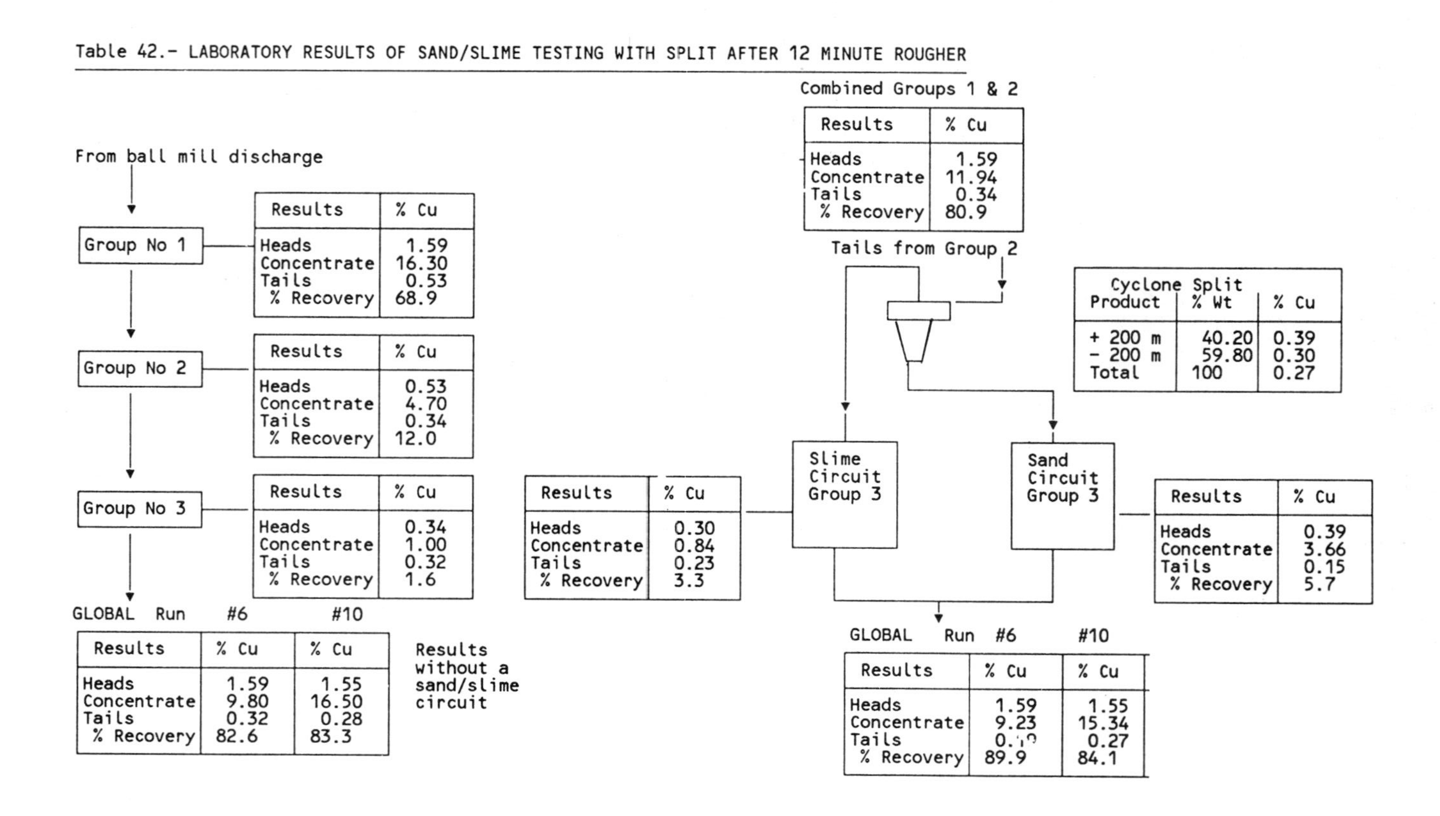

Group No 1

Results	% Cu
Heads	1.59
Concentrate	16.30
Tails	0.53
% Recovery	68.9

Group No 2

Results	% Cu
Heads	0.53
Concentrate	4.70
Tails	0.34
% Recovery	12.0

Group No 3

Results	% Cu
Heads	0.34
Concentrate	1.00
Tails	0.32
% Recovery	1.6

GLOBAL Run #6 #10 (Results without a sand/slime circuit)

Results	% Cu (#6)	% Cu (#10)
Heads	1.59	1.55
Concentrate	9.80	16.50
Tails	0.32	0.28
% Recovery	82.6	83.3

Combined Groups 1 & 2

Results	% Cu
Heads	1.59
Concentrate	11.94
Tails	0.34
% Recovery	80.9

Cyclone Split

Product	% Wt	% Cu
+ 200 m	40.20	0.39
- 200 m	59.80	0.30
Total	100	0.27

Slime Circuit Group 3

Results	% Cu
Heads	0.30
Concentrate	0.84
Tails	0.23
% Recovery	3.3

Sand Circuit Group 3

Results	% Cu
Heads	0.39
Concentrate	3.66
Tails	0.15
% Recovery	5.7

GLOBAL Run #6 #10

Results	% Cu (#6)	% Cu (#10)
Heads	1.59	1.55
Concentrate	9.23	15.34
Tails	0.[illegible]	0.27
% Recovery	89.9	84.1

To complete the sand/slime test series, a full plant test was carried out where the B section was modified and the ten existing cells repiped to conform to the flow diagram in Fig. 29. A series of eight runs were made in the plant during the A shift (8 am to 4 pm). Samples were taken from the points listed in Fig. 29, using specially design air lift pumps for the pulp samples and diversion launders for the concentrate samples. Simultaneously, the same samples were taken on the A section, to evaluate each experimental run.

The testing shown in Table 43 completed the non-sulphidisation part of the slime/sand study. The mill test showed quite clearly that a sand/slime separation followed with separate flotation circuits, and the use of an amyl xanthate kicker in the sands circuit, was a justified permanent investment, and certainly paid the cost of a mill test, as copper recovery was improved by over a point, and molybdenum more so. Examination of the distribution of the copper in the different size particles of the tails, shown in Table 44, suggests that a sand regrind could improve recovery; but with an underground concentrator, Andina did not have the space to add a ball mill without a very large capital cost.

The second part of this study was the possibility of avoiding the investment in a sand/slime modification of the plant to see if the non-sulphide copper can be floated by sulphidisation. The experimental design for the sulphidisation testing was based on preliminary trials to determine the range of the NaSH dose and the conditioning time required to complete the reaction. The laboratory flotations were made using the standard grind and flotation time. The responses measured were sulphide and oxide copper recovery. Another important variable was the optimum addition point for the NaSH (Table 45).

To determine the optimum addition point for NaSH in the plant, a set of runs were made in the laboratory with a plant pulp sample from the feed distributor box; another set with a tailings sample from the first group of cells (i.e., after 6 minutes of flotation); and finally a tailings sample from the second group of cells (i.e., after 12 minutes of flotation). The results of these experiments are shown in Table 46.

These tests did show a definite favourable effect of sulphidisation, but not enough to discourage the installation of a sand/slime separation and individual flotation circuits.

Table 43.- PLANT TEST OF THE EFFECTIVENESS OF SAND/SLIME SEPARATION

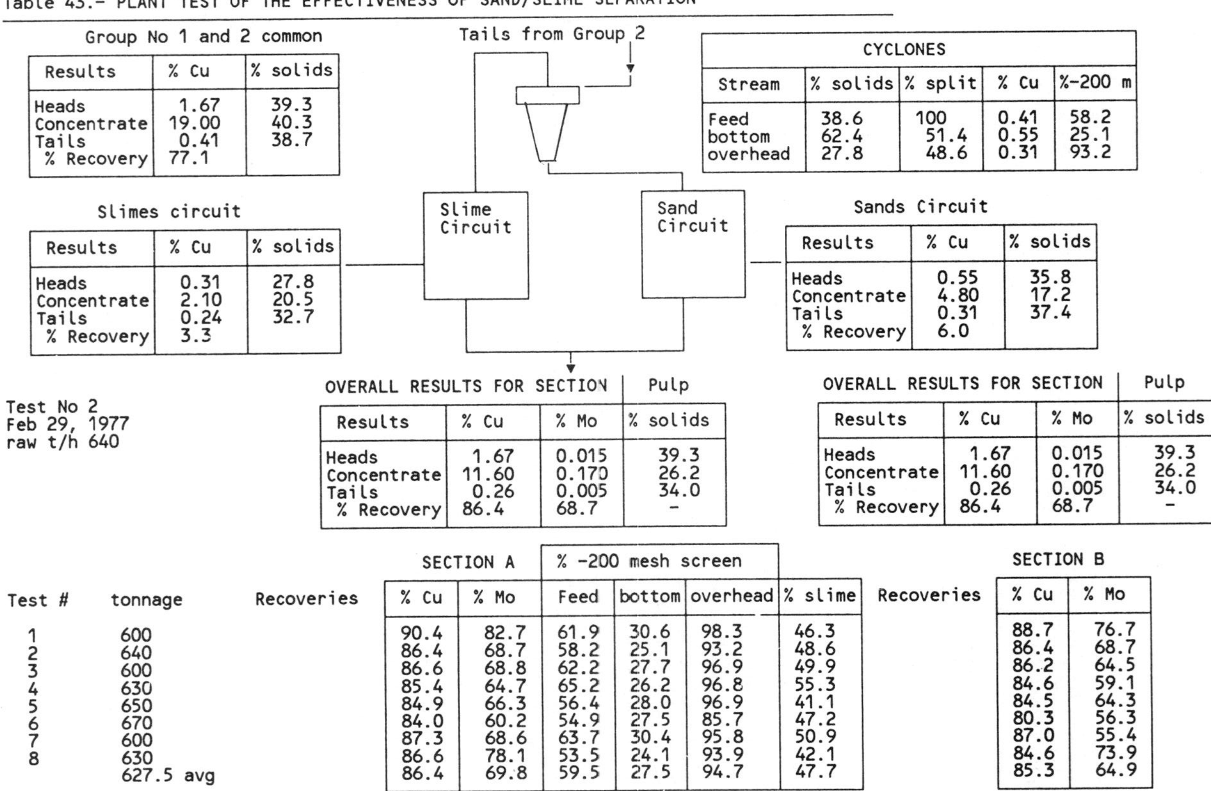

Group No 1 and 2 common

Results	% Cu	% solids
Heads	1.67	39.3
Concentrate	19.00	40.3
Tails	0.41	38.7
% Recovery	77.1	

CYCLONES

Stream	% solids	% split	% Cu	%-200 m
Feed	38.6	100	0.41	58.2
bottom	62.4	51.4	0.55	25.1
overhead	27.8	48.6	0.31	93.2

Slimes circuit

Results	% Cu	% solids
Heads	0.31	27.8
Concentrate	2.10	20.5
Tails	0.24	32.7
% Recovery	3.3	

Sands Circuit

Results	% Cu	% solids
Heads	0.55	35.8
Concentrate	4.80	17.2
Tails	0.31	37.4
% Recovery	6.0	

Test No 2
Feb 29, 1977
raw t/h 640

OVERALL RESULTS FOR SECTION			Pulp
Results	% Cu	% Mo	% solids
Heads	1.67	0.015	39.3
Concentrate	11.60	0.170	26.2
Tails	0.26	0.005	34.0
% Recovery	86.4	68.7	-

OVERALL RESULTS FOR SECTION			Pulp
Results	% Cu	% Mo	% solids
Heads	1.67	0.015	39.3
Concentrate	11.60	0.170	26.2
Tails	0.26	0.005	34.0
% Recovery	86.4	68.7	-

Test #	tonnage	SECTION A Recoveries % Cu	SECTION A Recoveries % Mo	% -200 mesh screen Feed	% -200 mesh screen bottom	% -200 mesh screen overhead	% slime	SECTION B Recoveries % Cu	SECTION B Recoveries % Mo
1	600	90.4	82.7	61.9	30.6	98.3	46.3	88.7	76.7
2	640	86.4	68.7	58.2	25.1	93.2	48.6	86.4	68.7
3	600	86.6	68.8	62.2	27.7	96.9	49.9	86.2	64.5
4	630	85.4	64.7	65.2	26.2	96.8	55.3	84.6	59.1
5	650	84.9	66.3	56.4	28.0	96.9	41.1	84.5	64.3
6	670	84.0	60.2	54.9	27.5	85.7	47.2	80.3	56.3
7	600	87.3	68.6	63.7	30.4	95.8	50.9	87.0	55.4
8	630	86.6	78.1	53.5	24.1	93.9	42.1	84.6	73.9
	627.5 avg	86.4	69.8	59.5	27.5	94.7	47.7	85.3	64.9

Table 44.- SCREEN ANALYSIS, COPPER DISTRIBUTION, ANDINA A 14,000 t/d, 1977

HEADS

Mesh	Wt % on	% Cu	Wt % of Cu	%cumm. Cu
+65	12.88	0.53	4.43	4.43
+100	10.44	0.65	4.40	8.83
+150	9.36	1.03	6.25	15.08
+200	10.00	1.57	10.18	25.26
-200	57.32	2.01	74.74	100.0
Total	100	1.54	100.0	

TAILS FIRST GROUP

Mesh	Wt % on	% Cu	Wt % of Cu	%cumm. Cu
+65	13.86	0.48	12.38	12.38
+100	9.60	0.48	8.58	20.96
+150	9.34	0.51	8.87	29.83
+200	8.60	0.50	8.00	37.83
-200	58.60	0.57	62.17	100.0
Total	100	0.54	100.0	

TAILS SECOND GROUP

Mesh	Wt % on	% Cu	Wt % of Cu	%cumm. Cu
+65	16.70	0.44	24.77	24.77
+100	10.98	0.41	15.18	39.95
+150	10.02	0.33	11.15	51.10
+200	8.80	0.25	7.42	58.52
-200	53.50	0.23	41.48	100.0
Total	100	0.30	100.0	

TAILS THIRD GROUP

Mesh	Wt % on	% Cu	Wt % of Cu	%cumm. Cu
+65	13.58	0.41	25.19	25.19
+100	11.22	0.37	18.78	43.97
+150	10.00	0.25	11.31	55.28
+200	10.26	0.16	7.43	62.71
-200	54.94	0.15	37.29	100.0
Total	100	0.22	100.0	

CONCENTRATE FIRST GROUP

Mesh	Wt % on	% Cu	Wt % of Cu	%cumm. Cu
+65	0.48	11.40	0.20	0.20
+100	2.14	16.10	1.25	1.45
+150	6.28	19.30	4.38	5.83
+200	11.18	22.10	8.93	14.76
-200	79.92	29.50	85.24	100.0
Total	100	27.50	100.0	

CONCENTRATE SECOND GPROUP

Mesh	Wt % on	% Cu	Wt % of Cu	%cumm. Cu
+65	0.90	5.50	0.46	0.46
+100	1.74	9.80	1.56	2.02
+150	3.68	11.90	4.02	6.04
+200	5.78	12.90	6.84	12.88
-200	87.90	10.80	87.12	100.0
Total	100	10.90	100.0	

CONCENTRATE THIRD GROUP

Mesh	Wt % on	% Cu	Wt % of Cu	%cumm. Cu
+65	0.88	1.30	1.00	41.00
+100	0.96	3.70	3.18	4.18
+150	1.50	3.60	4.83	9.01
+200	2.94	2.70	7.10	16.11
-200	93.72	1.00	83.89	100.0
Total	100	1.20	100.0	

The first of the test series is an evaluation of the effect of adding the NaSH to the flotation feed. For this, two pulp samples were obtained under normal plant operating conditions. One sample is floated under standard laboratory conditions that simulate plant conditions. The other sample is floated under the same conditions as the first sample, except that 0.10 lb/t is added to the flotation cell at the start of the test.

Table 45.- THE EFFECT OF CONDITIONING TIME, AND DOSAGE OF NASH, ON COPPER RECOVER

FIRST GROUP OF EXPERIMENTS - NaSH to PLANT FEED

NaSH lb/ton	Condition -ing min.	% recovery oxide Cu	% recovery sulph. Cu	% recovery Total Cu
0	0	2.7	84.7	87.4
0.05	1	3.0	84.7	87.7
0.20	1	3.5	84.9	88.4
0.12	2	3.7	85.0	88.8
0.05	3	3.4	84.2	87.6
0.20	3	3.5	85.4	89.0

SECOND GROUP OF EXPERIMENTS - NaSH to TAILS GRP 1 CELLS

NaSH lb/ton	Condition -ing min.	% recovery oxide Cu	% recovery sulph. Cu	% recovery Total Cu
0	0	2.9	84.6	87.5
0	0	2.7	84.7	87.3
0.05	1	2.7	85.1	87.8
0.25	1	3.2	86.8	90.0
0.15	2	3.2	85.0	88.3
0.05	3	2.7	85.1	87.7
0.25	3	3.5	86.4	90.0

THIRD GROUP OF EXPERIMENTS - NaSH to TAILS GRP 2 CELLS

NaSH lb/ton	Condition -ing min.	% recovery oxide Cu	% recovery sulph. Cu	% recovery Total Cu
0	0	1.3	84.0	85.3
0	0	1.4	83.9	85.2
0.05	1	1.7	85.4	87.1
0.25	1	1.7	88.0	89.7
0.15	2	1.6	85.3	86.9
0.15	2	1.6	85.0	86.6
0.05	3	1.8	85.2	87.0
0.25	3	1.9	86.3	88.2

The second series is intended to simulate the addition of NaSH in the plant, after 6 minutes of flotation (i.e., to the tails of the first group of 3 cells). Again two samples are taken of the pulp feed from the plant. One is floated under standard conditions, and the second has NaSH added to the flotation cell 2 minutes into the test, which corresponds to 6 minutes in the plant.

The third series simulates the addition of NaSH to the plant after 12 minutes of flotation (i.e., to the tails of the second group of 3 cells). Again two pulp samples of the plant feed were collected using the standard procedures. The first sample was floated under standard conditions, and the second sample had NaSH added to the laboratory flotation cell after 4 minutes of testing, which corresponds to 12 minutes of plant flotation.

The sulphidisation experiment in Andina was less successful than the sand/slime programme. Examination of the results in Table 45 indicates that, to be effective as a sulphidiser, NaSH must be added at a high enough concentration and given at least 3 minutes of conditioning prior to exposure to the oxygen in the aerated flotation cells. As a consequence, the series reported in Table 46 gives improved recoveries that are of the same order of magnitude as the sand/slime separation and separate flotation. The data is nonetheless interesting because of the extensive evaluation of the recoveries for each of the size fractions. Andina has since increased its flotation capacity to 40,000 t/d, and has been purchasing a large quantity of NaSH, so it is probable that an operable sulphidisation process has been worked out. It also employs the sulphhydrate in the depression of copper in the copper/moly separation plant.

The copper porphyry ores in Chile all have about 0.25 to 0.30% Cu as non-sulphide minerals. As a consequence, with the exception of Chuquicamata, whose ore is unique, all the major concentrators have copper recoveries of under 85%. The cerrusite content of the gangue contributes to this poor recovery because it generates colloidal slimes that adsorb on the copper mineral surface and retard flotation. In this case, if sulphides improve flotation, the effect may well be to peptise the cerrusite rather than to activate the oxide minerals or sulphides. There is clear evidence of this in the Andina data reported in Table 46, where the NaSH addition was more successful in activating sulphides, especially when added late in the flotation.

CODELCO El Teniente - molybdenite recovery improvement programme

A case history featuring a comprehensive data acquisition and test programme to improve molybdenite recovery at El Teniente was published by Beas et al (1991), from which these data are taken.

At El Teniente, partly because the declining ore grades and serious rock-burst problems required pushing equipment capacity beyond its design limits to keep on budget in copper output, molybdenum recoveries, which normally ran at about 50%, started drifting down from 48% in 1985, to a low of 34% in 1989 (Fig. 30). The programme undertaken by the in-house metallurgical staff to find improved reagents, and/or a more efficient processing scheme, included extensive ore sampling both for head grade scheduling and to smooth out processability upsets, and provided a good exercise in test planning and data evaluation. The period described in this section covers

Table 46.- ANDINA - RESULTS OF TESTS TO DETERMINE ADDITION POINT OF NASH FOR SULPHIDIZATION

HEADS

Mesh	% Cu	Mesh	% Cu
+150	1.05	+325	1.71
+200	1.52	-325	1.96
Total Cu 1.58%			

Flotation tests without addition of NaSH

Flotation Time 0 to 6 minutes

CONCENTRATE # 1

Mesh	% Cu	% of total Cu recov.	per size Cu recovery
+150	9.68	12.51	54.0
+200	14.87	5.54	61.7
+325	18.59	7.98	63.7
-325	21.93	33.64	60.8
Total	16.45	59.67	

Flotation time 6 to 12 minutes

CONCENTRATE # 2

Mesh	% Cu	% of total Cu recov.	per size Cu recovery
+150	6.24	5.37	23.2
+200	9.49	2.36	26.2
+325	11.89	3.41	27.2
-325	14.10	14.42	26.1
Total	10.57	25.56	

Flotation time 12 to 24 minutes

CONCENTRATE # 3

Mesh	% Cu	% of total Cu recov.	per size Cu recovery
+150	3.11	0.57	2.5
+200	3.72	0.19	2.1
+325	3.69	0.29	2.3
-325	3.83	1.97	3.6
Total	3.71	3.02	

TAILS

Mesh	% Cu	% of total Cu recov.	per size Cu recovery
+150	0.24	4.72	20.3
+200	0.17	0.89	9.9
+325	0.18	0.86	6.9
-325	0.21	5.28	9.6
Total	0.21	11.75	

Flotation test with NaSH added to feed

CONCENTRATE # 1

Mesh	% Cu	% of total Cu recov.	per size Cu recovery
+150	9.57	12.54	53.6
+200	14.88	5.61	61.3
+325	18.41	7.87	63.9
-325	22.03	33.58	61.0
Total	16.54	59.60	

CONCENTRATE # 2

Mesh	% Cu	% of total Cu recov.	per size Cu recovery
+150	6.19	5.32	23.2
+200	9.53	2.39	26.4
+325	12.07	3.37	27.2
-325	14.22	14.40	26.1
Total	10.48	25.48	

CONCENTRATE # 3

Mesh	% Cu	% of total Cu recov.	per size Cu recovery
+150	3.09	0.61	2.5
+200	3.75	0.22	2.2
+325	3.66	0.27	2.2
-325	3.87	1.94	3.6
Total	3.73	3.04	

TAILS

Mesh	% Cu	% of total Cu recov.	per size Cu recovery
+150	0.25	4.69	10.7
+200	0.17	0.91	10.2
+325	0.15	0.84	6.7
-325	0.21	5.44	9.2
Total	0.22	11.88	

CONCENTRATE # 1

Mesh	% Cu	% of total Cu recov.	per size Cu recovery
+150	9.58	12.57	54.1
+200	14.63	5.48	62.4
+325	18.53	8.03	62.8
-325	22.17	33.73	60.7
Total	16.49	59.81	

Flotation test with NaSH added to tails of first group

CONCENTRATE # 2

Mesh	% Cu	% of total Cu recov.	per size Cu recovery
+150	6.17	5.42	22.9
+200	9.45	2.19	25.9
+325	12.28	3.53	28.4
-325	14.33	14.35	27.2
Total	10.63	25.49	

CONCENTRATE # 3

Mesh	% Cu	% of total Cu recov.	per size Cu recovery
+150	3.15	0.48	2.2
+200	3.68	0.21	2.5
+325	3.71	0.31	2.4
-325	3.85	2.07	3.7
Total	3.69	3.07	

TAILS

Mesh	% Cu	% of total Cu recov.	per size Cu recovery
+150	0.23	4.80	20.9
+200	0.18	0.93	9.3
+325	0.14	0.75	6.5
-325	0.20	5.15	8.48
Total	0.21	11.63	

CONCENTRATE # 1

Mesh	% Cu	% of total Cu recov.	per size Cu recovery
+150	9.43	12.73	57.8
+200	14.55	4.68	61.3
+325	19.49	8.50	61.8
-325	22.80	34.22	60.5
Total	16.67	60.13	

CONCENTRATE # 2

Mesh	% Cu	% of total Cu recov.	per size Cu recovery
+150	6.10	5.47	24.9
+200	9.40	2.00	26.3
+325	12.53	3.63	26.4
-325	14.59	14.67	25.9
Total	10.71	25.77	

Flotation test wth NaSH added to tails of second group

CONCENTRATE # 3

Mesh	% Cu	% of total Cu recov.	per size Cu recovery
+150	2.03	0.86	3.9
+200	2.47	0.35	4.6
+325	2.28	0.35	2.6
-325	3.11	2.51	4.4
Total	2.73	4.07	

TAILS

Mesh	% Cu	% of total Cu recov.	per size Cu recovery
+150	0.16	2.96	13.4
+200	0.12	0.61	7.9
+325	0.20	1.27	9.3
-325	0.20	5.19	9.2
Total	0.18	10.03	

RUN # 1	OVERALL RECOVERY	88.3%	OVERALL RECOVERY	88.1%	OVERALL RECOVERY	88.4%	OVERALL RECOVERY	90.0%
RUN # 2	OVERALL RECOVERY	88.5%	OVERALL RECOVERY	88.4%	OVERALL RECOVERY	88.2%	OVERALL RECOVERY	89.6%
RUN # 3	OVERALL RECOVERY	85.6%	OVERALL RECOVERY	85.5%	OVERALL RECOVERY	85.6%	OVERALL RECOVERY	87.6%
RUN # 4	OVERALL RECOVERY	88.2%	OVERALL RECOVERY	88.3%	OVERALL RECOVERY	88.2%	OVERALL RECOVERY	89.0%

about 36 months of work between 1988 and 1991. First, a general review of the plant operations is given, to provide a base line for the studies.

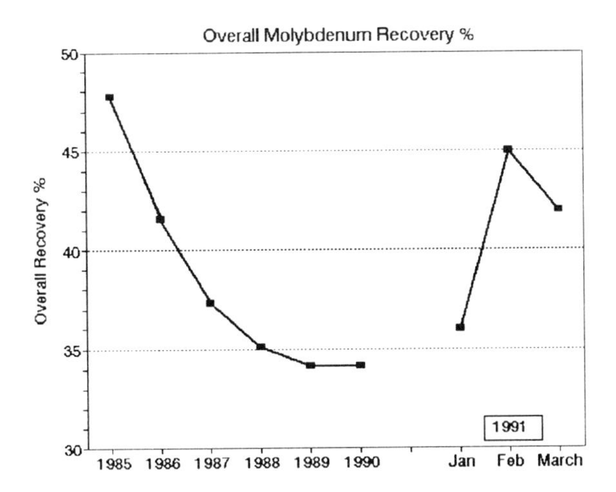

Fig. 30.- Overall molybdenum recovery at El Teniente

In 1990 the El Teniente Division of CODELCO produced 300,472 t of copper and 2,633 t of molybdenum. To obtain this output, 32,000 t/d of ore was mined from the weathered upper levels of the ore body, crushed and ground in the eighty year old Sewell mill (altitude 2,100 m), and, after conditioning with sulphuric acid to help recover copper contained in the non-sulphide minerals, the pulp was transported for 10 km in the old Sewell wooden tailings flume, which drops down the mountain 800 m to the Colon concentrator (altitude 1,800 m), where it was floated in a rougher and cleaner circuit kept at pH=4 with sulphuric acid. Under the same roof, an additional 54,000 t/d of ore from the primary area of the mine (Teniente levels 6 and greater), were crushed underground, transported by rail to Colon, and then ground and fed directly to a separate alkaline flotation circuit (maintained at pH 10.5 to 11.0 with milk of lime) in this mill. In 1990, all the ore treated at Sewell came from the weathered upper levels of the mine, while a little over 60% of the Colon feed was from primary ore sources.

Both Colon and Sewell's flotation processes start with a rougher equipped with 1,500 cu.ft WEMCO flotation cells, followed by one cleaning stage equipped with 1,000 cu ft WEMCOs. The concentrates from these cleaners are cycloned, the sands re-ground in open circuit ball mills, both cyclone slime and reground pulps are mixed (the pH obtained on mixing is 8.5 to 9), then fed to a common final cleaning and scavenging circuit, where the pulp undergoes a second cleaning (3 banks each with eight 500 cu. ft. cells), followed by a scavenger bank (one bank with 7 large 1,500 cu.ft. cells), and finally a re-scavenger (2 banks containing twelve modified 100 cu.ft. cells). The final tails from this section are returned to the Colon primary rougher head, and the concentrate is sent to the moly/copper separation plant. The primary tails from the acid flotation are mixed with the tails from the Colon circuit and fed to the tailings re-treatment plant, where they are scavenged using 3,000 cu.ft., 1,500 cu.ft., and 1,000 cu.ft. flotation cells without addition of fresh reagents.

Fig. 31 is labelled "simplified", as pipes and valving are in place to route all or a portion of the copper circuit tailings to the retreatment plant, as well as the tails from the first cleaner banks of the Colon and Sewell sections. The copper concentrate from the copper circuit is then processed in a moly/copper separation plant using Nokes Reagent (phosphorous pentasulphide plus caustic) to depress copper. The different streams are number coded, and their pulp tonnage and Cu assays are shown in Table 47. For the particular month documented (April 1989), global copper recovery was 79.8%. Colon's alkaline circuit recovered 82.27% of the gross copper input, and from the acid circuit (Sewell), copper recovery was 71.6%. The overall cleaning recovery was 96% for copper.

The current copper collector in both the alkaline (Colon) circuit, and in the Sewell acid flotation circuit, is diethyl xanthogen formate (now Shellflot 203; xanthogen formates used previously were Minerec A and TM-2001), and the frothers are Dowfroth 250 and MIBC. As a moly collector, diesel oil is added to the pulp canal prior to coming down the hill from Sewell, and to the conditioner in Colon. In both cases the dosage is in the 10 to 20 g/t range for the diesel oil, 15-20 g/t for the frother, and about 35 g/t for the xanthogen formate. Prior to 1989, 90 g/t of a mixture (designated T-3010 and patented by Teniente), consisting of Minerec A or Tecnomin 2001 (60 parts), gasoline (30 parts), and MIBC (10 parts), was the standard collector. The addition of gasoline was justified as a molybdenite collector, and the MIBC as an emulsifier. As well as improving recovery, the mixture resulted in a significant cost reduction as, prior to 1979, 90 g/t of unmixed Minerec A was the standard. As frothers, 25 g/t of Dowfroth 1012, or Powell accelerator or TEB, were used. Because of a very leathery and dry overflow from the flotation cells with DF-1012 after the changeover to the large WEMCO cells, Dowfroth 250 was substituted, and is used in the 15 to 20 g/t dosage range. The addition of MIBC as a modulator has resulted in a more manageable froth when ore changes occur.

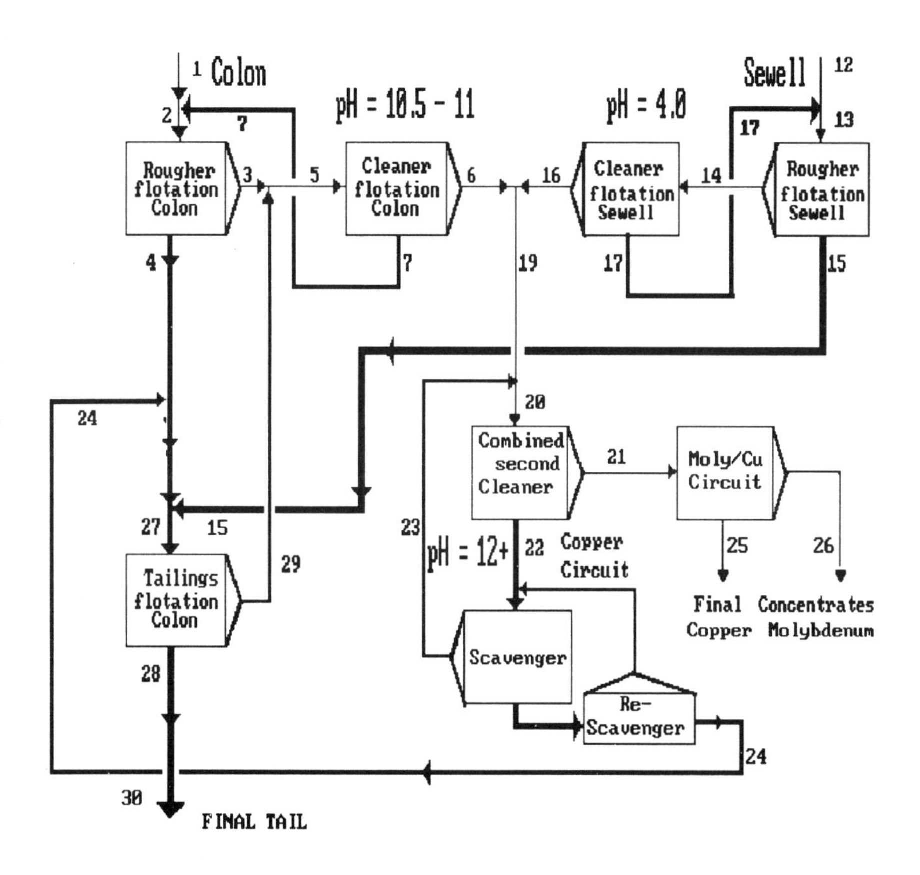

Note:Reagent T-3010 = 60% diethyl xanthogen formate mixed with 30% gasoline and 10% MIBC

Flotation Reagents: (1989) Colon - Collector Shell 203 at 25 g/t
Fuel Oil 20 g/t
Dowfroth 250 plus MIBC
(1989) Sewell - Collector T-3010 70 g/t
Fuel Oil 15 g/t
(1990) Sewell - Collector Shell 203

Fig. 31.- Simplified flotation schematic for El Teniente

Table 47.- MATERIAL BALANCES FOR THE FLOW SHEET - FIGURE 31

TENIENTE MATERIAL BALANCES April 1989

units:t/d	COLON	SEWELL	Retreat	Tailings	Overall	Primary	Rougher	Primary	Cleaners
PRIMARY FLOAT									
Stream number	(1)	(12)	(19)	(27)	(1)+(12)	(2)	(13)	(5)	(14)
Tonnage	53520	35389	4684	86593	88909	60599	37210	8342	3274
% Cu Tot	1.31	1.33	20.8	0.31	1.32	1.27	1.29	8.02	10.51
Cu content	701.1	470.7	974.5	266.9	1171.8	771	478.5	668.8	344.1
Concentrate									
Stream number	(6)	(16)	(21)	(29)	(25)	(3)	(14)	(6)	(16)
Tonnage	3231	1453	2717	401	2700	7942	3274	3231	1453
% Cu Tot	19.73	23.19	34.45	7.76	34.65	8.03	10.51	19.73	23.19
Cu content	637.5	337	935.9	31.1	935.5	637.7	344.1	637.5	337
Tails									
Stream number	(8)	(18)	(24)	(28)	(30)	(4)	(15)	(7)	(17)
Tonnage	52657	33936	1968	86192	96192	52657	33936	5111	1821
% Cu Tot	0.253	0.394	1.96	0.274	0.274	0.253	0.396	0.612	0.394
Cu content	133.2	133.7	38.6	235.9	235.9	133.2	134.4	31.3	7.2

RETREATMENT PLANT			
Overall balance			
Stream number	(19)	(21)	(24)
Tonnage	4684	2717	1968
% Cu Tot	20.8	34.45	1.96
Cu content	974.5	953.9	38.6
Copper circuit			
Stream number	(19)	(21)	(22)
Tonnage	4684	2717	4704
% Cu Tot	20.8	34.45	14.17
Cu content	974.5	935.9	666.5
Scavenger circuit			
Stream number	(22)	(23)	(24)
Tonnage	4704	2736	1968
% Cu Tot	14.17	22.95	1.96
Cu content	666.5	628	38.6
Moly/copper Plant			
Overall balance			
Stream number	(21)	(25)	(26)
Tonnage	2717	2700	17
% Cu Tot or MoTot	34.45	34.65	2.52
Cu content or Mo	935.9	935.5	0.4

Fraction of Recirculation retreated	
Colon scavenger tails	K0=1.00
Sewell scavenger tails	K1=1.00
Scavenger circuit tails to Colon rougher	K3=1.00
Sewell tail to Tailings retreat	K4=1.00
Combined Colon + scav. tails to Tailings retreat	K2=1.00

Copper recoveries	Percent
Overall Colon concentrator	82.69
Overall Sewell concentrator	71.59
Retreatment plant	96.04
Global for both mills	79.84
Rougher flotation Colon	82.72
Scavenger - Colon	95.32
Rougher flotation Sewell	71.92
Scavenger - Sewell	97.91
Copper circuit recovery	58.41
Scavenger circuit	94.21

Note: Colon recovery based on adding Colon + Tailings retreat concentrate + tails from retreat sections
"Stream number" = code shown by lines in Figure 31

The unique acid flotation circuit employed in Sewell since 1926 was imposed by the ore from the altered cap, which required an uneconomic amount of lime to produce a stable alkaline pulp. This inherent acidity has been a constraint on the choice of flotation reagents ever since. Due to the low natural pH, and the use of sulphuric acid as a modifier, the initial mill experience in the twenties was the need for 2 to 4 kilos of potassium ethyl xanthate per tonne of ore to obtain an acceptable recovery. This prohibitive collector consumption inspired the commercial development of acid stable xanthogen formate as the preferred collector (in 1928) by Dr. Arthur Fischer at Minerec. The acid circuit's extraordinarily high xanthate consumption justified the continued use of over 100 g/t of exotic Minerec A as the main collector for the next fifty years.

Molybdenum losses: The experimental programme devised to identify the cause of the drop in the overall molybdenum recovery, and to help define the steps necessary to improve it, concentrated on the following known trends:

- a progressively lower content of molybdenum in the ore treated
- an increase in the amount of oxidised molybdenum in the ore
- a reduced flotation time generated by an increased ore tonnage treated in the mill
- low floatability of colloidal molybdenite particles
- the need to operate with a very high alkalinity of the pulp in the copper cleaning circuit (pH >12), because excess pyrite in the concentrate was reducing the Caletones smelter copper capacity, requiring a more selective cleaning step.
- changes detected in the molybdenite rate of flotation

The laboratory aspects of the experimental programme concentrated on determining the grind required for good liberation of the valuable copper and molybdenum minerals, and the re-grind required to adequately reject pyrite; the skewed distribution of copper and moly as a function of particle sizes; and the effect of pH on floatability of the mineral species.

The process survey concentrated on identifying mechanical and equipment factors that could influence moly recovery.

Teniente ore mineralogy: The mineralogical trends in the average ore received at Colon reflect the reduced importance of the tonnage of ore from the upper part of the ore body which has traditionally supplied the old Sewell concentrator. In future, as mining levels above Level 6 are exhausted, both the copper oxides and the oxidised moly will decrease, reducing today's recovery problems. Fig. 32 shows total copper and oxidised copper trends for Sewell ores for the period after 1983, and the longer term trends of the non-sulphide content of the Sewell, Colon and composite Teniente ore are shown in Fig. 33.

As can be seen from Figs. 32 and 33, the copper and molybdenum grade is slowly decreasing, as is the oxide copper content. Partial data for the oxidised molybdenum content has been:

Proportion of total moly as oxide	1988	1989
Sewell	23.2%	17.9%
Colon	27.0%	20.7%

The higher oxidised moly values for Colon, as compared to Sewell, are due to feeding a significant amount of Teniente level 4 ore directly to Colon, rather than to the rougher flotation in the Sewell circuit. Mine levels above Teniente 6 are very high in oxides; in one case - Teniente 4 North Standard - they run close to 40% non-sulphide. To minimise the effect of this non-floating oxide, blocks with particularly refractive molybdenum (i.e. N-9, N-13, and N-15) were identified and were dosed to Colon to be diluted with primary ore.

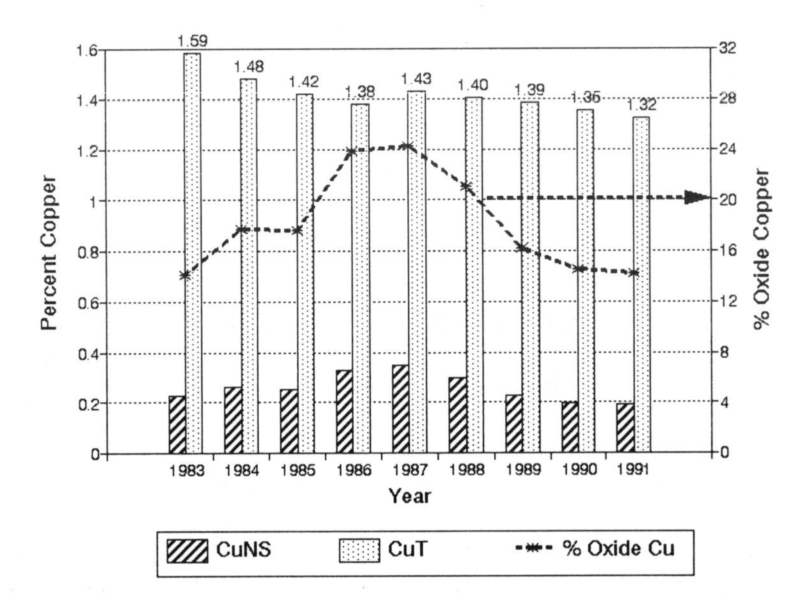

Fig. 32.- Sewell ore CuT, CuNS and % oxide copper content

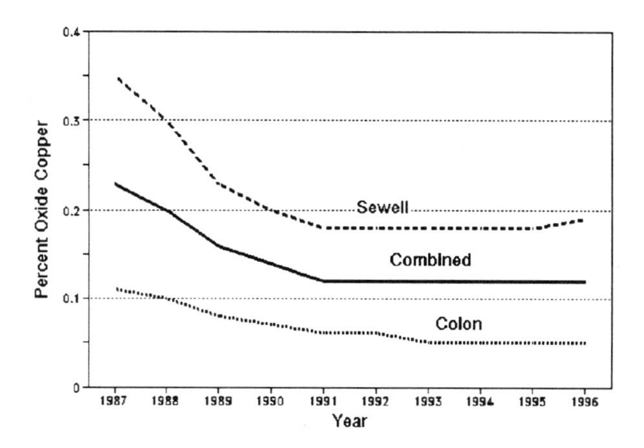

Fig. 33.- Sewell, Colon and combined ore non-sulphide assay 1987 to 1996

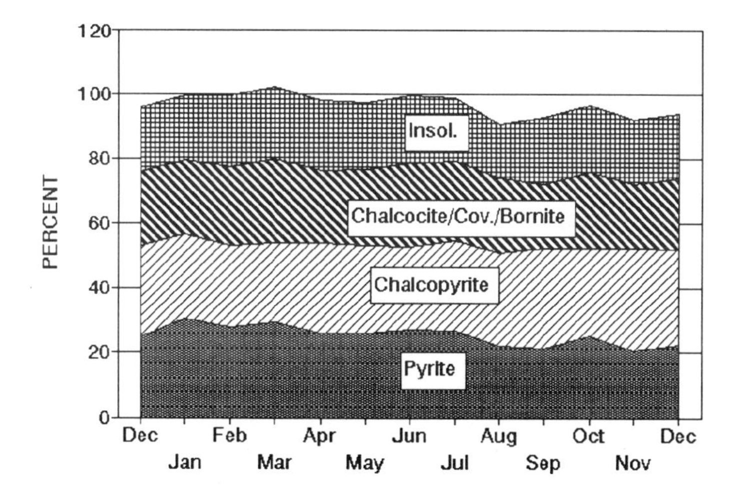

Fig. 34.- Mineral content of Teniente concentrate - 1984

The mineralogy of the final concentrate was studied using monthly samples that covered most of the year 1984. The main copper mineral recovered was chalcopyrite, which as can be seen from Fig. 34, was predominant over the secondary copper minerals: chalcocite, covellite and bornite.

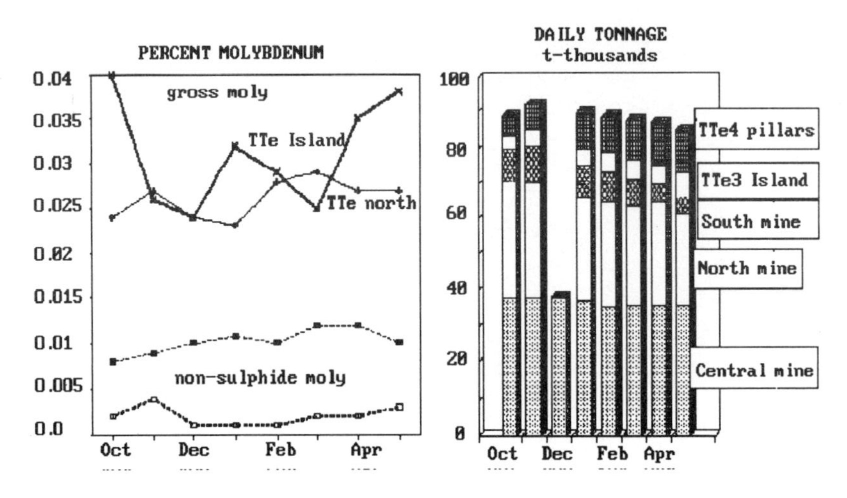

Fig. 35.- Monthly moly assays and tonnage produced by sectors - 1988/89

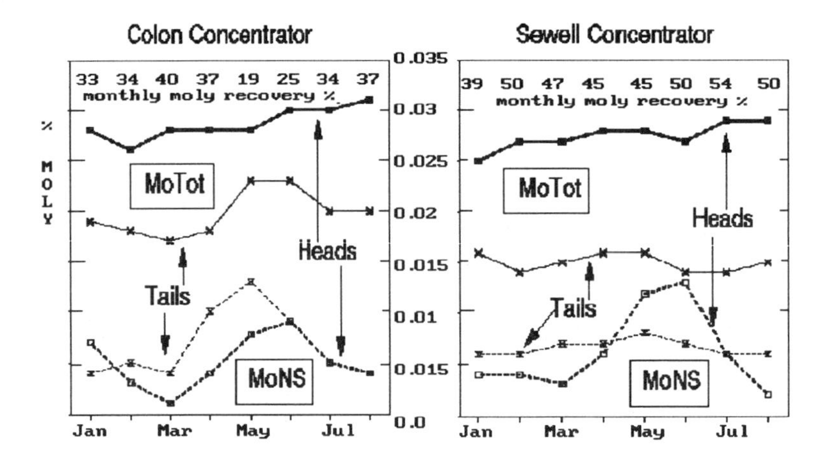

Fig. 36.- Monthly moly assays for 1989 for Colon and Sewell

To be able to correctly evaluate the laboratory data collected during 1988 and 1989, it was necessary to extensively sample the different areas being worked in the mine to obtain some idea of the short term variability in the minerals reaching the two concentrators. Fig. 35 shows composite monthly assays for two important sectors of the mine, and the monthly tonnage mined in 1988/89, from all important mine sectors, during the main course of this study. The average total molybdenum assays of the ore fed and tails obtained each month in Colon and Sewell, covering the same time period, as well as the non-sulphide moly concentrations, are shown in Fig. 36.

The effect of processability of different sectors of the mine was also tested with sufficiently large samples to perform normal industrial size laboratory flotations using Teniente's standard procedure. The raw data is shown in Table 48.

Table 48.- FLOATABILITY OF SAMPLES FROM DIFFERENT SECTORS OF THE MINE
% Recovery on Roughing

Mine Sector	Sample No	CuTot Acid	MoTot Acid	CuTot Alkaline	MoTot Alkaline	Acid - Alkaline CuTot	Acid - Alkaline MoTot
TTe 4 Fortuna	1	81.61	46.23	77.57	44.56	4.04	1.67
	2	67.80	47.15	59.05	31.57	8.75	15.58
	3	89.45	34.07	84.06	32.89	5.39	1.18
TTe 4 North Standard	4	67.63	69.47	62.65	65.82	4.98	3.65
	5	81.43	81.39	78.42	84.37	3.01	-2.98
	6	89.16	86.39	88.86	87.56	0.3	-1.17
	7	81.20	80.90	76.43	80.82	4.77	0.08
	8	78.84	75.01	74.33	79.99	4.51	-4.98
	9	51.86	28.44	48.62	24.29	3.24	4.15
	10	33.39	24.90	35.25	35.25	-1.86	-10.35
	11	33.08	24.58	36.47	30.28	-3.39	-5.7
	12	18.97	20.31	20.61	35.87	-1.64	-15.56
TTe3 Island	13	97.09	86.22	96.60	90.26	0.49	-4.04
	14	93.45	87.05	92.78	86.99	0.67	0.06
	15	92.95	82.34	93.12	81.12	-0.17	1.22
	16	96.84	92.18	97.24	90.21	-0.4	1.97
TTe5 Pillars	17	92.64	90.59	92.10	87.87	0.54	2.72
	18	94.71	81.82	92.54	81.80	2.17	0.02
	19	93.00	86.26	92.14	89.70	0.86	-3.44
	20	93.28	90.08	93.76	89.61	-0.48	0.47
Weighted avg. recoveries		88.2	86.769	86.2	84.7415	2.00	2.00

The most interesting data obtained from this very extensive floatability test programme based on mine sector samples is the difference between the recovery obtained from an acid pulp and an alkaline pulp, rather than the absolute value of the recovery, whose representativeness is difficult to evaluate a priori. These differences have been plotted in bar form in Fig. 37, which also shows the sector source for the different groups of samples. From these data it is obvious that Teniente level 4 ore contains extremely refractory copper and molybdenum, while Teniente levels 3 and 5 are normal. For these sectors, Table 48 shows absolute recoveries for copper above 90%, and for molybdenite above 80%. From the differential recovery values, the optimum process for the easily floated ore in levels 3 and 5 is not obvious, though for copper there is a slight tendency in favour of an acid environment.

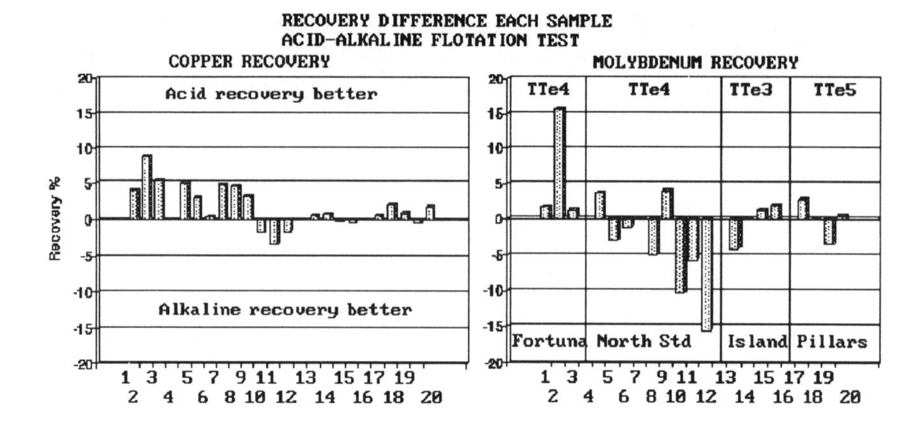

Fig. 37- Floatability of samples from different sectors of the mine

It is ironic that twenty years ago CODELCO management opted to design Colon as an alkaline flotation mill to avoid dependence on xanthogen formates as collectors because there was only one producer in the world, and because, supposedly, xanthogen formates were useful only in acid circuits. Today diethyl xanthogen formate is again the sole collector because it is unique in an acid circuit. The only selective competitor in alkaline flotation, isopropyl thionocarbamate (Z-200), was used for a number of years in the pH 10-11 Colon mill, but now has been judged not to match the recovery of xanthogen formate collectors (at least with Teniente ores) at any pH.

For Teniente level 4 samples, copper recovery is better in acid circuits, and a higher pH (10-11) favours moly recovery. The moly recovery problem at Teniente is losses in the cleaner circuits and not primarily molybdenite losses in the rougher flotation. The month by month moly recoveries for Colon and Sewell, shown in Fig. 36, suggest that moly recoveries from acid pulps could be better than standard basic pulp flotation.

From Fig. 37, note that certain mine sectors are not floatable, particularly in Teniente level 4.

Performance of the copper cleaner circuit: The final concentrate cleaning plant operating conditions are chosen to minimise moly losses, and maximise copper grade; but because of the design of the smelter, Teniente also has to have very tight specification on insol and pyrite content of the final copper concentrate, a problem that has become more serious because pyrite content of the ore has increased. Meeting these requirements has forced regrinding the primary concentrate to assure pyrite and insol liberation, and has had as an unavoidable side effect the over-grinding of the moly fraction (as can be seen in Fig.s 46 and 47). This operating procedure, causing over-grinding, has been continued because the value of the moly lost, at current prices, is lower than the potentially lost copper production. It has also resulted in raising the flotation pH to a maximum (pH greater than 12), to help depress pyrite, despite the possibility that the very high alkalinity of the concentrate triggers soluble moly losses in the tailings as well as providing more calcium to depress floatable moly (see Fig. 42). If the pH in the second cleaners is dropped from 12 to 10, the moly lost in the tails halves, but the iron in the final concentrate rises from 20 to 25%, and the copper grade drops from 32 to 28%.

To check the performance of the "copper circuit" whose flow diagram and equipment is shown in Fig. 38, particularly to obtain data on the rates at which moly and copper floats in the Colon cleaner circuits, plant tests were run in which samples of the concentrate from each of the flotation cells in the second cleaners, scavengers and re-scavengers were taken and analysed. The results are shown in Fig. 39. These data can be interpreted assuming that, as the residence time in each bank of cells is essentially the same for all the flotation cells, the copper and moly content in the concentrate samples is a measure of the flotation rates.

The first pair of graphs show the copper and insolubles assay in the concentrates in the first panel, and moly grades in the second panel, during normal operating conditions. From the shape of the copper curve, which does not show a drop in concentrate grade by the last cleaner cell, it is obvious that the rate of Cu flotation is too slow for the design cleaner flotation time, so a very significant amount of copper is carried through to the scavengers causing a large recirculating load (67% of the total recovered) from the scavengers and re-scavengers, though overall copper recovery is not adversely affected (94.2%), even when it continues to float vigorously through cell #5 of the scavenger bank. Molybdenum, on the other hand, is being severely depressed in all the cleaner bank, and only starts to float after the fifth cell in the scavengers. Note that the moly depression is such that the Mo grade of the cleaner tails (i.e. the feed to the scavengers) is 1%, nearly double the grade of the pulp fed to the first cleaner cell which contains 0.6% Mo. The fact that moly only starts to float in the scavenger when the amount of copper floating is dropping suggests that one factor contributing to moly depression is crowding out by the copper.

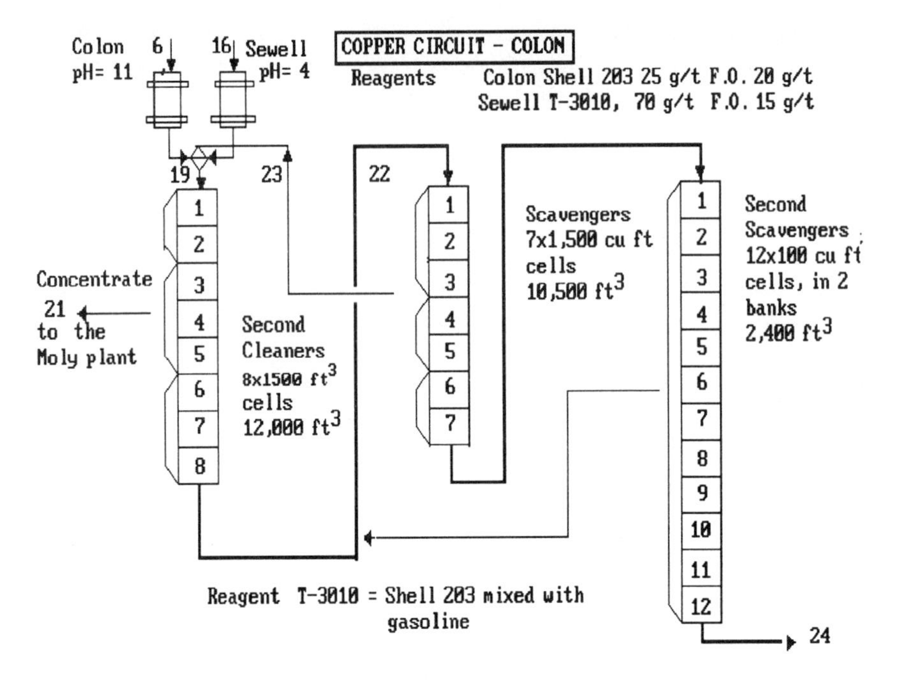

Fig. 38.- The alkaline cleaner flotation circuit at Colon

The second pair of graphs show the results of a plant test in which the Sewell regrind ball mill was stopped. The rate of flotation of copper increased, though the number of cleaner cells available appears to be barely adequate, as the first scavenger cell still produces a concentrate with a specification copper grade. The tail from the cleaners dropped from 18% to 11%, signalling the higher flotation rate obtained.

The shape of the molybdenum flotation curves are improved, but because of an ore change that doubled the grade fed to the second cleaners, the data cannot be compared with the next plant trial. As the copper concentrate coming off the head of the scavenger bank is nearly at specification, this suggested the second plant trial in which, as well as shutting down the Sewell regrind mill, the concentrate from the first three cells of the scavenger, instead of being recirculated to the cleaners, was piped directly to the P-4 thickener which feeds the copper/moly separation plant. Under these operating conditions the copper response is essentially normal, with a fully functioning second cleaner bank. Note though that the second cleaner tail now is at 4% Cu, as compared to 11% when cells 1 to 3 of the scavenger were being recirculated, and 18% when the Sewell regrind was operating. The moly flotation is marginally better, with recoveries improved as the scavenger

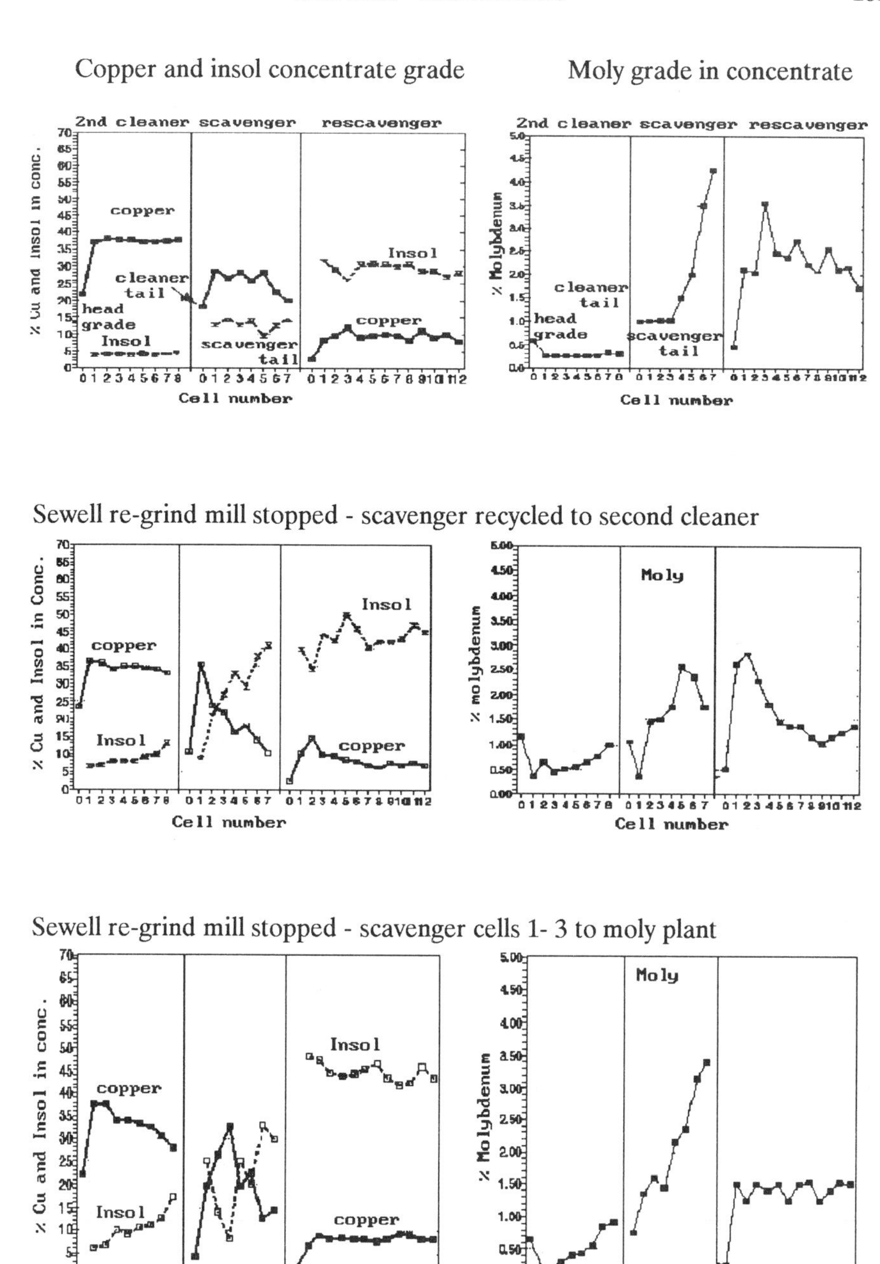

Fig. 39.- Performance of the copper cleaner circuit

tail now is at 0.3% Mo, compared to 0.45% at normal operations with both regrind mills going.

Effect of lime as a depressant of molybdenite and pyrite: It is well known that, particularly with by-product molybdenite, highly alkaline flotation conditions reduce recoveries. Shirley (1979), in a review article, points out: "The sensitivity of molybdenite to lime (CaO) usage was first noted at the huge Utah Division concentrators of Kennecott. Their operational procedure was modified to operate at the lowest pH possible consistent with maximum copper recovery. During a plant test at Chuquicamata conducted on half of the flotation sections, it was noted that when the normal operating flotation pH was lowered to 9.6 from 11.3 a considerably higher recovery of molybdenite was obtained."

Chander and Fuerstenau (1972) have shown that in an alkaline media, washed molybdenite has a high negative zeta potential (about -43 millivolts), with concomitant low flotation recovery (circa 30%), while reducing the pH to the 3.5 - 4.5 range will reduce the negative zeta potential by over 10 millivolts (to -30 mV), at which point the recovery jumps to 70-80%. The Teniente data in Fig. 36 confirms that lower pH during flotation improves moly recovery; recalculated and plotted versus the oxide moly contained in the ore (Fig. 40), it shows that oxide content reduces recovery. An increase in oxide reduces recovery more markedly in an alkaline environment, but the data show that moly recovery is better at all oxide levels in an acid circuit, presumably because the molybdenite surfaces are cleaner in a media consisting of very dilute sulphuric acid, and/or because adsorbed calcium ions are removed from the surface, and because moly solubility is reduced.

The strong effect of calcium ions on the depression of pyrite has been known for some time, but surprisingly little flotation data confirming this is available. Cea and Castro (1975) published Hallimond tube data that clearly shows the multiplying effect of using lime rather than caustic soda or soda ash as the source of alkalinity. This confirms Glembotskii et al's (1963) earlier observation "a comparison of the effect of $Ca(OH)_2$ with that of NaOH on pyrite shows that lime is the more powerful depressant". In Fig. 41, a comparison is shown between the effect of pH on pyrite recovery, as measured by Li Guongming et al (1989), with that of caustic and lime on molybdenite flotation, measured by Castro and Paredes (1979).

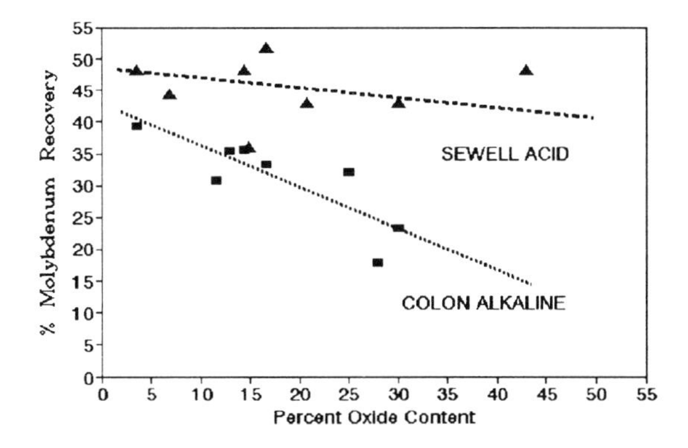

Fig. 40.- Moly recovery Sewell and Colon as a function of oxide content

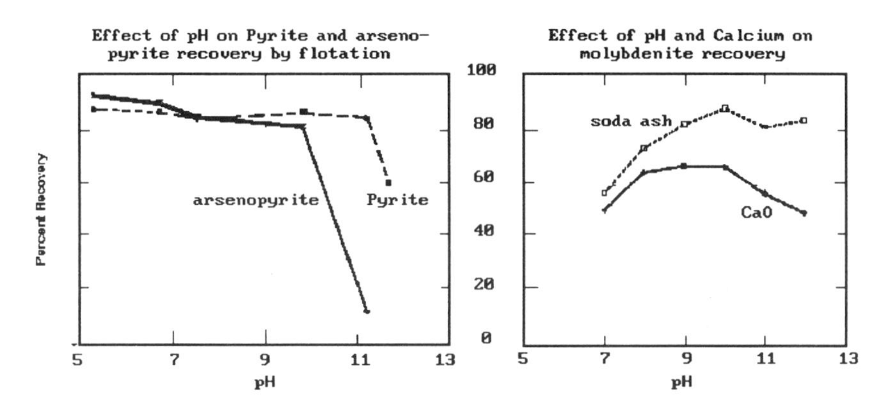

Fig. 41.- The effect of pH on pyrite and molybdenite recovery

The combined pulp feed to the copper circuit (which contains 25% chalcopyrite and 30% pyrite) was submitted to bench scale flotation tests performed at half pH unit intervals. Copper, moly, pyrite and insolubles assays were made for all tests. The results are shown in Fig. 42. They confirm that chalcopyrite is not significantly depressed up to a pH = 11. They also confirm that pyrite starts to depress at pH 11, but indicate that even at pH = 12 there is far from complete depression, and that depression of molybdenite, as a function of pH, in the presence of calcium ions, is greater than that for pyrite. The behaviour of the insolubles is indicated by the grade recovered in the concentrate rather than by recovery. From these figures one can conclude that there is no significant depression of insol in the pH range

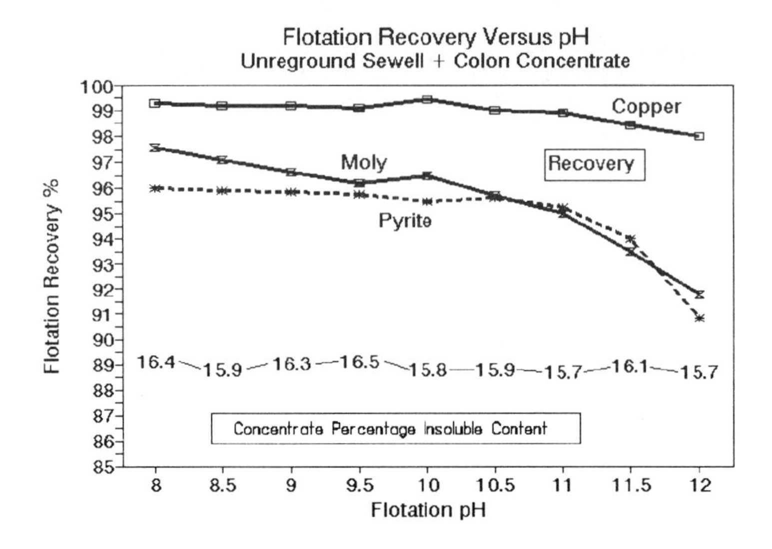

Fig. 42.- Effect of pH on Cu and moly recovery from second cleaner feed pulp

studied. The obvious alternative of testing soda ash or caustic as the modifier reagent, to eliminate the effect of the calcium ion, is impractical because of reagent carry-over from the rougher flotation at Colon where, with the high ore tonnage treated, the cost of substitution of lime as the pH regulator is not justified by the additional moly recovery. Similarly, opting to operate an acid rougher in a concentrator designed for alkaline flotation requires expensive acid proofing.

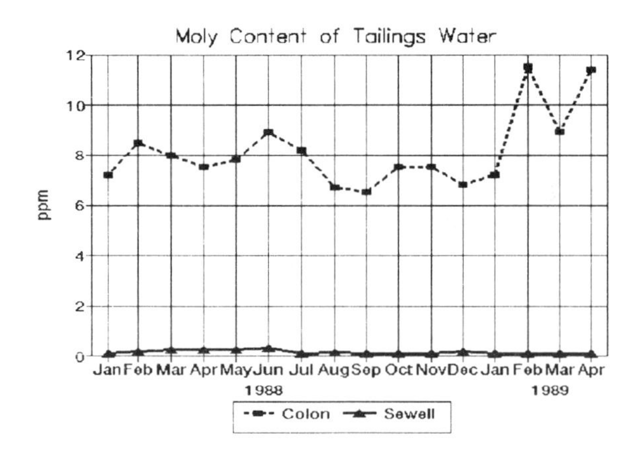

Fig. 43.- Molybdenum content of Teniente tailings water

Lime, and high alkalinity, also reduce moly recovery because oxidised molybdenite reacts at very high pHs with calcium ions to form powellite, or, in the presence of limonite, the surface compound formed could be ferromolybdite [3(FE_2O_3).MoO_3.8H_2O]. Both of these are water soluble and can be lost in the tailings water. Fig. 43 shows that this happened in 1989, when the moly content in the Colon tailings jumped up. At 12 ppm the losses in solution in 1989 would have amounted to about 300 tonnes of Mo.

CIMM (Chile's Mining and Metallurgy Institute) developed the electrochemical phase diagram (Fig. 44) for the system molybdenum - sulphur - water, which shows that pH 12 is beyond molybdenite's stability region. It suggests that above pH = 11, molybdenite is not stable, thus, partial oxidation at that pH can form a hydrophyllic surface making the moly particles unfloatable in the presence of calcium. The reactions are:

$$MoO_3 + Ca(OH)_2 \longleftrightarrow \underset{\text{Powellite}}{CaMoO_4} + H_2O \quad Kps = 4.2 \times 10^{-8}$$

for dissolution, and

$$MoS_2 + 9/2\ O_2 + 6\ OH^- \longleftrightarrow MoO_4^{=} + 2SO_4^{=} + 3H_2O$$

for the precursor reaction for the formation of ferromolybdite, the water avid compound on the molybdenite surface.

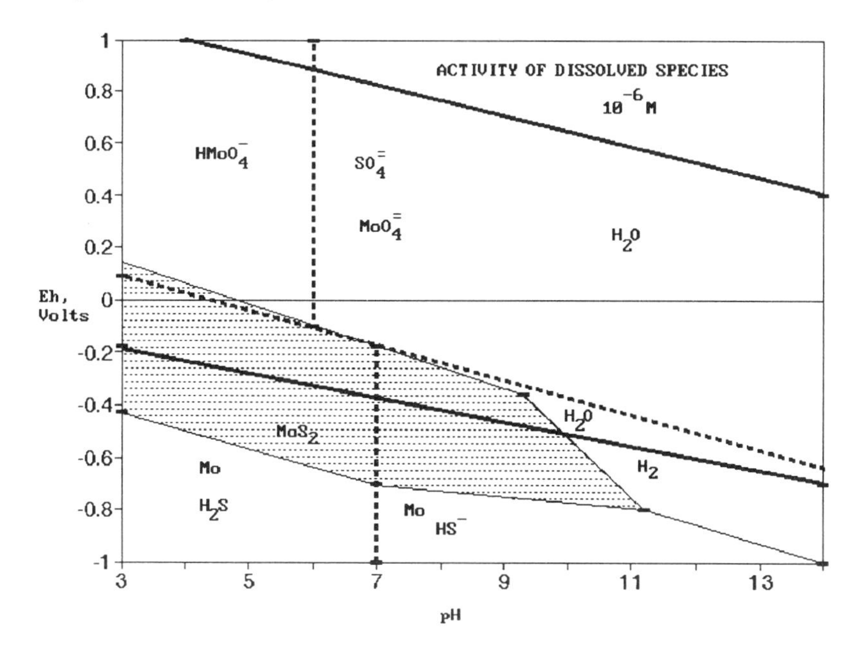

Fig. 44.- Potential-pH diagram for the moly-sulphur-water system

The CIMM report further suggests that the mechanism of calcium depression can be explained if the oxide layer on the pyrite consists of an undercoat of $Fe(OH)_2$ and $FeCO_3$, on which there is a coating of a porous layer of $Fe(OH)_3$. The calcium ions would adsorb on to this latter porous layer impeding adherence of the collector on the pyrite. A similar mechanism has been put forward by Tsai et al (1971) and Shimoiizaka (1972), in which only OH groups on the mineral surface are necessary for calcium ion adsorption. Presumably a ferromolybdite surface can undergo similar reactions and adsorb calcium ions.

Liberation of Copper and Moly: In June 1989, to check on possible over-grinding, the Colon and Sewell concentrates were carefully sampled, as well as all the major streams in the two regrind mills whose combined output feed the copper circuit (second cleaners et al). The standard cleaner circuit and operating conditions are shown in Fig. 45, and the results of the liberation tests in Table 49.

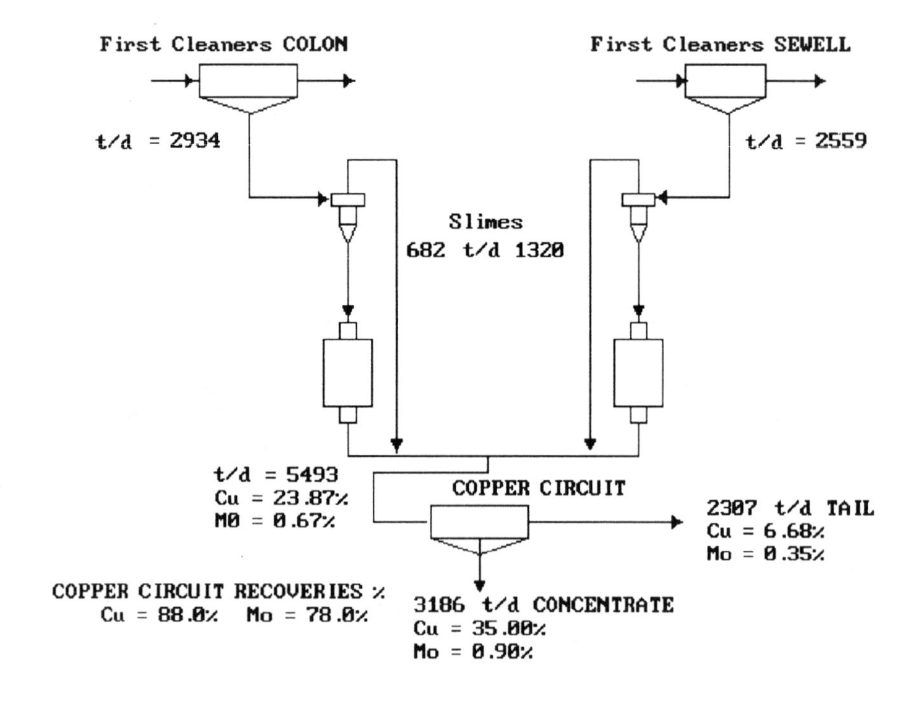

Fig. 45.- First cleaners and copper circuit flow diagram

For both Colon and Sewell liberation of the molybdenite prior to regrinding is practically complete - 99.0 and 99.4% respectively. In the case of the copper minerals, there is a moderate to marginal improvement in liberation, which could justify regrinding - 92.8% to 96.8% for Colon and 96.8% to 97.2% for Sewell - but the critical factor in concentrate specifications is pyrite and insolubles. The greater sensitivity of the Sewell ore to form colloidal slime, which are notoriously refractory to flotation, is due to the softer nature of the ore. This shows up in the power consumption figures of Table 49 - 4.88 kWh/t for a reduction ratio of 1.68 for Sewell, as compared to 8.11 kWh/t consumed for a reduction of 1.79 at Colon.

Table 49- MATERIAL BALANCE AND SCREEN ANALYSIS RETREAT REGRIND

	Colon Cleaner Concentrate				
	Feed to Cyclone	Cyclone underflow	Regrind Mill out	Cyclone overflow	Final Product
Dry Tonnes/day	2,934	1,614	1,614	1,320	2,934
Fv Pulp (m3/d)	14,821	1,247	1,247	13,574	14,821
% solids	17.32	67.11	67.11	9.08	17.32
% -325 mesh	57.35	26.64	51.37	90.77	69.09
D80 (micron)	93	143	80	33	59
% Liberated Cu	92.76				96.84
% Liberated Mo	98.86				99.45
Mill reduction ratio R80			1.79		
Mill Energy kWh/t			8.11		

	Sewell Cleaner Concentrate				
	Feed to Cyclone	Cyclone underflow	Regrind Mill out	Cyclone overflow	Final Product
Dry Tonnes/day	2,559	1,877	1,877	682	2,559
Fv Pulp (m3/d)	6,738	1,447	1,447	5,291	6,738
% solids	29.58	65.92	65.92	11.75	29.58
-325 mesh	44.79	29.62	54.39	86.31	62.90
D 80 (micron)	102	124	74	40	65
% Liberated Cu	95.75				97.21
% Liberated Mo	99.36				99.16
Mill reduction ratio R80			1.68		
Mill Energy kWh/t			4.88		

In late 1988, CIMM had performed a similar size distribution study as a part of the design of the moly plant rougher. The data showed that the moly fraction of the concentrate fed to the moly plant contained 41.9% -13 micronparticles, while the CIMM study showed that the extremely fine moly in the second cleaner concentrate amounted to 64% -13 micron particles and the copper fraction contained only 38% -13 micron. Based on these studies, in 1990, to ameliorate over-grinding, the ball charge was reduced from 40% to 30% of the volume in both mills.

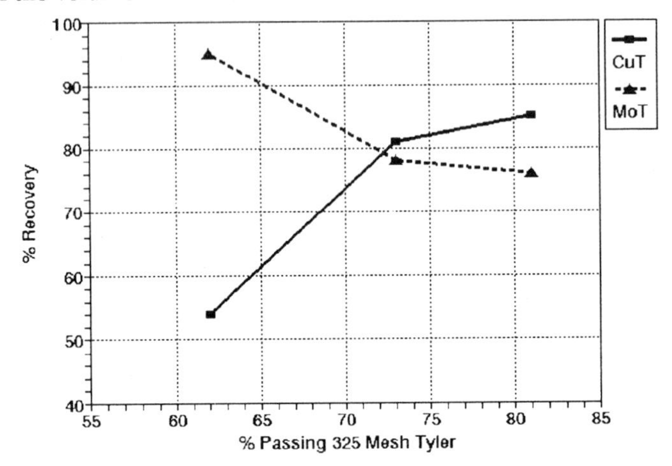

Fig. 46.- Effect of grind on copper and moly recovery

The data in Fig. 46 show that regrinding second cleaner feed can improve copper recovery, but at the expense of moly losses. Considering the depressant effect of any type of slime on sulphide mineral flotation, picking the optimum degree of comminution becomes a very difficult exercise. The problem in choosing a moly flotation system is seen in the studies made of the distribution of the molybdenite as a function of particle size. The result for Sewell and Colon ores is shown in Figs. 47 and 48.

Sand slime separation: An obvious alternative studied in the flotation laboratory, using pulps and cyclone overflows sampled during plant operations, was simulation of a sand slime circuit. The reground sands, whose assay was 21.14% Cu and 0.75% Mo, on flotation in the 2 litre laboratory machine yielded a concentrate grading 33.3% Cu and 2.09% Mo, and tails with 14.8% Cu and only 0.054% Mo; i.e. a Cu recovery of 54%, and a 95.28% moly recovery. The slimes, which assayed 23.86% Cu and 0.94% Mo, yielded a concentrate with 29.55% Cu and 1.06% Mo, and tails with 11.7% Cu and 0.68% Mo; i.e. a Cu recovery of 84.37% and moly at 76.9%. Globally these recoveries add up to only 95.94% on copper and 87.72% moly. Compared to the standard circuit (Fig. 45) the sand slimes process improved moly recovery significantly, but showed much poorer copper recoveries because of losses in the sands cleaner. Regrinding the sands was not followed up because of the rapid drop off in fine moly recovery.

Post moly/copper separation pyrite scavenger: A serious look has been given to removal of the excess pyrite after recovering the molybdenite at an optimum pH throughout the Sewell and Colon flotation process. The proposed final cleaner circuit is shown in Fig. 49. The advantage in moly recovery of operating the second cleaners at the rougher pH, or lower, is obviously a greatly enhanced molybdenite recovery. The disadvantage is also obvious - the need for a considerably expanded moly/copper separation plant, as there are already problems of moly losses due to overloading the existing plant.

Moly/copper separation plant studies: To compensate for the reduced flotation time generated by the increased ore tonnage treated in Colon, an additional circuit (A'), consisting of two banks of twelve 100 ft^3 cells, was added to the moly/copper separation rougher circuit (A). Studies were made by outside consultants (CIMM) to check the operation of individual cell banks in the moly plant. It was found that, in the roughers, 26 to 32% of the fine particles of molybdenite were short-circuiting the flotation zone due to the fact that the tank of the original 60 ft^3 cells adapted for use in the moly plant had been modified to 100 ft^3, after which they were prone to sanding. Increased cell maintenance cleaning and raising the impeller speed from 230 rpm to 300 rpm improved the operation. A greater impeller speed, as suggested in a paper by Malhotra et al (1977), improves flotation of fine moly particles. Increased processing time was also obtained by the addition of two square 36"x36" column cells, and a 36" diameter round cell to the molybdenite circuit.

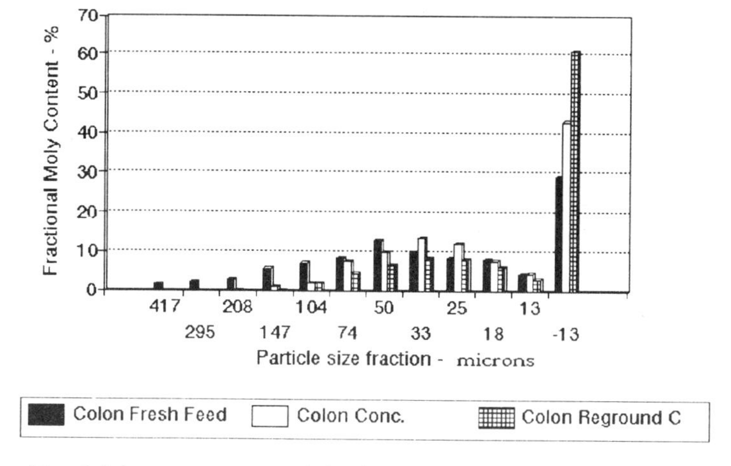

Fig. 47.- Moly content vs particle size - Colon before and after regrind

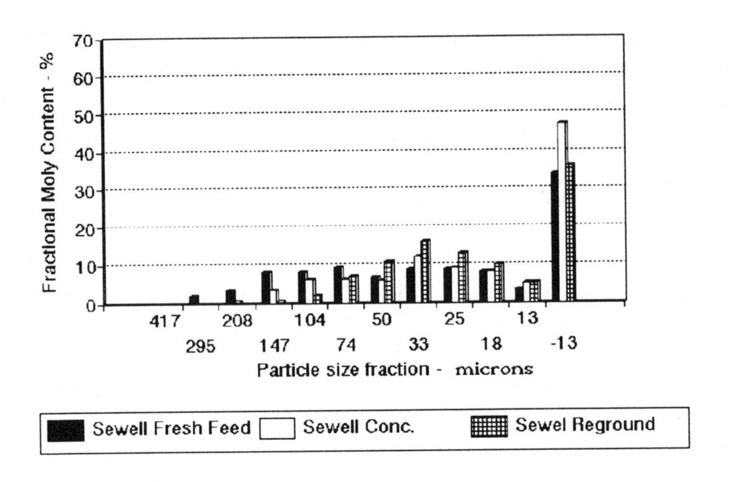

Fig. 48.- Moly content vs particle size - Sewell before and after regrind.

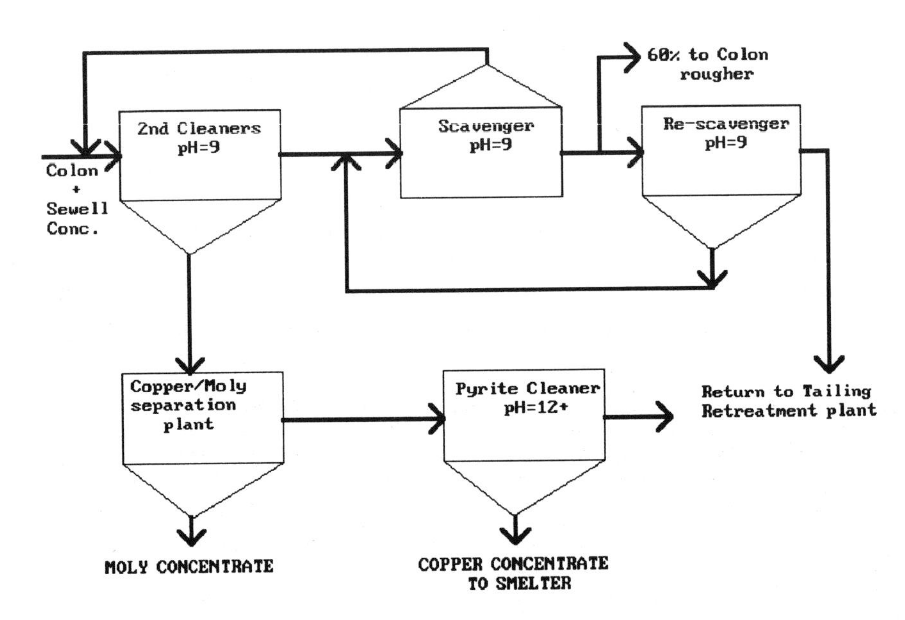

Fig. 49.- Proposed new Colon cleaner circuit

Depressant studies - Nokes reagent optimization: The reagent employed in the Colon moly plant is designated LR-744. It is made on site by the reaction of phosphorous pentasulphide with caustic soda in a water solution, in the ratio of 10 parts by weight pentasulphide to 13 parts NaOH. The mechanism of depression of Nokes reagents and NaSH was studied by Castro and Pavez (1977), who found that the direct effect of these reagents is a large reduction of the pulp potential, Eh.

A May 1989 Teniente report on the effect of reagent dosage on moly recovery and copper depression can be summarised as follows:

Moly plant rougher flotation - increased dosage of LR-744 (1,500 g/t to 2,000 g/t) will improve total molybdenum flotation slightly. In laboratory tests, with higher levels (2,500 to 4,500 g/t), moly flotation is slightly impaired, possibly because there is physical occlusion of moly with the increased depression of the copper minerals. Moly selectivity is directly proportional to LR-744 dosage, though the depressive effect of high doses restricts this route to the production of a molybdenite concentrate with a low Cu content. Typical laboratory data show:

Table 50.- EFFECT OF NOKES REAGENT ON ROUGHER RECOVERY

Feed rate LR-744	Rougher Recoveries	
grams/tonnes	CuT	MoT
2,500	20.9%	91.1%
3,500	18,5%	89.5%

Moly plant first cleaner stage - up to 300 g/t of LR-744 can be fed without reducing MoT recovery, but these amounts of depressant provide only marginal improvements in CuT rejection. Greater additions of LR-744 increase selectivity, but at the cost of a significant drop in MoT recovery, viz:

Table 51.- EFFECT OF NOKES REAGENT ON FIRST CLEANER RECOVERIES

Feed rate LR-744	First cleaner Recoveries	
grams/tonnes	CuT	MoT
0	37.1%	81.8%
300	30.6%	81.2%
800	26.9%	79.3%

These data underline that excess depressant should be avoided in flotation stages where the primary function is molybdenum recovery.

Use of flotation columns in the moly plant - laboratory tests of columns in the cleaner stages gave the following results:

Table 52.- MOLY/COPPER SEPARATION CLEANER TESTS WITH COLUMNS

	First Cleaner		First Cleaner	
	%CuT	%MoT	%CuT	%MoT
Feed	37.0	1.44	34.3	2.92
Concentrate	24.2	15.2	17.1	24.9
Recovery MoT	74.6%		93.3%	
Depression CuT	95.4%		94.5%	

	Second Cleaner		Third Cleaner	
	%CuT	%MoT	%CuT	%MoT
Feed	20.35	23.32	7.21	43.32
Concentrate	7.7	42.40	3.98	49.83
Recovery MoT	92%		83%	
Depression CuT	81%		60%	

Plant tests also have confirmed that the use of column cleaner stages is economically very attractive in the moly/copper separation process. Teniente is therefore going to modify the Cu/Mo separation plant.

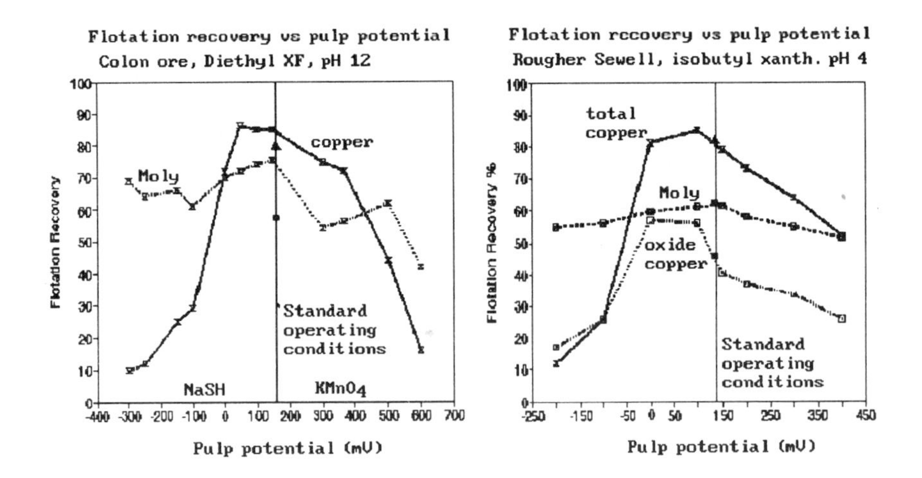

Fig. 50.- Redox measurements of rougher flotation - Teniente

As part of this molybdenum flotation improvement programme, a study of the effect of adding a reagent to change redox potentials of the pulp has also been carried out. Results for rougher flotation of Colon and Sewell ore are shown in Fig. 50. The collector used for the Sewell ore at pH = 4 was isobutyl xanthate, as there were problems adjusting the redox potential values with the normal xanthogen formate. In light of the similar shape of the curves, the acid results should also apply to the xanthogen formates.

This work is being used to screen sulphidisation, as well as other inorganic reagents for which there is no prior experience with Teniente ores. In addition, a screening programme is on-going to evaluate pyrite selective flotation reagents.

Chapter 11

SULPHIDE MILL PRACTICE

Reagent consumption in the flotation industry

The statistics and other data quoted in this section provide practical target performance figures from operating mines (i.e., ore grades, concentration ratios, concentrate grade, average mineral grind, power and water consumption per tonne processed for the different types of ores, etc.), which can be used as a guide in setting parameters for laboratory studies. The very extensive average operating data, contained in Tables 56 and 57, should be used, based on the predominant mineral of the ore to be tested, to select the initial system conditions, and starting reagents.

The practical importance of the reagent families, and their evolution in flotation, as well as the total volume of the reagent market in the U.S.A., can be obtained from Bureau of Mines statistics, some of which are summarised in Tables 52 and 53. Table 54 shows how the economic recessions have affected the volume of ores processed by the mineral industry. Depletion of mineral resources can be deduced from the drop in the grades treated, and the evolution of the concentration ratios with this drop. Certainly the catastrophic drop in metallic mining with the 1985 crisis is obvious; as is the ecological pressure to clean up coal, reflected in the ore tonnage upgraded by flotation, which has more than doubled.

The observed concentration ratios are a simple way to distinguish between different types of flotation. If the concentration required to attain a commercially saleable product is 30:1, the flotation regime is nascent bubble capture. In the coal industry, and in the recovery of industrial minerals, where the concentration ratio is also very low, the flotation regime is bubble capture. When a mineral exhibits an intermediate concentration ratio, either flotation mechanism may dominate, or both may be present simultaneously, as is the case with complex metallic ores such as silver-zinc-lead, where the task is to upgrade a 7 - 9% ore to a 25% concentrate to feed a smelter; i.e., a ratio of roughly 3:1 to 4:1. The low concentration ratio in Table 53 for metallic carbonate and oxide ores, and the increase in tonnage treated since 1960, is due to iron ore, which the USBM puts in this classification.

Table 53.- FROTH FLOTATION IN THE U.S.A.; source: U.S. Bureau of Mines

Year	1960	1970	1980	1985
Sulphide Ores				
Ore Treated tonnes	141,022,000	256,545,000	254,455,000	176,520,000
Concentration ratio	26.5:1	31.8:1	38.1:1	33.3:1
Metallic Carbonate and Oxide Ores				
Ore Treated tonnes	2,594,000	20,193,000	39,003,000	22,264,000
Concentration ratio	3.0:1	1.7:1	1.8:1	1.7:1
Nonmetallic Ores				
Ore Treated tonnes	32,901,000	73,603,000	178,545,000	124,865,000
Concentration ratio	3.0:1	3.4:1	3.5:1	4.2:1
Anthracite and Coal				
Ore Treated tonnes	3,738,000	11,824,000	11,728,000	25,200,000
Concentration ratio	1.5:1	1.5:1	1.7:1	1.4:1
Grand Total ore treated by flotation	80,255,000	362,165,000	483,723,000	348,849,000

The quantities of collectors consumed in 1975 and 1985 (the only two years reported in comparable detail in the Minerals Year Books), are shown in Table 54. Note that the fatty acids dominate the oily collector market, and the xanthates the metallic sulphide market.

The sulphide and non-sulphide ore tonnage treated is about equal, but the oily collector consumption is thirty times sulphhydryl reagents because industrial minerals require a unit dosage that much higher. The average calculated reagent consumption is 589 g/t for the oily collectors, and only 23.5 g/t for the sulphide minerals. This is one reason why the inorganic and oxide minerals' collector and frother market is financially much more attractive to the reagent suppliers, despite the lower unit value of the fatty acids and organic amines.

The breakdown by ore type for frother usage and consumption is less precise, because the same reagents are used with all the different ore types. In addition, there does not seem to be a consistent reporting basis between the two latest years, as can seen from Table 55. Fuel oil and kerosene have been separated in this table, because they modify frothers, and act as mild collectors. But they can be a primary collector if the mineral is naturally floatable, so are difficult to classify, and should be allocated to both the frother and the collector tables.

Table 54.- U.S. CONSUMPTION OF COLLECTORS IN 1975 AND 1985

Sulphide Collectors	Consumption, tonnes 1975		1985	
Alkyl xanthates	3,699	58.6%	1,647	57.1%
Dithiophosphates	1,060	16.8%	611	21.2%
Xanthogen formates	507	8.1%	22	0.8%
Thionocarbamates	485	7.7%	586	20.3%
Mercaptobenzothiazol	185	2.9%	17	0.6%
	5,936		2,883	
Avg dosage g/t ore	23.5		16.3	
Oily Collectors	**Consumption, tonnes 1975**		**1985**	
Fatty acids	59,509	77.0%	76,264	79.1%
Tall oil	10,953	14.2%	*	
Amines	4,276	5.5%	14,301	14.8%
800 series (sulphonates)	2,146	2.8%	5,411	5.6%
Petroleum sulphonates	347	0.5%	484	0.5%
	77,231		96,460	
	589.3		559.7	

* included in fatty acids

In 1975, frother consumption of 12,100 t compares with 83,167 t of sulphide and oily collectors. This ratio widened considerably in 1985, reflecting either improved frother chemistry, or more probably a shift to frothing collectors, such as amines and sulphonates, in the industrial mineral applications. The overall drop in both figures for 1985 confirms the recession that hit the mining industry in the early eighties.

Copper mill data

The material covered in Table 56 is a composite of a 1975 copper mines survey (Crozier(1977)), up-dated by data published in the SME Handbook, miscellaneous plant visits, and reports in the trade magazines covering mainly the Pacific basin. It is therefore more representative of western porphyry copper technology than the design norms followed in Europe and Africa, where the geology of the deposits are somewhat different. The reagent use and operating conditions reported also tend to be more representative of the larger and newer mines.

Table 55.- CONSUMPTION OF FROTHERS IN THE TOTAL U.S. MINING INDUSTRY

1975	Tonnes	Percent
MIBC	5,401	44.7%
DF 250	1,130	9.3
Barrett Oil	1,078	8.9
Pine Oil	721	6.0
Cresylic Acid	583	4.8
Aerofroth 71	345	2.9
Other Aerofroths	369	3.1
Other Dowfroths	92	0.8
Nalco frother	233	1.9
UCON	778	6.4
All others	1,367	11.2
Total frothers	12,099	
Fuel Oil	93,249	
Kerosene	3,157	
1985	**Tonnes**	**Percent**
Aliphatic alcohols	2,997	44.1%
Polyglycol ethers	763	11.2
Petroleum based blends	1,057	15.6
Pine oil	495	7.3
Phenol	353	5.2
Unspecified polyols	1,125	16.6
Total frothers	6,791	
Fuel Oil	66,265	
Kerosene	4,246	

At the time the SME Handbook was published, over a million tonnes of copper capacity had been removed from the market, mainly in the U.S., because of the mining industry crisis. Thus, temporarily or permanently, most of the mines treating the lower ore grades were shut down, or shifted to leaching processes combined with low cost direct electrowinning of the copper. All these factors must be taken into account when generalising from the data.

Nevertheless, it is interesting to see how predominant chalcopyrite and chalcocite are in the ores of commercially important mines. But the types of reagents employed, and their point of addition to the flotation circuit, certainly are not very standardised.

Tables 56 and 57 are both divided into ore characteristics, head grades, and concentrate properties; and separately the collectors, frothers and modifiers (where available) are shown with the quantities employed, and an attempt is made to indicate the point in the flotation process where the different collectors are added. The completeness of the statistical data for the different mines varies considerably. As there have been significant ore changes in the older mines over the years, the reagents reported are consistent with the ore description, but do not necessarily reflect current use.

Looking over Tables 56 and 57, it is interesting to note that there are operating conditions and reagents listed that reflect quite arbitrary regional preferences. Most noticeable is the tendency of U.S. designed plants (North and South America) to grind their ore slightly finer and to operate at higher alkalinity than the Pacific (mainly Philippines) mines. The only two mines listed that employ an acid circuit, El Teniente in Chile and Lepanto in the Philippines, do so because they have very specific ore characteristics that require low pH operation. Lepanto uses an acid circuit because its ore is very high in arsenic (enargite), while the El Teniente ore not only has a relatively high non-sulphide content but also has acidic components in the gangue that in the past have required very large amounts of lime to raise the alkalinity to the habitual pH 9-11 (this may not be true in future when harder, primary, ore will predominate).

From the mines listed in Tables 56 and 57, which represent over half of the world's copper production (4,500,000 tonnes) in 1975, we can list average operating conditions and performance for the different producing areas. As can be seen in Table 58, between 1965 and 1975 the average copper concentrator treated ores with 0.782% Cu, of which 85.35% was recovered. Regionally, head grades range from a low of half a percent copper in Canada, to a high of 1.455% in Chile, while recoveries go from a high of 87.7% in Canada, to a low of 80.4% in Europe.

Table 56.- INDUSTRIAL COPPER MILL FLOTATION DATA -

Principal Mineral CHALCOPYRITE

No.	Location	Daily t milled	% Oper time	Mill Feed % Cu	% Mo	% Fe	% Zn	Secondary Minerals	Gangue Composition
1B	Twin Buttes, AZ	32,000	91	0.83	0.003		-	Cu sulph	Arkose, silicated limestone
1C	Yerington, NV	13,500	98	0.60	-	2.8	-	Cc	Quartz, monzonite, granodiorite
3A	Mission, AZ	23,000	96	0.66	0.024	3.7	0.07	Cc	Quartz; Ca, Al silicates
3B	Silver Bell, AZ	11,000	97	0.64	0.008	3.4		Cc	Quartz, pyrites
4	Highland Valley, BC	16,000	97	0.51	-	-	-	Bo	Quartz, plagioclase, mico, clays
5	Bagdad, AZ	5,500	98	0.81	0.030	-	-	Cc	Pyrites, silicates
7	Aitik, Sweden	6,500	95	0.49				Py	Quartz, feldspars, garnet, mica
8	Bouganville	90,000		0.48	0.010			Bo, Mag	Quartz, feldspars
9	Merritt, BC	5,000	98	1.00	0.029			Mag, Hem	Quartz, feldspars, calcite, kaolinite
10	Pinto Valley, AZ	40,000		0.44				moly	
10A	Miami, AZ	17,500		0.50	0.005			Cs	
13	Sierrita, AZ	80,000	96	0.30	0.029			moly	Monzonite, diorite
13A	Mineral Park, AZ	17,000		0.40	0.04			moly	
13B	Esperanza, AZ	15,000		0.40	0.03			moly	
14	Erstberg, W. Irian	7,500		2.80	-	35.0		Bo,Cov	Magnetite, hematite
15	McLeese Lake, BC	30,000		0.40	0.020			Cc	Quartz, sericite
16	Tide Lake, BC	7,000		1.50				Bo	
17	Nairne, S. Australia	2,400	95	1.00				Cc	Andalusite, biotite, schist
18	Lakeshore Casa Grande	8,200		1.50	0.001	5.0			
20B	McGill, NV	22,000	97	0.73				Cc	Silica, alumina
20C	Ray, AZ	25,000	98	0.03	5.400			Cc Cup	Quartz, sericite, monzonite, porphyry
20D	Utah Mines	108,000		0.68	0.027	3.1		Cc,Bo,Cov	Quartz, feldspars, biotite
21	Noranda, Quebec	1,500	99	2.54				Sp	Quartz, pyroxene, feldspars
21A	Logan Lake, BC	47,500	93	0.43	0.016			Bo	Quartz, diorite, hornblende
22A	Superior, AZ	3,600		4.50		32.0		Bo	
23	Marinduque, Phil.	19,000	97	0.70				Bo	Quartz, calcite, feldspars
25	Hokkaido, Japan	1,100	97	2.35			1.10	Sp	Quartz, calcite, chlorite
26	Mount Isa, Australia	16,800	95	3.00		7.0			Quartz, dolomite
27C	Rokana, Zambia	17,000	92	1.95		3.2		Bo, Car	Quartz, dolomite
28A	Geco, Ontario	5,125	96	2.57			5.90	Sp, Ga	Silicates
28B	Horne, Quebec	1,860		2.45					Silicates
28C	Brenda	26,000	95	0.19	0.051			Moly	Quartz-diorite, feldspars, biotite, epidote

Table 56a.- INDUSTRIAL COPPER MILL FLOTATION DATA (continued) -

Principal Mineral CHALCOPYRITE

		Copper Concentr %						Moly Concen			Precious
		%Cu	%Mo	%Fe	%Zn	Insol	%Recov	%Mo	%Cu	%Recov	Met. Rec
1B	Twin Buttes, AZ	32	0.26				87	51	1.40	30	Ag
1C	Yerington, NV	25		29		8.7	83				
3A	Mission, AZ	28	0.29	28	1.7	7.7	90	52	0.66	43	Ag
3B	Silver Bell, AZ	29	0.15	24		7.0	84	51	0.70	21	Ag
4	Highland Valley, BC	30	0.49	26		9.7	82				
5	Bagdad, AZ	32	1.45	22		15.5	74	54	0.75	64	Ag
7	Aitik, Sweden	29					90				Au, Ag
8	Bouganville	30	0.13				85				Au, Ag
9	Merritt, BC	27									
10	Pinto Valley, AZ	27	0.05	29		5.6		53	1.25		Au, Ag
10A	Miami, AZ	27	0.06				87		1.25	54	
13	Sierrita, AZ	25					90	53		75	Ag
13A	Mineral Park, AZ	20	0.1				76		0.30	62	
13B	Esperanza, AZ	25	0.12				87			74	
14	Erstberg, W. Irian	28					95				Au, Ag
15	McLeese Lake, BC	29					81	53	0.60	30	
16	Tide Lake, BC	28					95				
17	Nairne, S. Australia	24					92				Au, Ag
18	Lakeshore Casa Grande	27	0.03	27		13.5	91				
20B	McGill, NV	18	0.30	30		12.5	76				Au, Ag
20C	Ray, AZ	21	0.41	25		15.2	85	49	1.40	37	
20D	Utah Mines	27	0.25	26		14.5	89	55	0.50	66	Au
21	Noranda, Quebec	23			5.8		95				Au, Ag
21A	Logan Lake, BC	33					90	56	0.65	75	
22A	Superior, AZ	25		30		2.5	96				Au, Ag
23	Marinduque, Phil.	27				6.0	87				Au, Ag
25	Hokkaido, Japan	22			4.0		94				Au, Ag
26	Mount Isa, Australia	25		29		8.5	96				
27C	Rokana, Zambia	32		18		9.6	89				
28A	Geco, Ontario	27			8.0		94				Au, Ag
28B	Horne, Quebec	16					96				Au, Ag
28C	Brenda	30	0.20				86	55	0.30	78	

Table 56.- INDUSTRIAL COPPER MILL FLOTATION DATA (continued) -

Principal Mineral CHALCOPYRITE

		Daily t milled	% Oper time	Mill Feed % Cu	% Mo	% Fe	% Zn	Secondary Minerals	Gangue Composition
28D	Gaspe, Quebec	34,000		0.60	0.015			Cc	
30A	O'Okiep, S. Africa	2,100	98	1.30				Bo	Plagioclase, biotite, hypersthene
30B	Nababeep, S. Africa	3,150	98	1.20				Bo	Plagioclase, biotite, hypersthene
30C	Carolusberg, S.Africa	4,200	98	1.25				Bo	Plagioclase, biotite, hypersthene
31A	Keretti, Finland	1,600	95	3.30		23.0	0.60	Py,Sp	Quartz, serpentine
31B	Virtisalmi, Finland	750	95	0.65					Amphibibole, pyroxene, garnet
31C	Pyhasalmi, Finland	?	95	0.88		29.6	4.20	Sp, Py	Quartz, feldspars, sericite
32	Phalaborwa, S. Africa	53,000		0.55				Bo,Cc,Cup	Carbonatite, phoscorite, dolomite
33A	New Cornelia, AZ	33,000	99	0.68		2.6		Cc,Bo	Quartz, feldspars
34A	Sto Tomas, Phil	5,000	97	0.39		3.2		Mag	Andesite, diorite
34B	Benguet, Philippines	16,000	97	0.39		3.4		Mag	Andesite, diorite
35	Sahuarita, AZ	58,000	96	0.50	0.017	3.6	0.07	Cc	Arkose, garnet, hornfels
35A	Bagdad, AZ	40,000		0.70	0.03			Cs	
37A	Chambishi, Zambia	3,000	95	2.73				Cc,Bo	Quartz, mica, feldspars
37B	Luanshya, Zambia	16,000	96	1.78				Cc,Bo	Quartz, mica
40	Princeton, BC	15,000	90	0.45			-		Albitized Feldspars, pyrites
41	Niihama, Japan	1,400	98	1.53		17.0		Py,Prh	Chlorite, albite, quartz
42	Kidd Creek, Ontario	9,000	97	1.90			8.90	Sp,Ga	Quartz, feldspars, talc
43A	Kipushi, Zaire	3,000	95	6.00			12.00	Bo,Sp	Limestone, dolomite, schist
44	Fierro, NM	3,200	95	2.00		2.7	0.70	Cc,Car,Het	Limestone, dolomite, schist
46	Island Copper, BC	38,000	90	0.48	0.017	3.7		Moly	Andesite, quartz, porphry
50	El Teniente, Chile	105,000		1.48	0.047	2.2		Cc,Moly	
52	Andina, Chile	30,000		1.75	0.015	2.0		Cc,Moly	
53	Los Bronces, Chile	15,000		1.55	0.020	1.8		Cc,Moly	
54	El Cobre, Chile	12,000		2.11	0.025	2.0		Cc,Moly	

Table 56a.- INDUSTRIAL COPPER MILL FLOTATION DATA (continued) -

Principal Mineral CHALCOPYRITE

		Copper Concentr %						Moly Concen			Precious
		%Cu	%Mo	%Fe	%Zn	Insol	%Recov	%Mo	%Cu	%Recov	Met. Rec
28D	Gaspe, Quebec	30	0.06				88		0.50	56	
30A	O'Okiep, S. Africa	27		31							
30B	Nababeep, S. Africa	35		18		30.2					
30C	Carolusberg, S.Africa	29		16		42.8					
31A	Keretti, Finland	21		34	1.4	7.0	97				Au, Ag
31B	Virtisalmi, Finland	24				12.5	97				
31C	Pyhasalmi, Finland	23		29	6.1		94				Au, Ag
32	Phalaborwa, S. Africa	34					84				Au, Ag
33A	New Cornelia, AZ	29		26		8.5	87				Au, Ag
34A	Sto Tomas, Phil	25		2		22.1	85				Au, Ag
34B	Benguet, Philippines	27		1		20.2	88				Au, Ag
35	Sahuarita, AZ	27	0.24	27	2.0	8.5	87	40	1.90	31	Au, Ag
35A	Bagdad, AZ	32	0.15				88		1.00	64	
37A	Chambishi, Zambia	46				20.0	92				
37B	Luanshya, Zambia	29		20		21.5	91				
40	Princeton, BC	27	-	28		7.0	87			-	Au, Ag
41	Niihama, Japan	22		37		2.0	92				
42	Kidd Creek, Ontario	24			6.3		95				Ag
43A	Kipushi, Zaire	28			8.5		91				Ag
44	Fierro, NM	25		27	7.5	7.0	95				Au, Ag
46	Island Copper, BC	23	0.80	25			88	42	0.50	30	Au, Ag
50	El Teniente, Chile	34	0.20	20		5.0	86		0.50	54	Au, Ag
52	Andina, Chile	30	0.21	15		5.0	87		1.00	54	
53	Los Bronces, Chile	31	0.30	15		4.0	82			52	
54	El Cobre, Chile	30	0.35	10		5.0	84			50	

Table 56.- INDUSTRIAL COPPER MILL FLOTATION DATA (continued)

Principal minerals Chalcocite, Malachite, and Bornite

		Daily t milled	% Oper time	Mill Feed % Cu	% Mo	% Fe	% Zn	Secondary Minerals	Gangue Composition
1A	Butte, Montana	44,000	95	0.69				Cp,Bo	Quartz, complex silicates
1D	Cananea, Mexico	21,000	83	0.00	3.500			Cov,Cp	Quartz, clays, sericite
3C	Sacatan, AZ	11,000	93	0.72		3.0		Ma	Silicates, iron oxides, clays
19A	Inspiration, AZ	20,000	94	0.50	0.008	1.4		Cp,Bo	Silica
19B	Christmas, AZ	5,100	96	0.65					Andesite, diorite, limestne
20A	Chino, NM	23,000	94	0.91	0.020	13.2		Cp	Ca silicates, quartz, monzonite
22B	San Manuel, AZ	43,000	98	0.75	0.015	3.5			
27A	Chingola East, Zambia	16,500	92	1.10				Cp,Bo,Ma	Quartz, dolomite, feldspars, mica
27B	Chingola West, Zambia	25,000	97	1.50		2.2			
27D	Konkola, Zambia	5,100	90	3.00				Bo,Ma,Az	Quartz, Ca, Mg, Fe, Mn carbonates
33B	Copper Queen, AZ	20,000	90	1.10		16.6		Py	
33C	Morenci, AZ	61,000	99	0.85	0.008	3.7		Ma,Cov	Quartz, feldspars
33D	Tyrone, NM	29,000	99	0.85		2.6		Cov	Quartz, feldspars, sericite
33E	Metcalf, AZ	30,000	97	0.80	0.010	4.5		Cp,Bo,Cov	
36	Gibralter, BC	40,000		0.40	0.01			Cp	
39	Toquepala, Peru	44,000	98	1.15	0.025	3.7		Cp	Silica, Alumina
45	White Pine, Mich	23,000		1.00					Silicates
48	Chuquicamata, Chile	153,000		1.45	0.030	1.5		Cp,Bo	
49	El Salvador, Chile	28,000		1.44	0.024	1.3		Cp,Moly	
	Principal Mineral MALACHITE								
43B	Kamoto, Zaire	5,000	94	4.20				Cc,Car,Het	Limestone, dolomite, schist
43C	Kakanda, Zaire	2,400	96	5.00				Chr,Het	
	Principal Mineral BORNITE								
37C	Mufulira, Zambia	24,000	98	2.50		1.3		Cp,Cc	Silicates

Table 56a.- INDUSTRIAL COPPER MILL FLOTATION DATA (continued) -

Principal minerals Chalcocite, Malachite, and Bornite

		Copper Concentr %						Moly Concen			Precious
		%Cu	%Mo	%Fe	%Zn	Insol	%Recov	%Mo	%Cu	%Recov	Met. Rec
1A	Butte, Montana	26		19	-	3.7	75				Au, Ag
1D	Cananea, Mexico	32	0.14	25		4.7	72				Au, Ag
3C	Sacatan, AZ	32		20		11.7	84				Ag
19A	Inspiration, AZ	42	0.70	2		1.5	76	53	0.50	30	Au, Ag
19B	Christmas, AZ	23		21		10.0	72				Au, Ag
20A	Chino, NM	21	0.50	32		5.4	77	50	1.15	29	Au, Ag
22B	San Manuel, AZ	28	0.54	28		7.0	86	56	0.90	69	Au, Ag
27A	Chingola East, Zambia										
27B	Chingola West, Zambia	33		8		1.4	77				
27D	Konkola, Zambia	39					85				
33B	Copper Queen, AZ	13		33		7.0	80				Au, Ag
33C	Morenci, AZ	24	0.09	28		8.4	79				Au, Ag
33D	Tyrone, NM	23	0.15	30		8.0	79				Au, Ag
33E	Metcalf, AZ	26	0.15	27		10.0	61				Au, Ag
36	Gibralter, BC	30	0.15				85		1.00	45	
39	Toquepala, Peru	28	0.40	27		6.0	86	53	1.20	30	
45	White Pine, Mich	31					83				Ag
48	Chuquicamata, Chile	45	0.15	10		7.0	80		0.10	54	
49	El Salvador, Chile	52	0.35	14		3.0	84		0.20	56	
Principal Mineral MALACHITE											
43B	Kamoto, Zaire	43					65				
43C	Kakanda, Zaire	24					80				
Principal Mineral BORNITE											
37C	Mufulira, Zambia	46		11		16.0	91				

Table 57.- INDUSTRIAL COPPER MILL FLOTATION REAGENT DATA - COLLECTORS

Principal Mineral CHALCOPYRITE

		Reag.	g/t	POA	Reag.	g/t	POA	Reag.	g/t	POA	Reag.	g/t	POA
1B	Twin Buttes, AZ	Z-6	16.5	RF	Z-200	1.7	CF						
1C	Yerington, NV	Z-6	7.8	RF	Z-11	3.7	RF	Z-6	3.3	SdF	Z-11	2.5	SdF
3A	Mission, AZ	Z-6	10.3	G, RF	238	2.1	G, RF	97A	6.2	G, RF			
3B	Silver Bell, AZ	Z-10	8.3	G	238	8.3	G	97A	3.3	RF			
4	Highland Valley, BC	NEX	12.4	RF	KAX	15.3	Sc						
5	Bagdad, AZ	Z-4	20.6	RF	Z-6	4.1	RF	404	12.4	Sc			
7	Aitik, Sweden	KAX	18.6	C	NIX	0.8	C	NIX	4.1				
8	Bouganville	3501											
9	Merritt, BC	242	13.6	RF,Sc	Z-200	0.8	RF						
10	Pinto Valley, AZ	Z-14	12.4	RF	238	12.4	RF						
10A	Miami, AZ	Z-14											
13	Sierrita, AZ	KAX	2.1	RF	3302	6.2	G						
13A	Mineral Park, AZ	Z-200			3302								
13B	Esperanza, AZ	Z-6			3302								
14	Erstberg, W. Irian	Z-200	16.5	G,RF	Z-14	16.5	RF						
15	McLeese Lake, BC	NIX	8.3	RF,Sc	Z-200	8.3	RF,Sc						
16	Tide Lake, BC	Z-200	14.9	RF	Z-200	1.7	CF						
17	Nairne, S. Australia	Z-200	24.8	RF									
18	Lakeshore Casa Grande												
20B	McGill, NV	Z-200	6.2	G	Z-4	20.6	G						
20C	Ray, AZ	Rac	41.3	RF									
20D	Utah Mines	Reco	8.3	RF	Oil	6.2	Sc						
21	Noranda, Quebec	404	792.4	G	343	466.4	RF	CuSO4	24.6	RF			
21A	Logan Lake, BC	NIX	24.8	G,RF	KAX	28.9	Sc						
22A	Superior, AZ	404	24.8	RF	NIX	4.1							
23	Marinduque, Phil.	Z-200	16.5	G	Z-6	20.6	RF,Sc	Z-11	20.6	RF,SC			
25	Hokkaido, Japan	3501	11.1		3477	11.1		KAX	92.5	SC	KEX	13.6	CF
26	Mount Isa, Australia	NIB	206.4	G									
27C	Rokana, Zambia	NIX	177.5		PKD	132.1	C						
28A	Geco, Ontario	350	24.8	RF	NIX	26.4	RF						
28B	Horne, Quebec	KAX	44.2	PA	KAX	102.4	CF	NIX	69.8	PC			
28C	Brenda	KAX	31.8	G	FO	12.8	C						

Table 57.-INDUSTRIAL COPPER MILL FLOTATION REAGENT DATA - COLLECTORS (continued)

Principal Mineral CHALCOPYRITE

	Reag.	g/t	POA	Reag.	g/t	POA	Reag.	g/t	POA	Reag.	g/t	POA
28D Gaspe, Quebec	Z-6	24.8		3302	11.2							
30A O'Okiep, S. Africa	238	24.8	RF	NIX	24.8	RF						
30B Nababeep, S. Africa												
30C Carolusberg, S.Africa	NIX	12.4	G	NEX	4.1	RF						
31A Keretti, Finland	NAX	144.5	G									
31B Virtisalmi, Finland	NAX	165.1	RF									
31C Pyhasalmi, Finland	NAX	66.0	C									
32 Phalaborwa, S. Africa	KAX	33.0	G									
33A New Cornelia, AZ	Z-11	4.5	G	TP	1.7	RF	238	0.8	RF	25	0.8	RF
34A Sto Tomas, Phil	NIB	14.9	G	3501	4.5	G	Z-14	111.4	SlF	2501	33.0	SLF
34B Benguet, Philippines	Z-14	12.4	G	Z-200	0.8	G						
35 Sahuarita, AZ	Z-6	7.4	RF	238	2.5	RF						
35A Bagdad, AZ	Z-6			Z-6								
37A Chambishi, Zambia	NEX	8.3	RF	KAX	4.1	RF						
37B Luanshya, Zambia	NIX	28.9	G,RF									
40 Princeton, BC	NEX	20.6	RF	KAX	24.8	Sc						
41 Niihama, Japan	KA	18.2	C	Z-200	27.2	C	238	18.2	C			
42 Kidd Creek, Ontario	208	78.4	C	NIB	12.4	Sc						
43A Kipushi, Zaire	KEX	177.5	RF									
44 Fierro, NM	Z-11	74.3		Z-200	18.6	G						
46 Island Copper, BC	KAX	3.3	RF									
50 El Teniente, Chile	MA	60.0	G									
52 Andina, Chile												
53 Los Bronces, Chile												
54 El Cobre, Chile												

Table 57.- INDUSTRIAL COPPER MILL FLOTATION REAGENT DATA - COLLECTORS (continued)

Principal minerals Chalcocite, Malachite, and Bornite

		Reag.	g/t	POA	Reag.	g/t	POA	Reag.	g/t	POA	Reag.	g/t	POA
1A	Butte, Montana	Z-6	2.1	RF	Z-200	3.3	RF	1331	10.7	RF	MA	16.5	RF
1D	Cananea, Mexico	238	7.8	RF	343	1.2	RF						
3C	Sacatan, AZ	Z-6	4.1	G	Z-6	0.8	RF	Z-6	0.8	Sc			
19A	Inspiration, AZ	898	12.4	RF									
19B	Christmas, AZ	Z-6	10.3	RF	404	7.8	G						
20A	Chino, NM	Rac	28.9	RF									
22B	San Manuel, AZ	Z-11	0.4	G	3302	0.4	G	SM8	4.5	G			
27A	Chingola East, Zambia												
27B	Chingola West, Zambia	NIX	177.5	C	PDK	132.1							
27D	Konkola, Zambia	NIX	239.4	RF									
33B	Copper Queen, AZ	Z-200	3.3	G	Z-11	0.8	Sc						
33C	Morenci, AZ	Z-200	21.9	G									
33D	Tyrone, NM	Z-11	18.6	G									
33E	Metcalf, AZ	Z-200	6.2	G	NIX	3.7	Sc						
36	Gibralter, BC	Z-11											
39	Toquepala, Peru	Z-11	10.3	RF	3302	8.3	G						
45	White Pine, Mich	NIB	90.8		DAP	9.9	Sc						
48	Chuquicamata, Chile	SF-323											
49	El Salvador, Chile	SF-323											
	Principal Mineral MALACHITE												
43B	Kamoto, Zaire	KEX	66.0		KAX	66.0		NaSH	1238.2				
43C	Kakanda, Zaire	P	920.0	RF									
	Principal Mineral BORNITE												
37C	Mufulira, Zambia	NEX	17.3	RF	KAX	17.3	Sc						

Table 57a.- INDUSTRIAL COPPER MILL FLOTATION REAGENT DATA - FROTHERS

Principal Mineral CHALCOPYRITE

		Reag.	g/t	POA	Reag.	g/t	POA	Reag.	g/t	POA	pH	Reag.	g/t	POA
1B	Twin Buttes, AZ	R-23	10.3	RF							11.0	Lime	3,880	G
1C	Yerington, NV	A-65	2.9	RF	A-65	5.8	SdF	A-710	3.7	SdF	11.0			
3A	Mission, AZ	PO	2.9	RF							11.5	Lime		G
3B	Silver Bell, AZ	PO	37.1	G							11.0	Lime	825	G
4	Highland Valley, BC	MIBC	14.0	RF	PPG	4.5	RF					Lime	351	G
5	Bagdad, AZ	1639	24.8	RF	PO	28.9	RF				11.5	Lime	2,064	G
7	Aitik, Sweden	PO	18.6		41G	8.3						Lime	289	G
8	Bouganville	MIBC		RF	250		RF	A-65			8.5	Lime		G
9	Merritt, BC	PO	5.8	RF								Lime	413	
10	Pinto Valley, AZ	250	16.5	RF										
10A	Miami, AZ	250			PO						10.5			
13	Sierrita, AZ	A71	8.3	G	MIBC	33.0	M				11.0	Lime	1,032	G
13A	Mineral Park, AZ	MIBC									11.5			
13B	Esperanza, AZ	MIBC									11.5			
14	Erstberg, W. Irian	PO	8.3	G,RF	MIBC	24.8	G,RF	250	28.9	G,RF		Lime	619	G
15	McLeese Lake, BC	MIBC	8.3	RF,Sc	250	4.1	RF,Sc					Lime	413	G
16	Tide Lake, BC	PO	14.0	RF	PO	6.6	CF					Lime	825	G
17	Nairne, S. Australia	M300	2.5	RF								Lime	495	G
18	Lakeshore Casa Grande											Lime		G
20B	McGill, NV										10.8			
20C	Ray, AZ	1639	74.3	RF							11.5	Lime	2,683	G
20D	Utah Mines	MIBC	22.3	RF							8.5	Lime	743	G
21	Noranda, Quebec	TEB	26.0	G							8.5	Lime	1,486	G
21A	Logan Lake, BC	250	11.6	RF	PO	G,Sc						Lime	908	G
22A	Superior, AZ	MIBC	20.6	RF								Lime	2,064	G
23	Marinduque, Phil.	MIBC	20.6	G	A65	16.5	RF,CF	250	16.5	RF,CF	9.0	Lime	310	G
25	Hokkaido, Japan	PO	47.9									Lime	2,542	G
26	Mount Isa, Australia	TEB	82.5	RF										
27C	Rokana, Zambia	TEB	12.4	RF	TEB	6.2	C							
28A	Geco, Ontario	PO	27.2	RF								NH3	118	G
28B	Horne, Quebec	PO	14.0	PA	PO	38.4	PC					Lime	2,022	G
28C	Brenda	MIBC	12.4	C							8.0			

Table 57a.- INDUSTRIAL COPPER MILL FLOTATION REAGENT DATA - FROTHERS (continued)

Principal Mineral CHALCOPYRITE

	Reag.	g/t	POA	Reag.	g/t	POA	Reag.	g/t	POA	pH	Reag.	g/t	POA
28D Gaspe, Quebec	MIBC									10.0	Lime		G
30A O'Okiep, S. Africa	CA	16.5	RF								Lime	1,238	G
30B Nababeep, S. Africa													
30C Carolusberg, S.Africa	CA	16.5	RF								Lime	619	G
31A Keretti, Finland	TEB	7.4	C								Lime	2,064	G
31B Virtisalmi, Finland	PO	6.2	RF								Lime	2,683	G
31C Pyhasalmi, Finland	421	20.6	C								Lime	1,403	G
32 Phalaborwa, S. Africa	TEB	8.3	G							8.0			
33A New Cornelia, AZ	R-55	10.7	RF								Lime	702	G
34A Sto Tomas, Phil	250	13.2	G	65	12.8	G					Lime	248	G
34B Benguet, Philippines	250	10.7	G								Lime	318	G
35 Sahuarita, AZ	MIBC	10.3	RF								Lime	1,156	G
35A Bagdad, AZ	MIBC									11.5			
37A Chambishi, Zambia	MIBC	16.5	RF										
37B Luanshya, Zambia	F1	4.5	RF	38Y	3.7	RF							
40 Princeton, BC	PO	24.8		41-G	8.3						Lime	1,032	G
41 Niihama, Japan	PO	5.4	C	MIBC	12.8						Lime	1,362	G
42 Kidd Creek, Ontario	MIBC	14.9	C								Lime	244	G
43A Kipushi, Zaire	RO	99.1	RF								Lime	2,270	G
44 Fierro, NM	250	8.3	RF	MIBC	8.3	RF					Lime	1,238	G
46 Island Copper, BC	71R	22.7	RF	1012	1.7	RF				10.5	Lime	619	G
50 El Teniente, Chile										4.2	H_2SO_4		G
52 Andina, Chile										9.0	Lime		G
53 Los Bronces, Chile										10.5	Lime		G
54 El Cobre, Chile											Lime		G

Table 57a.- INDUSTRIAL COPPER MILL FLOTATION REAGENT DATA - FROTHERS (continued)
Principal minerals Chalcocite, Malachite, and Bornite

		Reag.	g/t	POA	Reag.	g/t	POA	Reag.	g/t	POA	pH	Reag.	g/t	POA
Principal M CHALCOCITE														
1A	Butte, Montana	MIBC	4.1	RF							10.5	Lime	3,880	G
1D	Cananea, Mexico	250	13.6	RF								Lime	2,518	G
3C	Sacatan, AZ	MIBC	20.6									Lime	1,238	G
19A	Inspiration, AZ	MIBC	12.4	RF							10.5	Lime	1,238	G
19B	Christmas, AZ	MIBC	22.7	RF							9.5	Lime	3,302	G
20A	Chino, NM	PO	51.2	RF	MIBC	44.6	RF				11.0			
22B	San Manuel, AZ	MIBC	23.5	RF	250	0.4	RF				10.5	Lime	702	G
27A	Chingola East, Zambia											Lime		G
27B	Chingola West, Zambia	TEB	12.4	RF	TEB	6.2	C					Lime		G
27D	Konkola, Zambia	PO	27.2	RF								NH_3	118	G
33B	Copper Queen, AZ	250	9.1	RF							11.5	Lime	2,889	G
33C	Morenci, AZ	CA	21.9	RF							10.5	Lime	2,229	G
33D	Tyrone, NM	250	16.5	RF							11.4	Lime	1,651	G
33E	Metcalf, AZ	250	8.3	RF								Lime	3,508	G
36	Gibralter, BC	MIBC									10.5			
39	Toquepala, Peru	PO	8.3	RF	73	8.3	RF				11.5	Lime	1,238	G
45	White Pine, Mich	250	5.0	F							9.5	Lime	660	G
48	Chuquicamata, Chile										11.0	Lime		G
49	El Salvador, Chile										11.0	Lime		G
Principal M MALACHITE														
43B	Kamoto, Zaire											Lime	908	G
43C	Kakanda, Zaire	250	8.2	RF	MIBC	8.2	RF					Na_2CO_3	1,068	G
Principal M BORNITE														
37C	Mufulira, Zambia	250	21.8								10.7	Lime	781	G

Abbreviations for Tables 56 and 57

Symbol identification

C	conditioner	OxRo	oxide rougher	REG	regrind
CF	cleaner feed	PA	primary aerator	Sc	scavenger
CRT	copper rougher tailings	PC	pyrite circuit	SdF	slime feed
G	grinding circuit	POA	point of addition	SRT	sulphide rougher tailings
M	middlings	Reag	reagent		

Depressants

CN	cyanide
GG	guar gum
Nig	nigrosine and dextrin

Reagent Abbreviations

Collectors

3302	Aeropromotor 3302	NAX	sodium amyl xanthate
3501	Aeropromotor 3501	NEX	sodium ethyl xanthate
DTP	dithiophosphate	NIB	sodium isobutyl xanthate
DAP	di-isoamyl DTP	NIX	sodium isopropyl xanthate
KAX	potassium amyl xanthate	Rac	Raconite
KEX	potassium ethyl xanthate	SM8	Minerec SM8
25	Aerofloat 25	238	Aerofloat 238
MA	Minera A	242	Aerofloat 242
244	Aerofloat 244		
Z-200			
343	Aeroxanthate 343		
350	Aeroxanthate 350		
898	Minerec 898		
1331	Minerec 1331		
FO	fuel oil		
P	palm oil		

Frothers

R55	Union Carbide frother	CA	cresylic acid
R23	Union Carbide frother	FI	Flotol B
RO	resin oil	73	Aerofroth 73
PO	pine oil	M300	PPG derivative
PA	Powells accelerator		
Que	quebracho		
250	Dowfroth 250		
1638	Shell MIBC + 10% stillbottoms		
1639	Shell MIBC + 15% stillbottoms		

Flocculants and modifiers

GPG	Aero-dri 100	202	Superfloc 202	TP	tarpine
41G	Palabora stabilizer	206	Superfloc 206	WG	waterglass
60	complex 60	M-59	Polyhall M-59	330	Superfloc 330
97A	Stepanflote 97A	Sep	Separan	673	Nalco 673
127	Superfloc 127	SPP	sodium tripolyphosphate	1801	Nalco 1801
				2299	Nalco 2299

Table 58.- AVERAGE PORPHYRY COPPER MINE PERFORMANCE IN 1975

Region	Number of mines	Copper Grade	Copper Recovery	Production Capacity, t/yr
USA	29	0.695%	84.2%	1,758,000
Canada	11	0.504%	87.7%	373,310
Chile	5	1.455%	86.3%	1,059,500
Australasia	8	0.663%	86.1%	508,780
U.S.S.R	7	0.815%	83.4%	391,270
Europe	3	0.632%	80.4%	72,000
Africa	2	0.939%	85.1%	195,000

The flotation data being analysed are biased towards U.S. design preferences, as the statistical information represents very nearly a 100% of the U.S. mines, 90% of Chilean mines, only half the mines in Canada, U.S.S.R. and Pacific area, and less than half of the European and African mines.

Fig. 51 shows four frequency bar charts listing flotation recovery for copper and moly, ore treated grade, concentrate grade, grind minus 200 mesh percent, flotation pH and collector consumption for all the mines in Tables 56 and 57. From these we see that copper recovery shows two distinct peaks which probably reflect the difference in treatability of primary sulphide ores, which even for very low copper grades show 90+% recovery, and that of partially weathered, high grade ores whose mean recoveries trend in the mid eighties range. Mines with recoveries below 80% are obviously involved with ores with special treatability problems. The greater complexity of the mixed molybdenum ores is reflected in the very broad scatter in the moly recoveries, which range rather indiscriminately between 30 and 80%. The copper concentrate grades produced, bunch in the 25 to 34% range. Mines with higher concentrate grades are treating relatively unusual ores, like those found at Chuquicamata and El Salvador, where the weathered cap has rarer mineralisation, such as atacamite. The numbers confirm that for most ores, dissemination of the mineral is fine, and that a grind that gives 50 to 65% minus 200 mesh is required for proper liberation.

As noted above, only two mines in the list operate with an acid pulp, but at normal recoveries. Otherwise the listing shows that flotation at a neutral pH is not attractive. There is no good explanation for this, other than to speculate on the depressant effect of ions that are normally present in solution in natural mine waters, but that could be precipitated by the lime used to increase the pulp alkalinity. The scatter in the last graph underlines that there are no generalisations that can be made on what sort of a collector consumption should be considered normal, though not unexpectedly, there is some correlation between collector consumption and copper grade. Also noticeable is the use of multiple collectors to try and optimise by-product recoveries. The best correlation between copper grade and collector consumption is obtained with the data from the U.S. mines, where the linear correlation coefficient is about 83%.

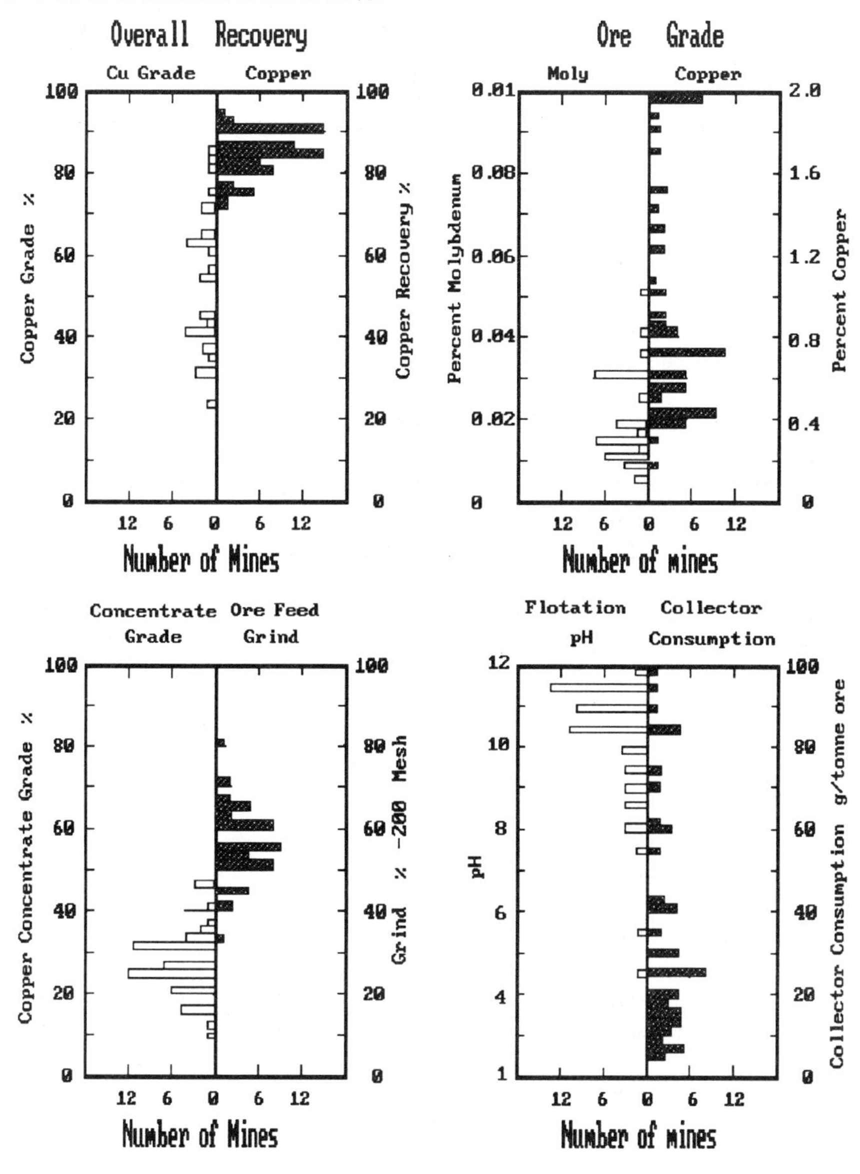

Fig. 51.- Overall mill performance, Tables 56 & 57

Table 59 shows the frequency of use of single collectors and mixes as practised in 1975. The bar graphs in Fig. 52a illustrates this. The trend has not changed significantly, with xanthates being, by far, the most popular thiol collectors. An attempt to correlate the 1975 collector data with flotation showed a significant correlation between the use of xanthic esters and enhanced recovery of molybdenite, and also a direct correlation between the use of the xanthogen formate structure and acid pulps. There is some support for the use of dithiophosphates as supplementary collectors in

Table 59.- COLLECTOR USE FREQUENCY COMPARISON FOR COPPER MILLS % OF TOTAL MILLS SURVEYED

	Collector usage as		
	Sole Collector or	Primary collector	Gross Total
Xanthates	30.4%	78.3%	78.3%
Dithiophosphates	4.3%	5.8%	27.5%
Xanthogen Formates	1.4%	7.3%	7.3%
Thionocarbamates	5.8%	8.6%	18.8%
Xanthic Esters	0	0	8.7%
Mercaptobenzothiazol	0	0	5.8%
Fuel oil	0	0	17.4%
Total	41.9%	100%	163.8
<u>Two or more collectors mixed</u>:			
Xanthate plus ...			
Dithiophosphates			15.9%
Xanthogen Formates			1.4%
Thionocarbamates			7.2%
Xanthic Ester			5.8%
Fuel oil			11.6%
Dithiophosphate plus ...			
Fuel oil			2.9%
Fuel oil plus Xanthic Ester			1.4%
Fuel oil plus Mercaptobenzothiazol			1.4%
Thionocarbamate plus MBT			1.4%
Xanthogen Formate plus ...			
Fuel oil			4.3%
Thiophosphate plus MBT			1.4%
Thionocarbamate plus ...			
Xanthic Ester			1.4%
Dithiophosphate			1.4%

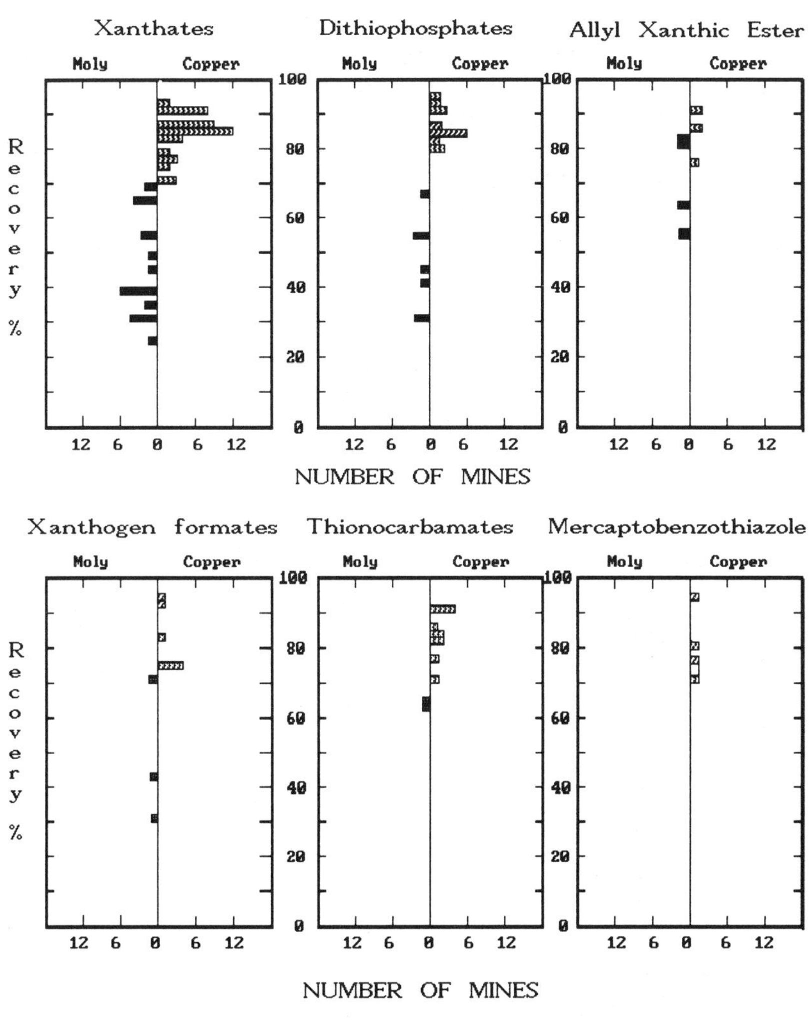

Fig. 52a.- Collector frequency bar graphs

cleaning circuits. Cyanamid has always pushed their thiophosphate and mercaptobenzothiazole as good enhancers for the recovery of precious metals and other secondary values in copper ores. It also introduced AP 3302 (xanthic ester) as a special collector for molybdenum, and the data bears this claim out, though the ester is too weak to be used as a primary collector for copper. Fuel oil is the other collector that is universally used to enhance molybdenum recovery. The quantity used affects the selection of the frother.

There is an inadequate amount of frother data to try to detect recovery or selectivity trends, but some intuitive conclusions can be reached from the bar graphs in Fig. 52b as to the way frothers are being applied. The data is summarised in Table 60. Two trends are obvious from this table: an increased tendency to fine-tune the froth by employing multiple frothers, and a distinct increase in the popularity of MIBC. The driving force in these changes between 1960 and 1975 has primarily been improved operating stability, and enhanced recoveries; only secondarily have the changes come because of reduced reagent costs.

Table 60.- FROTHER USE FREQUENCY COMPARISON FOR COPPER MILLS
% OF TOTAL MILLS SURVEYED

	From Table 55 1975, 55 mills	From Booth (1962) 1960, 66 mills
Single frother		
Pine oil	15%	17%
Cresylic Acid	2%	11%
MIBC	29%	14%
Other Higher Alcohols	0	3%
Polyglycol ethers	11%	21%
TEB	2%	8%
Subtotal - Single Frothers	59%	74%
Two or more frothers mixed		
Polyglycol ethers plus....		
Pine oil	13%	5%
MIBC plus ...		
Pine oil	5%	5%
Cresylic acid	4%	3%
Polyglycol ethers	11%	0
Pine oil plus ...		
Cresylic acid	0	5%
Polyglycol ether plus MIBC	4%	0
Subtotal Mixed Frothers	37%	18%
Miscellaneous Frothers (Pyridine, etc)	5%	8%

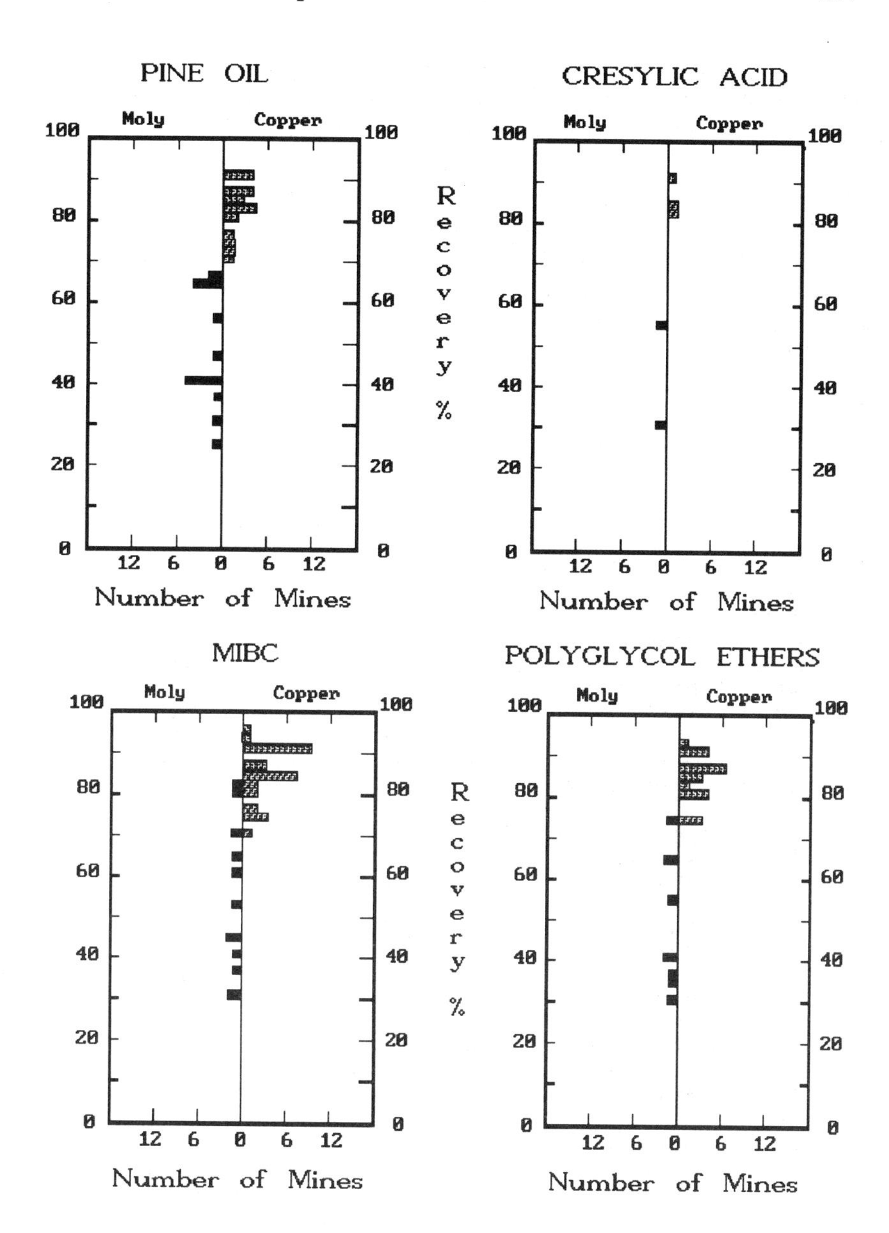

Fig. 52b.- Frother frequency bar graphs

Table 61.- COLLECTOR CONSUMPTION IN COPPER SULPHIDE ORE FLOTATION
metric tonnes

Collector	Total 1975	Canada Mexico 1975	South America 1975	Pacific area 1975	USA 1975	USA 1965	USA 1960
Isopropyl xanthate	4,054	69.5	2,156	369	1,459	1,082	823
Amyl xanthate	1,549	214	11	628	696	273	192
Sec Butyl xanthate	450	-	-	-	450	-	380
Ethyl xanthate	251	106.8	-	99	45	370	396
Total xanthate	6,304	390.4	2,167	1,096	2,650	1,726	1,790
Dicresyl DT	829	9.5	78	383	358	341	330
Disec butyl DT	579	234.1	118	87	140	13	-
diamyl DT	112	-	-	-	120	59	-
Diethyl DT	120	-	-	-	120	466	516
Total dithiophosphate	1,640	243.6	196	470	730	880	846
Diethyl XF	1,286	-	1,286	-	-	32	232
Ethyl isobutyl XF	105	-	-	-	105	89	118
Ethyl methylamyl XF	40	-	-	40	-	-	-
XF mixtures	98	-	-	-	98	68	-
Total xanthogen formate	1,528	-	1,286	40	203	189	350
Ethyl isopropyl TC	983	119.1	164	299	400	254	258
Methyl butyl TC	153	-	-	-	153	102	-
Total thionocarbamate	1,136	119.1	164	299	553	356	258
Allylamyl xanthic ester	371	218.2	-	-	153	-	-
Mercaptobenzothiazole	104	-	-	17	87	7	51
Sub Total Collectors	10,628	971.4	3,814	1,922	4,813	3,330	3,132
Ore treated dry t '000	194,741	36,318	28,704	34,268	95,414	73,926	3,290
Copper recovered, t '000	2,858	330	843	391	1,267	1,02	811
% of reported Cu production	**45%**	**40%**	**89**	**?**	**82%**	**83%**	**83%**

Table 61 shows Bureau of Mines data for sulphhydryl collector consumption in the U.S.A. for the period 1960 to 1975, and is the only consistent published data on the tonnage of individual collectors used in 1975 in copper flotation world wide. It covers approximately the same mines as those canvassed in Tables 56 and 57. The world wide data shows some regional preferences, or possibly availability of the different types of collectors, while the comparison on collector usage in the U.S., between 1960 and 1975, underlines how small

Table 62.- POWER AND WATER USE IN U.S. FLOTATION

Mineral	Plants Number	Capacity t/day	Ore - t Treated	Conc tonne	Conc Ratio	Energy kw-hr tonne	Water gal/ tonne
1975							
Copper	18	293,636	83,809,091	1,967,769	42.6	17.8	935
Copper-moly	15	467,273	132,209,364	2,400,717	55.1	19.4	842
Cu-Pb-Zn	13	26,636	6,658,300	500,620	13.3	18.2	743
Cu-Zn-Fe	3	7,727	1,966,700	735,889	2.7	22.3	671
Gold-silver	3	764	65,900	433	152.3	30.3	814
Lead-zinc	21	28,091	7,510,000	865,993	8.7	24.0	924
Zinc	10	17,818	3,139,800	193,066	16.3	20.4	682
Barite	3	W	W	W	W	W	W
Feldspar-mica-quartz	13	9,455	2,331,091	1,032,900	2.3	27.5	2,640
Fluorspar	3	1,291	249,545	77,[illegible]85	3.2	75.1	1,870
Glass sand	22	36,364	6,680,818	5,516,727	1.2	na	na
Iron ores	5	75,455	26,017,273	13,579,672	1.9	22.0	1,870
Kyanite	3	W	W	W			
Limestone-magnesite	5	1,773	365,727	309,909	1.2	22.8	594
Mercury	1	W	W	W	W	W	W
Molybdenum	2	W	W	W	W	W	W
Phosphate	19	51,091	68,251,182	16,360,336	4.2	9.8	3,465
Potash	8	44,545	12,600,909	3,045,875	4.1	18.3	319
Talc	2	W	W	W	W	W	W
Tungsten	2	W	W	W	W	W	W
Coal	78	58,455	11,890,000	7,435,000	1.6	na	na
Grand Total	252	1,465,273	384,113,636	54,788,472	7.0	17.4	1,397
1985							
Copper	8	145,805	62,559,090	2,519,090	24.8	16.8	762
Copper-moly	7	474,200	114,332,730	2,790,910	41.0	14.5	357
Cu-Pb-Zn	3	W	W	W	W	W	W
Cu-Zn-Fe	1	W	W	W	W	W	W
Gold-silver	3	470	118,182	4,550	26.0	72.6	616
Lead-zinc	9	12,530	3,254,550	672,720	4.8	19.5	572
Zinc	4	14,715	3,009,090	109,090	27.6	14.7	639
Barite	2	W	W	W	W	W	W
Feldspar-mica-quartz	10	7,584	1,894,550			49.0	4,872
Fluorspar							
Glass sand	8	11,600	2,521,180	2,162,730	1.17	NA	1,610
Iron ores	4	103,870	22,004,550	17,661,820	1.25	44.1	4,360
Kyanite							
Limestone-magnesite	4	960	176,360			NA	610
Mercury	1	W	W	W	W	W	W
Molybdenum	4	95,900	21,867,270	60,910	359.0	17.6	323
Phosphate	22	415,800	109,715,450	24,619,090	4.5	12.4	3,480
Potash	5	41,170	10,090,900	1,516,360	6.6	12.0	78
Talc	1	W	W	W	W	W	W
Tungsten	1	W	W	W	W	W	W
Coal	77	93,800	25,150,900	17,731,820	1.4	7.1	359
Grand Total	179	1,403,540	384,423,820	72,818,180	5.3		

W = withheld, but included in total; na = not available

the shifts in collector use were during those years. The numbers do underline the domination of this market by the xanthates. The frothers, shown in Table 60, and analysed graphically in Fig. 52b, also indicate that there was relatively little change in frother technology. Unfortunately there is no more recent data which is cohesive enough to update the trends. Certainly the copper industry took a bath in the early eighties, and, after numerous bankruptcies, it has been reborn with new owners, a different investment philosophy, and a fundamentally changed view on the most economic technology. Flotation per se has not seen changes other than the trend to reduced capital intensiveness through the use of giant flotation cells and columns for upgrading rougher concentrates in cleaners circuits, or to reduce the capital investment in moly-copper separations. Flotation is still the lowest cost method for partition of two similar solids, but heap, or in-situ leaching, followed by selective extraction of the metals from the solution, and then direct electro-winning, has rescued from the scrap heap a number of mines which had by-passed oxidised cap ores. The quantities of these reserves are limited, so should become exhausted in a relatively short time, but have given the industry a breathing spell to find new technology to treat the massive, low grade, disseminated deposits that are now idle. The crisis also spawned a gold-rush, which now has gradually tapered off.

The SME Handbook quotes comparable U.S. Bureau of Mines statistics on power and water use in flotation mills in the U.S.A., covering all types of minerals. These are reproduced in Table 62. These data are important as they provide a broader picture of the whole mineral industry, and how flotation is used as a unit operation.

Flotation process design

In order to design an efficient laboratory test programme for a new ore, it is important to visualise the type of flotation circuits that would normally be employed in treating a particular type of ore.

Circuit design normally starts with comminution. All floatable ores must first be crushed and ground to a particle size that will liberate the valuable component(s) from the gangue. The liberation grind will depend on the dissemination of the ore components, but usually the optimum from the recovery point of view is the 100 to 350 mesh range. At less than 50 mesh, the particle is usually too heavy to remain attached to the bubbles, and at grinds greater than 400 mesh the pulp will contain slime which can interfere with efficient flotation. The equivalent micron sizes for the nominal screen sizes normally found in a metallurgical laboratory are listed in Table 63. From this table we see that the approximate optimum particle size range found in a normal flotation pulp is between 30 to 150 microns.

Run of mine ore is crushed in jaw or gyratory primary crushers, followed by secondary and tertiary cone crushers, to make a grinding circuit dry feed. Normally each stage of crushing is scalped with a vibratory screen. The feed size to the grinding process depends on design. Autogenous grinding mills are justified because of capital savings in the crushing circuit. A two stage rod mill/ball mill sequence also will tolerate a coarser feed. Modern grinding circuits usually consist of an autogenous mill followed by one or more ball mills. Both stages are operated in a closed circuit with a scalping cyclone returning the oversize back to each stage of grinding. One equipment supplier is recommending the installation of a "flash" flotation cell into the cyclone circuit to recover larger particle ore which contains malleable components such as native gold or copper. Process water is added to each stage of grinding. The final pulp can be discharged directly into the first flotation cell, or more usually to an agitated conditioning tank where the pulp density is adjusted to 30 to 40% solids, by addition of process water, and the lime, collector, and part of the frother are added. For most ores the pH adjusting reagent, conditioners, activators and depressants are preferably added to the final ball mill feed. Modern plants control the ore feed rate to the mill based on a particle size analyser located on the flotation pulp feed line and the mill drive power demand. This method tends to provide uniform tonnage and particle size distribution of the feed.

Table 63.- SCREEN SIZE EQUIVALENTS

microns	Tyler Mesh	A.S.T.M Standard	British Standard	microns	Tyler Mesh	A.S.T.M Standard	British Standard
2,000		10	8	250		60	60
1,680	10	12	10	210	65	70	72
1,410		14	12	177		80	85
1,190	14	16	14	149	100	100	100
1,000		18	16	125		120	120
840	20	20	18	105	150	140	150
710		25	22	88		170	170
590	28	30	25	74	200	200	200
500		35	30	62		230	240
420	35	40	36	53	270	270	300
350		45	44	44		325	
297	48	50	52	37	400	400	

A major technology change during the past few decades is the great increase in the size of the individual flotation cells, and the use of high capacity non-mechanical flotation devices, such as columns. One consequence is that the traditional pulp distribution tank, which doubled as a conditioning tank, is now no longer included in the flotation circuit. This trend in the size of the equipment has not changed the basic unit operations in the pulp flow. All mills still have flotation banks dedicated to: 1) an initial rougher bank; 2) scavengers; followed by 3) cleaners, and 4) re-cleaners. And the pulp is: 1) conditioned with flotation reagents and air; 2) sometimes parted into sands and slimes in cyclones or classifiers for separate treatment; 3) reground at the cleaner and re-cleaner stage to further liberate the valuable minerals; and 4) thickened to recover process water.

The simplest flow scheme used with bulk sulphides and simple industrial mineral upgrades is the rougher-cleaner arrangement:

Rougher-Cleaner Flotation Circuit

Final Conc.
Feed Pulp
1
2
3
4
5
6
7
8
tails
CLEANER
ROUGHER FROTH
water

A more conventional circuit includes a scavenger after the rougher:

Rougher-Cleaner-Scavenger Flotation Circuit

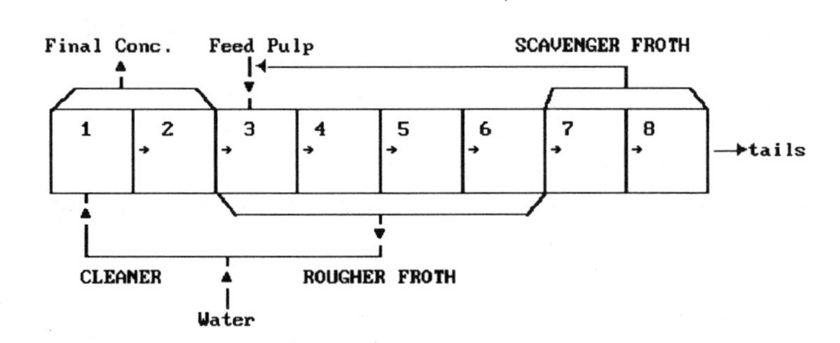

Rougher bank pulp solids are normally maintained at 35% plus or minus 5%, while the cleaner feed is normally diluted to around 25% to improve selectivity.

If the mineral contains native gold, as is not uncommon with some copper ores, the feed and tailings launder can be lined with sacking or corduroy, to recover coarse gold particles that do not float. The cleaner can be cascaded with re-cleaners. In a moly/copper separation, these may reach 15 stages.

The residence time in the rougher and cleaner varies according to the ore involved. With typical copper ores, roughers in the Pacific basin are designed for 8 to 12 minutes residence, while European and African mines will provide up to 40 minutes of roughing time. Cleaners typically are half to two thirds of the rougher time. Complex ores (Zn, Cu, Pb, Ag mixtures) normally involve European type flotation times.

All flotation processes involve differential flotation, as in all cases the separation is valuable mineral/gangue, or, in the case of copper ores, valuable mineral/pyrite plus gangue; but the term *differential flotation* usually is reserved for complex ores, where more than one valuable mineral is recovered, and mutually separated from each other and the gangue tailings. The tonnage and monetary value of complex ores commercially floated is considerably lower than the tonnage of bulk metallic and industrial minerals milled, but technically it is more challenging as it normally involves multiple flotation steps, multiple collectors and frothers, in addition to conditioners, activators and depressants; while bulk ores typically only require conditioners, frothers and collectors. All mills employ flocculants and drying agents in their tailings disposal and process water recovery systems, so this technology is more generally applicable, though different for water soluble and water insoluble minerals.

Complex ore processing schemes can become very involved, but they all can be simplified to a general set of circuits, consisting of a primary rougher-scavenger flotation bank for the feed and each separable mineral, followed by a cleaner-scavenger and recleaner - scavenger bank. The scavenger concentrates are reground and returned to the primary feed, while the recleaner scavenger concentrate is recycled. In each case the recleaner concentrate is the final concentrate. The same circuit is used for the recovery of the lead and zinc sulphides. The tails from the previous circuit is the feed for the downstream systems.

The reagent suite used is different for each mineral recovered. Not shown in Fig. 53 is a dewatering and reconditioning step placed between the copper-lead and lead-zinc circuits to remove the collector, frother and other reagents adhered to the ore in the previous circuit, and to add new conditioners, collectors, frothers, depressants and activators for the next separation. Reagent selection for the flotation of complex ores is dependent on the ores

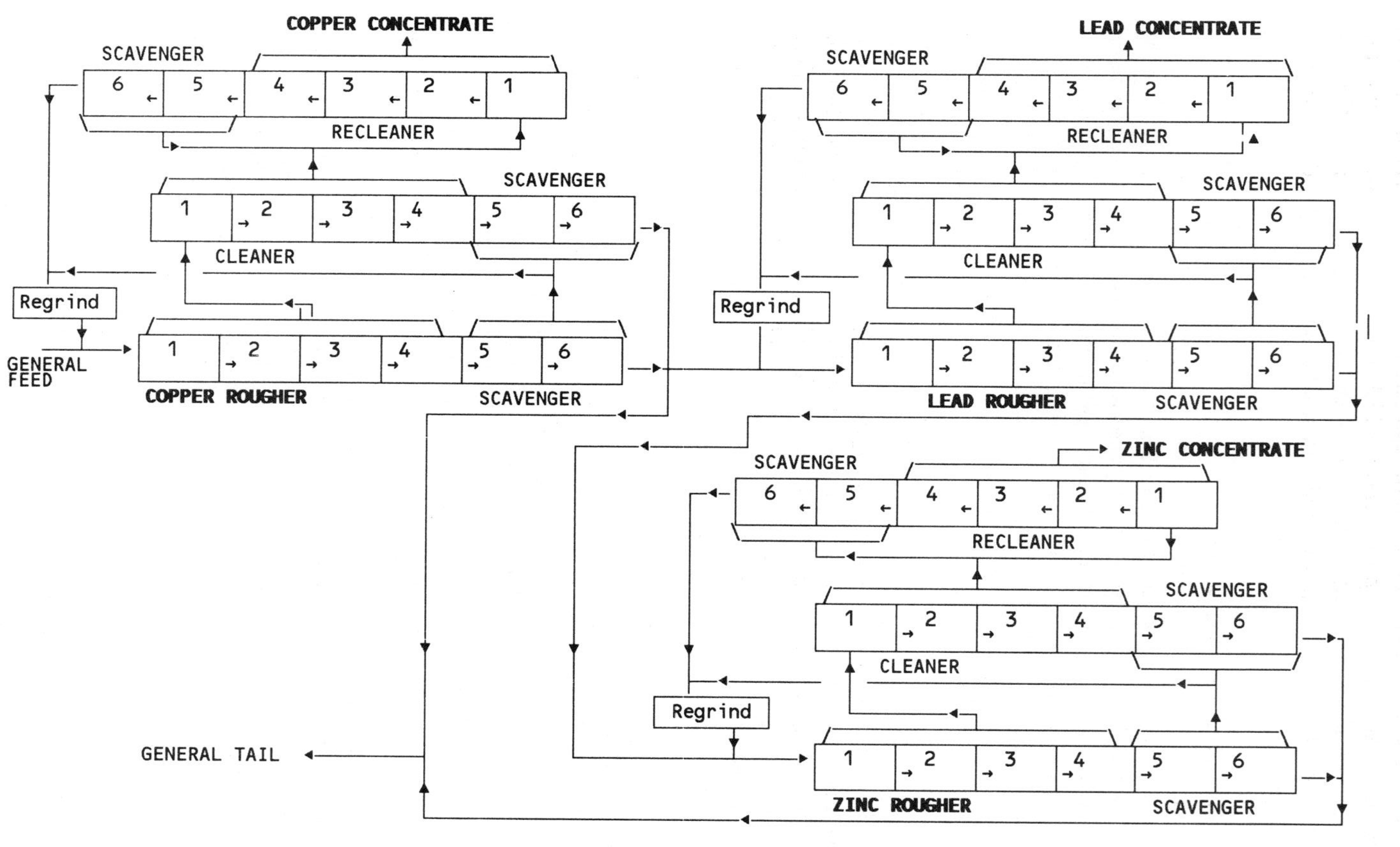

Fig. 53.- Complex ore differential flotation circuit

and the exact mill circuit, and flotation conditions favoured by the designer. As there are interactions between the inorganic and organic reagents, there is little that can be said, a priori, about which set of reagents is an optimum. Prediction of the circulating load from the scavengers is also difficult to determine in the laboratory, and essentially impossible to do from first principles.

Copper mill flotation practice

As we have seen, by far the most common copper mineral treated by flotation is chalcopyrite, followed by chalcocite, which is the primary mineral in a quarter of the mills; but essentially all the copper ores contain pyrite as a significant secondary sulphide, and in some cases other iron sulphides, such as pyrrhotite and magnetite are also present. The other copper minerals accompanying chalcopyrite as the major component are: chalcocite, bornite, covellite, cuprite, and native copper, more or less in order of importance. When chalcocite is the major component, which occurs in one out of five copper ores, the secondary copper minerals present are: chalcopyrite, covellite, bornite, malachite, and azurite. In a significant number of copper minerals, sphalerite is present as a less valuable secondary mineral. In Chile and the Philippines, where molybdenite, gold and silver are important minor minerals, arsenical ores, such as enargite and luzonite, are present. In African mines, cobalt is an important by-product. Chilean porphyries tend to contain around 0.3% copper as non-sulphide ores that are not readily soluble in acid.

Table 64 summarises the most important sulphide ores, their nominal formula, density and metal content. The copper minerals are listed in the approximate order of their commercial importance, and the constituents of complex ores are separated into primary and associated minerals.

The largest tonnage copper flotation mills treat low grade porphyry ores, with chalcopyrite as the predominant mineral. As chalcopyrite contains less than 36% Cu, concentrate grades obtained contain about 30% copper, and low head grades (0.4 to 0.9% Cu) mean that tailings with less than 0.1% Cu are unusual. To maximise grade and recovery, the flotation plant should be operated with powerful collectors having a high degree of selectivity. As the chemists have not developed a collector that combines these two properties, in most flotation plants a high recovery is obtained using a strong collector in the roughers, followed by a selective collector in the cleaners. If, as is usual in low cost and low capital mills, the pulp fed to the roughers has a coarse grind, a kicker (strong collector) is added to the main scavenger bank to recover mineralised coarse particles, and the scavenger concentrate is re-ground in a

Table 64.- PRINCIPAL SULPHIDE ORE MINERALS

Primary	Secondary	Composition	Metal %	Sp. grav.
Copper Minerals				
Chalcopyrite		$CuFeS_4$	34.6	4.1/4.3
Chalcocite		Cu_2S	79.8	5.5/5.8
	Bornite	Cu_5FeS_4	63.3	5.1
	Covellite	CuS	66.4	4.6
	Malachite	$Cu_2(OH)_2CO_3$		3.6/4.1
	Azurite	$Cu_3(OH)_2(CO_3)_2$		3.77
	Cuprite	Cu_2O		6.1
	Native copper	Cu	100.0	8.8
	Enargite	$Cu_2S \cdot AS_2S_5$	48.3	4.4
	Tennantite	$4Cu_2S \cdot As_2As_3$	57.5	4.4
	Tetrahedrite	$4Cu_2S \cdot Sb_3S_3$	52.1	4.8
	Chrysocolla	$CuSiO_3 \cdot 2H_2O$		2.0/2.4
	Digenite	Cu_2S	79.8	5.7
	Cobaltite	$CoAsS$	35.5	6.2
	Smaltite	$CoAs_2$	28.2	6.5
	Molybdenite	MoS_2	60.0	4.6/4.7
Other Sulphides				
Pyrite		FeS_2	46.6	4.8/5
Pyrrhotite		Fe_7S_8	60.4	4.6/4.7
Arsenopyrite		$FeAsS$	34.3	6.0
Galena		PbS	86.6	7.6
Pentlandite		$(Fe,Ni)S$	variable	4.6/5
Millerite		NiS	64.7	5.5
Sphalerite		ZnS	67.1	4.1
Marmatite		$ZnS(FeS)_x$	variable	4.0

small ball mill to liberate copper and help reject gangue, combined with the cleaner tails, and then returned to the rougher bank.

The flotation reagents employed in a specific operation will depend on the primary and associated minerals, and the gangue contained in the ore. For the great majority of Cu mills, pyrite must be rejected, so flotation is carried out at a pH above 9, generated with the addition of lime to the ball mills. But if the ore contains recoverable gold, consideration should be given to the use of soda ash or caustic soda for alkalinity control, as the calcium ion is known to depress Au. When the important by-product is molybdenite, its recovery is encouraged by adding hydrocarbons (kerosene or fuel oil), and possibly a moly specific collector, such as Cyanamid Aero 3302 or long chain mercaptans. The usual strong collector, habitually added to the ball mill, is amyl xanthate or if more selectivity is desired, isopropyl xanthate.

In simple circuits, xanthate addition to the ball mill will result in the recovery of a relatively low grade rougher concentrate. This concentrate is upgraded in the cleaners by diluting the pulp down to around 25% solids, and adding a more selective Cu collector, such as ethylisopropyl thionocarbamate (Z-200), in combination with a mild depressant, which can be more lime added to the cleaners to increase the pulp pH. When the copper minerals present contain a large proportion of chalcocite, and bornite, the Z-200 in the cleaners can be replaced by cheaper collectors - dialkyl dithio-phosphates. When high grade concentrates are obtained, it is always an indication that the ore being processed is from the alteration zone, and contains chalcocite, covellite and bornite. Another sophistication is to separate the rougher launders into sections, and stage-add the collectors: a dialkyl dithiophosphate or alkyl thionocarbamate in the ball mill, combined with separate collection of the concentrate from the first few rougher cells, followed by a selective xanthate in the second half of the roughers, and a strong xanthate in the scavenger section. Outokumpu Oy's "flash flotation" cells are a variant of this scheme, where cyclone bottoms are scalped prior to return to the ball mill feed.

Selectivity, i.e. activation of the flotation of a copper sulphide mineral before the promotion of the iron sulphides and gangue material, may be a function of the chelant properties of the collector molecule. Attempts to use molecular models and relate the length of the collector bite to the interatomic distances on the mineral surface have not been fruitful, but the rate of formation of a ligand bond may well affect the rate of flotation of the mineral species. The formation of metal cation - collector complexes is certainly important in the evaluation of depressants and pulp contaminants.

Alkaline circuits are practically universal in rougher flotation of porphyry copper ores. The exceptions are the acid circuits of El Teniente in Chile, Lepanto in the Phillipines, and Flin Flon in Canada. The ore of the latter two contains a significant amount of copper in arsenic minerals, such as enargite. The standard collector has been a xanthogen formate, usually ethyl butyl XF. This is very stable in acid media and only decomposes at pH 10 or greater. It also has been used in alkaline circuits successfully (Inspiration in the U.S.), and is a good collector in sea water pulps.

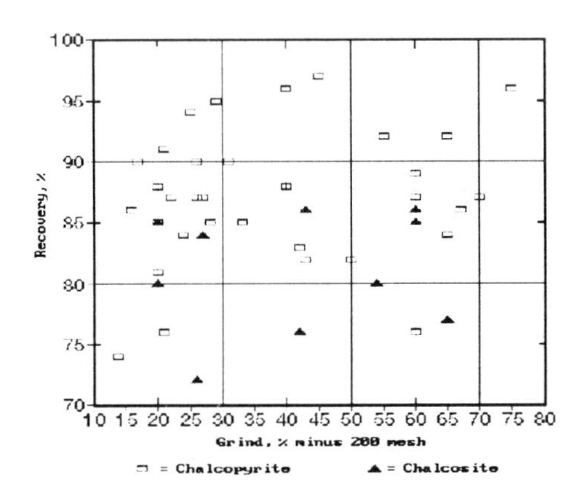

Fig. 54.- Copper recovery for chalcopyrite and chalcocite ores - Effect of grind on recovery

Fig. 54 plots copper recovery as a function of grind for the mines in Tables 56 and 57. The lower recovery of copper from ores where chalcocite predominates is obvious. Of interest is that, at least for gross statistical relations, the correlation between grind and recovery is tenuous; possibly this is due to the difficulty in floating fines, as undoubtedly liberation should be improved. The probable explanation of the lower recovery of copper from chalcocite ores is a greater fraction of the copper content corresponding to non-sulphide copper minerals, which in turn implies the presence of sericite and clays.

Non-sulphide copper ores

Sulphidisation, in the laboratory, is a viable process to increase recovery of oxidised copper minerals with flotation. This has very rarely been borne out on an industrial scale. The exception that comes to mind is Morenci, where at one time the ball mill feed was treated with ammonium sulphide which did improve recovery in an erratic manner. Later a modified LPF process was tried. In this process the ball mill feed was contacted with sulphuric acid in a trommel which fed into the ball mills. In the ball mill the acid pulp was neutralised with home made calcium sulphide, manufactured on-site by calcining limestone with pyrite. The crude calcium sulphide contained significant quantities of metallic iron which precipitated out cement copper from the liquid phase of the pulp. The sulphide mineral and the cement copper were then floated with ethyl isopropyl thionocarbamate. A similar process, but using iron powder, was used at a mine in Baja California where smelter slag was floated with sea water and Minerec A, after an acid leach. If liberation problems have required a very fine grind which has resulted in the formation of slime, a common solution used to improve overall recoveries has been to cyclone the ball mill output and separate the pulp into sand and slime circuits. The sand circuit is handled in the normal way, and the slimes are agglomerated with additives, such as water glass (silicates).

Malachite, azurite and cuprite can be concentrated by flotation, though the production of colloidal fines during crushing and grinding is a major problem. Fatty acid flotation was once a standard process, but today sulphidisation with NaSH and the use of benzothiazole as a collector is more in vogue. Cyanamid recommends their moly collector, Aero 3302, for copper carbonate flotation. The sulphhydrate is the preferred sulphidising agent as depression occurs at a pH of over 9.2 and sodium sulphide can contribute sufficient alkalinity to push pulp pH above this critical level. The new popularity of sulphidisation is due to the availability of reliable electrode probes that can assure the availability of sufficient sulphhydrate ion concentration for surface sulphidisation, while keeping the total NaSH concentration lower than the depression range.

Chrysocolla, and other copper silicates, do not float efficiently. They have been recovered by acid leaching and the LPF process, but if available in an economic quantity, the most profitable method of recovery now is heap leaching followed by liquid-liquid extraction and direct electrowinning.

Flotation of complex ores

The terminology complex ores generally is given to mixed sulphide ores containing copper, lead, zinc and silver. Setting up a processing scheme for this type of ore is complicated and empirical, as all of these ores contain non-floatable sphalerite and most of them have significant non-sulphide lead and silver present. The flow diagram for the flotation of a simple complex ore has been shown at the beginning of this section. The basic design consists in a sequential flotation of the different minerals employing depressants, activators, and selective collector-frother combinations to inhibit the flotation

Table 65.- COMMONLY OCCURRING MINERALS IN COMPLEX ORES CONTAINING GALENA (PbS) AND SPHALERITE (ZnS)

Secondary (oxidized) Minerals		Associated Sulphide Minerals		Gangue Minerals	
name	composition	name	composition	name	composition
Smithsonite	$ZnCO_3$	Pyrite	FeS	Calcite	$CaCO_3$
Calamine	$Zn_4Si_2O_7(OH)\cdot H_2O$	Chalcopyrite	$CuFeS_2$	Dolomite	$CaMg(C)_3)_2$
Hydrozincite	$Zn_5(OH)_6(CO_3)_2$	Marmatite	$ZnS(FeS)_x$	Ankerite	$Ca(Fe,Mg)(CO_3)_2$
Willemite	Zn_2SiO_4	Argentite	Ag_2S	Siderite	$FeCO_3$
Cerussite	$PbCO_3$	Native Silver	Ag	Rhodochrosite	$MnCO_3$
Anglesite	$PbSO_4$	Stromeyerite	$CuAgS$	Quartz	SiO_2
Pyromorphite	$Pb_5(PO_4)_3Cl$	Marcasite	FeS	Fluorite	CaF_2
Wulfenite	$PbMoO_4$	Arsenopyrite	$FeAsS$	Barite	$BaSO_4$
Vanadinite	$Pb_5(VO_4)_3Cl$	Tetrahedrite	$(Cu,Fe,Ag)_{12}Sb_4S_{13}$		
Plumbojarosite	$Fe_6(SO_4)_4(OH)_{12}$	Tennantite	$(Cu,Fe,Ag)_{12}As_4S_{13}$		
Cerargyrite	$AgCl$	Enargite	Cu_3AsS_4R		
Embolite	$Ag(Cl,Br)$	Pyrrhotite	FeS		
Bromyrite	$AgBr$	Stibnite	Sb_2S_3		
Iodyrite	AgI				
Argentojarosite	$AgFe_3(SO_4)_2(OH)_6$				

of each species. Native copper, as well as native gold and silver, can be easily floated at slightly above a natural pH with a xanthate and a collector-frother such as pine oil or cresylic acid. The dithiophosphates and benzothiazole mixtures are also good collectors. The minerals normally associated with lead-zinc ores are listed in Table 65. They must be taken into account in the selection of reagents, as many contribute soluble cations to the flotation pulp.

Chapter 12

OXIDE AND NON-METALLIC MINERAL FLOTATION

Non-sulphide, carbonate, and metallic oxide mineral flotation

So far, we have dealt with minerals in which the flotation mechanism involved is the selective adhesion of sulphhydryl collectors to sulphide surfaces; the technology that has been described is restricted to nascent bubble flotation (operative at concentration ratios above 15-30:1) where bubble attachment occurs through bridging to frother/collector complexes chemisorbed on active sites. As we saw, the principal minerals with this behaviour are the porphyry copper sulphides and some complex sulphide ores, such as high grade zinc-lead-silver sulphides, and sulphidised metallic non-sulphides. This section will review the behaviour of non-sulphide metallic minerals, industrial minerals and salts (insoluble, soluble and semi-soluble). And in contrast to nascent bubble flotation, data taken using experimental procedures and designs based on measuring contact angles, or using Hallimond tubes, and/or mechanical flotation machines, give acceptable and credible results with these minerals.

The reason for this change in viewpoint is that in bubble capture flotation, or oily flotation, there is no need for the transfer of an electron from the air/water interface to the mineral surface to trigger the deposition of a collector, and to provide the bubble particle bond. For non-metallics, adhesion between the particle and bubble is only (or possibly mainly) due to the hydrophobic nature of the final surface of the particle encountering the bubble.

Due to this, the choice of reagents is antipodal for non-metallic flotation - depressants and activators are chosen for their effect on the environment and not on the mineral surface. For example, cations depress when they react with the collector and take it out of the process by making it insoluble, and not by changing the nature of the mineral surface. Thus the properties and condition of the water phase are of the greatest importance in these flotation processes.

The nature of oily collector flotation

Concentration of non-metallic ores involves a completely different mechanism of selective particle attachment - bubble capture by collision - where the mineral surface must have natural or induced hydrophobicity to allow particle capture by bubbles. Ores involved are generally non-sulphides. They contain minerals which range from metallic oxides, carbonates, sulphates, or exotics such molybdenites, and tungstates, to mixtures of inert and semi-soluble ionic salts, and pure elements such as sulphur or iodine, and impure carbon, such as graphite and coals. The distinguishing characteristics of the bubble capture flotation process is that it involves concentration ratios in the 2:1 to 5:1 range. The economic importance of the reagent market for each of these types of minerals is illustrated in Tables 54 and 55 (page 176/77), which also show the tonnage treated by flotation by the different mineral industry sectors in America from 1960 to 1985, and the average concentration ratios which were required in winning the different types of minerals.

When concentration factors of two or three to one are involved, processes that replace or supplement flotation include gravity, electrostatic, magnetic, or heavy media separation processes. These are currently still popular in processing heavy metals, such as gold and members of the platinum group, titania sands, iron ores, etc. In the US flotation industry, only 40% of the tonnage floated is industrial minerals, but the volume of the flotation reagents employed is 80% of the market total, as ten times the grams per tonne are used when compared to selective sulphide flotation. Of the industrial minerals processed, the most important are phosphates (45%) and iron ore (15.5%), followed by potash (7%), coal (6.7%), glass sands (4%) and feldspars (2%).

The industrially important non-sulphide copper ores are malachite and azurite (carbonates), cuprite (oxide), chrysocolla (silicate), atacamite (chloride), brochantite and chalcanthite (sulphates). The last three are water soluble. At one time these were recovered by the LPF process, now they usually are heap leached, so are of importance in flotation only as contaminants during the early life of a mine when processing the mixed minerals in the ore cap, where they can provide undesirable activating cations in the pulps. For example, atacamite refractory surface properties are such that, during the Spanish colonial period, prior to the existence of blotters, the mineral was exported from northern Chile for use as ink-drying sands.

Industrially important non-sulphide lead ores include cerussite (carbonate), and anglesite (sulphate); and for zinc, smithsonite (carbonate). Iron ores of value are haematite and magnetite (oxides), goethite (hydroxide), and siderite (carbonate). This last can be a source of limonite, a bad actor in secondary enriched copper ore flotation, where it acts as a colloidal depressant for the copper minerals, and can form a coating of ferromolybdite on by-product molybdenum at high flotation pHs, cutting recovery significantly (see Beas et al (1991)).

Carbonate ores, and oxides such as cuprite, can be successfully recovered by sulphidisation; i.e., by artificially coating the valuable mineral surface with the metal sulphide produced by reacting the pulp with an alkali sulphide or sulphhydrate. The mechanism of flotation, and the reagents employed as collectors, modifiers and frothers, in this case, are similar to those used with copper sulphides, with the caveat that sulphidised surfaces are labile and must be kept in the presence of an active sulphidising agent during the whole flotation process. If the ores are high grade, and the required concentration ratios to obtain an acceptable concentrate are low, these can be treated as non-metallics, and economically floated with oils, fatty acids, amines, carboxyl collectors, long chain xanthates, etc. Thus, when sulphidised, they act in the flotation process as nascent-bubble capture minerals, and un-sulphidised or otherwise activated, they act as bubble capture minerals.

Note that when sulphhydryl collectors are tested with metallic minerals that exhibit oily flotation characteristics, the quantities of xanthates et al needed for successful recovery is such as to provide more than monomolecular surface coverage (i.e., at least a fivefold increase in dosage is required as compared to nascent or normal flotation). Laboratory tests are cited which show that synthesised dixanthogens act as collectors on sulphide and non-sulphide minerals; misinterpretation of these data can lead to erroneous mechanistic conclusions, as when minerals have surfaces fully coated with sulphhydryl collectors the dithiolate is adsorbing in the same way as a hydrocarbon oil, while when dixanthogen is detected as the active species in nascent bubble flotation, the xanthate has chemisorbed and polymerised in situ.

Polar, non-sulphide minerals and ores that are preferentially upgraded by flotation include alkali earth compounds, such as calcite, fluorite, apatite, phosphates, barite, scheelite, powellite, magnesite, dolomite, etc. Their ionic lattice provides the energy for hydration. Weak acids, such as xanthates or thiophosphate collectors, are useless in their flotation because of the soluble salts formed with the salts' cations. Soaps, which are unreactive, and fatty acids or other high molecular weight acids with a carboxyl group, can act as collectors for these minerals, but in the case of fatty acids it is because they will react with cations in solution to form soaps. Selective flotation, as known with the sulphide minerals, is not possible due to the generally undifferentiated surface chemistry of the group. This lack of selectivity is a general characteristic of oily flotation, where the concentration ratio is under 5:1; here depressants and activators are the more important tools that the metallurgist can use in concentrate upgrading. Because of the sedimentary genesis of these ores, one crucial step that is difficult to scale up from laboratory testing is the degree of comminution and the washing intensity required to remove clays and other colloidal slimes. This difficulty is compounded because only clean samples are usually sent in from the field for testing, so it is nearly impossible to obtain representative results without truck-load quantities of ore for a pilot scale test.

Fortunately, at concentration ratios below 5:1, usually the valuable mineral needs to be separated only from a homogeneous gangue, so the degree of rougher selectivity required is technically simpler, as the desired product and the reject have a different chemical composition. A key factor, though, is the mechanical removal of clays prior to the flotation step, as these will blanket any selectivity available from collectors, as well as providing unacceptable contamination of the concentrate. The roughing step, for a normal or reverse flotation, consists of selectively coating either the valuable mineral particles or the gangue, but not both, with a hydrophobic reagent so only the coated particles will float. The process sophistication then comes in the cleaning steps, where pH and selective regulators such as sodium silicate, or organics such as tannins, dextrin, starch, etc., can generate differential recovery.

Flotation of soluble salts has become a serious alternative to differential crystallisation in the chemical industry only in the post-World War period. In nearly all large scale applications the working fluid used is water, or, for soluble salts such as KCl from NaCl, the media is a saturated brine, so the overall process can be classified as semi-soluble salt flotation.

Glembotskii (1963) has a short chapter at the end of his book on flotation which summarises a Russian publication, "Concentration of Non-metallic Minerals by Flotation", that suggests there has been much wider use of flotation in the purification of soluble salts in Russia than in the West. He provides sparse details of the separation of nitrates from chlorides, and sulphates from chlorides, as well as the well known technology of sylvite purification. He underlines the use of lead nitrate as an activator for sodium chloride in some fatty acid reverse flotation systems. When it acts as a depressant, he suggests this is mainly with chlorides, where there is the formation of lead chloride which is relatively insoluble, and thus can coat crystal surfaces.

He also summarises a rule proposed by S.A. Kuzin which postulates that in a mixture of two soluble salts with greatly different solubilities, the most soluble can easily be floated off from the brine; but when the second salt's solubility approaches that of the floatable salt, it is depressed until neither salt can be floated if their solubilities are equal. The example cited is polyhalite - K_2SO_4-$MgSO_4$-$2CaSO_4$-$2H_2O$ - which is relatively insoluble in water, but its solubility increases in a potassium sodium chloride brine; concurrently its floatability increases with a change in composition of the brines which increases the polyhalite solubility. On the other hand its solubility is reduced in magnesium and potassium sulphate brines, which reduces its floatability accordingly.

Structure of mineral surfaces

In thiol collector adhesion to mineral particles, the key factor, which mandates the strength of reagent adhesion to the surface and the rate of capture of particles by bubbles, is the active sites where electron transfer to a conductive matrix is thermodynamically possible. To understand and quantify this phenomena, it is necessary to become conversant with the literature on catalysis. When a reagent adsorbs rather than chemisorbs on a particle surface, the macro properties of the structure are controlling, and bulk measurement of the mineral properties gives meaningful data which can be used to predict selectivity.

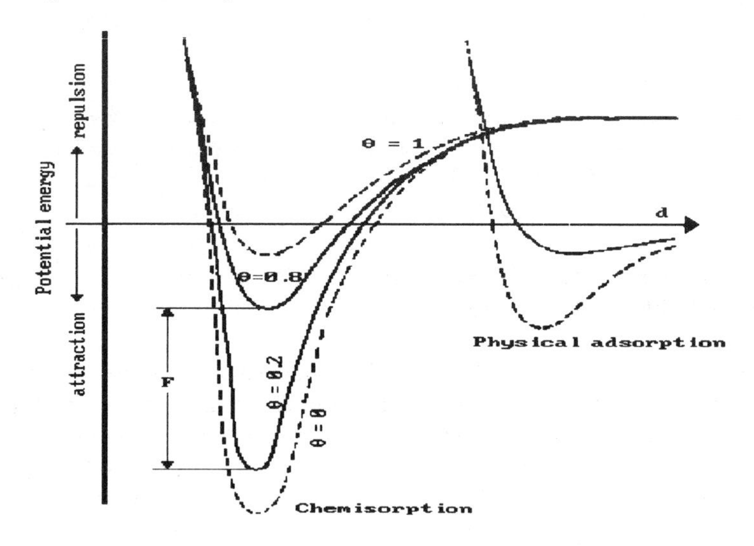

Fig. 55.- Forces in adsorption and chemisorption

For an in-depth survey of pertinent adsorption phenomena, the reader is referred to Leja's (1982) treatise on the surface chemistry of flotation, which explains how to estimate, and legitimise, the forces involved in adsorption and chemisorption. The latter is covered well by Clark (1974). Fig. 55 highlights two important adsorption energetic factors in flotation: the potential energy curve versus distance from the surface, with minima at r_p and r_c, shows that for chemisorption, both the repulsive forces and the attractive forces are greater and closer to the particle surface than is the case in adsorption; and that for adsorption the total adherence energy has a lateral component, which increases as the area of the individual particle covered increases.

The key difference in morphology between minerals which float by bubble capture and those that nascent bubble float, is their surface uniformity. The mineral particle can be conductive and still react in the bubble capture mode if its surface does not have active catalytic sites. We can thus generalise that the cations of these types of minerals must not be predominantly multivalent. Ionic and covalent crystal lattices will exhibit two distinct energetic levels. Ionic minerals will have charged surfaces that hydrolyse, and covalent minerals will be naturally floatable because they are hydrophobic.

Mechanism of collector adhesion

In a simplistic bubble adhesion model for a naturally floatable mineral, the particle surface, in an aqueous pulp, is completely unhydrated so that an approaching collector molecule will not be impeded from reaching the crystal surface. But with a hydrated solid, the surface is covered with water molecules and cannot be approached without displacing water molecules energetically attached to the solid, i.e. activation energy is required to displace water of hydration. This surface is usually shown as packed with spheres with dipole moments in a semi-oriented order (Fig. 2, page 21). A more modern picture of a hydrated mineral is to postulate that its surface is covered with water clathrates.

There has been considerable research on the crystal structure of water. Post-Pauling (1960), the molecular clusters postulated in liquid water at ambient temperature generally are pictured as having the shape of a distorted basket formed by an outer shell of five- and six-membered rings held together by hydrogen bonding, with two or three water molecules and a proton in the centre. Castleman et al (1991,1989) have studied both gas phase and liquid phase water clusters. In the gas phase they have found that the cluster which can be identified as either $H^+(H_2O)_{21}$ or $H_3O^+(H_2O)_{20}$ has a very stable form, and described it as weakly bound aggregates whose properties lie between those of water vapour and liquid water. Here the cage structure consists of 12 five-membered rings, where the oxygen of the water molecules forms the vertices. In the liquid phase they have identified anionic water cluster ions with the formulae $OH^-(H_2O)_n$ and $O^-(H_2O)_n$ which can show a different n, whose value is more diffuse than the gas phase n=20.

More interesting from the flotation point of view is that , with n=20, these clathrates contain 40 hydrogen atoms, 30 of which are involved in hydrogen bonding (forming O-H-O bonds), which means they have only 10 surface hydrogens available for hydrogen bonding (if an eleventh is associated with the proton in the centre of the protonated water formula). This structure was detected as quite stable in the vapour phase. In liquid water the picture is more diffuse, as it appears that n=20 is a preferred charge in the cluster, so then the available surface hydrogen bonds are ten. When n is less than 20, available hydrogen bonds are less than 10.

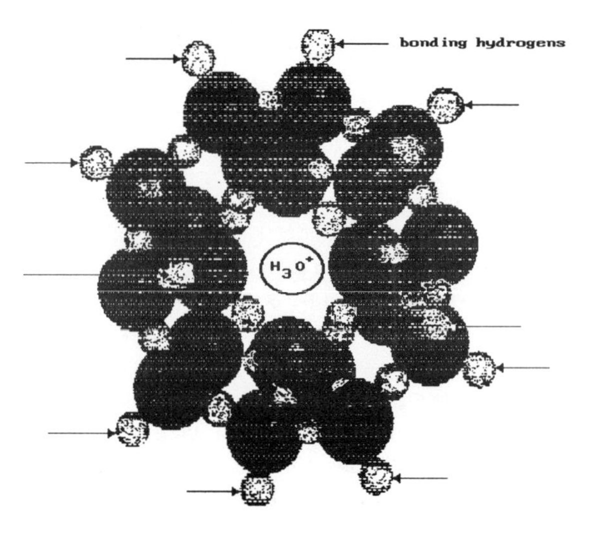

Fig. 56.- Water clathrates, maximum 11 hydrogen bonds available

Returning to our description of the collector molecule approaching the hydrated mineral particle, the obstacle that needs to be surmounted for adhesion is water that must be displaced with the collector's polar head. If the hydrated particle is scaled with clathrates, then more than one water molecule necessarily must be displaced by the collector. More than that, the number of water molecules per collector molecule should be multiples equal to the clathrate structure present. It is interesting that in a very recent paper by Yehia (in press), he measured heats of adsorption on apatite which suggested that 8 water molecules were being displaced per collector molecule adsorbed. He has interpreted this number to be related to the size of the area covered by the polar group of the Aerosol, but it is interesting to speculate if it is not more likely to be related to the hydrogen bonds available on the clathrate structures, and that this number could be fixed by the surface charge originally on the unhydrated mineral.

With non-sulphide metallic minerals, and with non-metallic minerals, bubble attachment does not depend on the transfer of electrons; rather, large areas of the desired constituent of the ore must be hydrophobic, either because the crystal lattice is naturally hydrophobic (covalent bonding), or because the hydratable surface is coated with a hydrophobic reagent. In this case, collector adhesion strength, and the induction times for coursing bubble attachment, can be calculated using classical Gibbs thermodynamics, and the energies involved can be unambiguously measured using bubble angles. Reagent quantities employed need to be orders of magnitude greater because selective complete coverage of the target mineral surface is required to attain high recoveries.

To determine if a specific collector will adhere promptly to a hydrated mineral, the electric charges on the surface must be determined, as must the effect of changes in the environment on these charges. Zeta potential, pH, electro-scans of the surface, etc., all provide pertinent information on surface properties that affect reagent selection. Concepts that are not pertinent to this type of flotation include dithiolate formation, frother-collector interactions, selection of modifiers, activators and depressants based on the physico-chemical effects on the mineral particles' surfaces, etc. These exception are important to keep in mind, as the same reagents normally will have contradictory action in the two types of flotation; so those metallurgists involved in non-metallic mineral flotation should try to compartmentalise their tools of the trade to avoid confusion.

The fundamental factor to keep in mind in evaluating experimental data, and applying these to selection of flotation reagents and operating conditions, is the difference in the bubble attachment process. In non-metallic bubble collision flotation there is no need for the transfer of an electron from the air/water interface to the mineral surface. Adhesion between the particle and bubble is only due to the hydrophobic nature of the surface encountering the bubble, while with sulphide minerals there is an electron bond which can be as strong as a covalent one.

Collectors for industrial minerals

These are reagents or oils that coat and/or react with mineral surfaces and make the mineral selectively water repellent and attachable to air bubbles. Those that react with the mineral surface can be classified as surfactants that chemisorb, such as hydroxamic acids on haematite, or oleates on fluorite or calcite, or long chain ionic surfactants which absorb physically as counterions in the electrical double layer, such as dodecyl sulphonate on alumina or haematite, and dodecylammonium chloride on quartz; but over 90% of the

Table 66.- COLLECTOR REAGENTS FOR NON-METALLICS

Type	Formula	Cation	Anion	Electrolyte
ANIONIC COLLECTORS				
Soaps and fatty acids	RCOOM	Na^+, K^+, H^+	$RCOO^-$	weak
Alkyl sulphates	RSO_4M	Na^+, K^+	RSO_4^-	strong
Alkyl sulphonates	RSO_3M	Na^+, K^+	RSO_3^-	strong
Alkyl hydroxamates	RCONHOM	Na^+, K^+	$RCONHO^-$	weak
Alkyl phosphates	RPO_4M_2	Na^+, K^+	$RPO_4^=$	weak
CATIONIC COLLECTORS				
Primary amine salts	RNH_3X	RNH_3^+	Cl^-,	weak
Alkyl ether amines	$RONH_3X$	$RONH_3^+$	Cl^-	weak
Quaternary ammonium salts	$RN(CH_3)_3Cl$	$RN(CH_3)_3C^+$	Cl^-	strong
Sulphonium salt	$RS(CH_3)_2Cl$	$RS(CH_3)_2C^+$	Cl^-	strong
Alkyl pyridinium salt	R–⟨pyridine ring⟩–NHCl	R–⟨pyridine ring⟩–NH$^+$	Cl^-	strong
Morpholinium salts	$RNH(CH_2CH_2)_2OCl$	$RNH(CH_2CH_2)_2O^+$	Cl^-	
Sulphonium salts	RR'R"SCl	RR'R"S^+	Cl^-	
Alkyl xanthates	ROCSSM	Na^+, K^+	$ROCSS^-$	strong
AMPHOTERIC COLLECTORS (zwitterion form)				
N-alkyl-2-aminopropionic acids	$RNH_2(CH_2)_2COOM$	$RNH^+{}_2(CH_2)_2COO^-$	Na^+, K^+	
n-alkyl-N,N-dimethyl glycines	$R(CH_3)_2NCH_2COOM$	$R(CH_3)_2N^+CH_2COO^-$	Na^+, K^+	
Alkyl betaines	$(CH_3)_3NRHCCOOM$	$(CH_3)_3N^+RHCCOO^-$	Na^+, K^+	

reagents used as collectors for industrial minerals are soaps and fatty acids (including tall oil from the naval stores industry).

Because the raw materials for the manufacture of these reagents are generally natural oils and fats, products from different manufacturers are not necessarily interchangeable. It should also be emphasised that, as is the case for detergents, cationic and anionics should not be used simultaneously, as they can react and neutralise each other. Water quality, which is a problem in sulphide ore flotation if heavy metal ions are in solution, is more of a problem in industrial mineral flotation, which is normally associated with soluble inorganic salts, as all cations and ions that are in solution in the flotation pulp can interfere with the reagent action.

As a large number of anionic and cationic collector molecules used in inorganic flotation are derived from natural oils and fats, an overview of fat chemistry is helpful. Table 67 lists the approximate fatty acid content of a few typical natural oils and fats. This assay cannot be taken as a specification, as the use of caustic to remove acids from fats is ancient and continues in the production of soaps; steaming to deodorise, and hydrogenation to control the unsaturated acid content, are processes now used for nearly a century, and other purification processes are common, so the selling specification for derivatives of these natural oils will vary for each manufacturer. Caveat emptor!

Fatty acids

A typical process for obtaining useful collectors is treating crude vegetable oils by alkali refining, water washing and drying, bleaching, hydrogenation, and deodorising. The free fatty acids neutralised with caustic can be recovered by centrifuging; bleaching and winterising results in removal of stearins, while hydrogenation can go as far as a solid shortening. Animal fats and oils are processed by rendering and hydrogenation, plus inter-esterification and isomerization, to improve their quality. Classic detergents, whose chemistry is similar to that of the flotation oils, consist of a hydrophobic hydrocarbon chain containing 8 to 18 carbon atoms and a hydrophyllic functional group which may be:

anionic, e.g., $-OSO_3^-$ or $-SO_3$;
cationic, e.g., $-N(CH_3)_3^+$ or $C_5H_5N-^+$;
zwitterionic, e.g., $-N^+(CH_3)_2(CH_2)_2COO^-$;
semipolar, e.g., $-N(CH_3)_2O$;

or

nonionic, e.g., $-(OCH_2CH_2)_nOH$.

Petroleum based linear alkylbenzene sulphonates, alkylbenzene-ether sulphonates, alkylglycerol-ether sulphonates, alkyl esters of isothionate, methyl alkyl laurates, fat based alkyl sulphates, fatty alcohol-ethylene oxide sulphates and soaps are the most common anionic detergents. Cationic surfactants include quaternary alkyl ammonium halides and other amines.

Table 67.- FATTY ACID CONTENT OF OILS AND FATS (WEIGHT %)

No. Carbon atoms	Acid	Formula	Soy bean oil	Fish oil	Coco nut oil	Tal- low	Lard
8	Caprilic	$C_7H_{15}COOH$	...	...	8.0	...	...
10	Capric	$C_9H_{19}COOH$	...	...	7.0	...	...
12	Lauric	$C_{11}H_{23}COOH$	...	...	48.0	...	...
14	Myristic	$C_{13}H_{27}COOH$	...	7.0	17.5	1.0	1.0
14	Myristoleic*	$C_{13}H_{25}COOH$	...	...	...	...	0.2
16	Palmitic	$C_{15}H_{31}COOH$	8.3	16.0	8.8	21.0	28.0
16	Palmitoleic*	$C_{15}H_{29}COOH$	...	17.0	...	...	3.0
18	Stearic	$C_{17}H_{35}COOH$	5.4	1.0	2.0	30.0	13.0
18	Oleic*	$C_{17}H_{33}COOH$	24.9	27.0	6.0	43.0	46.0
18	Linoleic*	$C_{17}H_{31}COOH$	52.7	...	2.5	5.0	6.0
18	Linolenic*	$C_{17}H_{29}COOH$	7.9	...	...	...	0.7
20	Arachidic	$C_{19}H_{39}COOH$	0.9	...	...	...	2.0
20	Arachidonic*	$C_{19}H_{31}COOH$	...	20.0			
22	Clupanadonic*	$C_{21}H_{35}COOH$	...	12.0			

* = unsaturated acids

Predicting the behaviour of a particular reagent suite in non-metallic flotation is complex because the reagents ionise, dissociate, and hydrolyse, so modifying agents become more important than in sulphide flotation, and their action less easy to forecast; thus, the description of the activity of a particular collector must always be qualified in terms of the environment. Laboratory adsorption data, electrophoresis measurements, zeta potential, etc., all seem, and are, very pertinent, but can be misleading because of the interference of other ions.

The tendency for a collector to adsorb onto a mineral surface will depend on:

1.- The length of its hydrocarbon chain
2.- The number of chains in the structure
3.- The number of ionic groups in the structure and the type of associated ions

4.- The position of the ionic groups on the chain and the distance separating the groups
5.- The nature of the ionic groups and the cross-sectional area of the groups
6.- The nature, position, and number of nonionic hydrophyllic groups

These factors affect the properties of the collector:

1.- The base strength, which can be quantified by the pK_a values
2.- The solubility of the collector
3.- The critical miscelle concentration (CMC)
4.- The Krafft point
5.- The tendency of the collector to adsorb onto a specific mineral

Table 68.- THE STRUCTURE OF SOME ANIONIC COLLECTORS

Surfactant group	Structure		pK_a
Carboxylic soap	$R-C(=O)O^-$	H^+ or Na^+/K^+	4.7 ± 0.5
Sulphate	$R-O-S(=O)_2-O^-$	Na^+ or K^+	
Sulphonate	$R-S(=O)_2-O^-$	Na^+ or K^+	≈ 1.5
Hydroxamate	$R-C(=O)-N(H)-O^-$	Na^+ or K^+	≈ 9
Phosphoric acid	$R-P(=O)(HO)-O^-$	H^+ or Na^+/K^+	7 ± 2.5

Anionic collector chemistry

The most important anionic collectors are shown in Table 68. Mishra (1988) points out that carboxylates are fully ionised at pH values above 10; the alkyl sulphates do so at a pH above 3 (at a lower pH they hydrolyse to alcohol and bisulphate); while alkyl sulphonates are stable even in strong acids. The long chain fatty acid carboxylates go through an ionisation process between pH 4 to 10. At pH around 8, the ionisation is to a 1:1 ion complex, $RCOOH\text{-}RCOO^-$. Similar intermediate species are present in the case of alkyl sulphates and sulphonates in the half-ionisation pH range; as a consequence, these compounds can act as a dual collector/frother at those pHs. In fatty acid flotation it has been noted that adsorption, and recovery, improve with the degree of unsaturation in the alkyl group. Collector properties also change if aryl groups are substituted for some or all of the hydrocarbon groups.

As in the case of sulphhydryl collectors, over-oiling will also convert a collector into a depressant, so with the above surfactants there are concentrations above which they will not act as collectors. The phenomena in this case can be explained by miscelle formation and the adsorption of these macro-molecules on the mineral surface. A list of critical miscelle concentration (CMC) values are shown in Table 69.

Note that, within the same species, the CMC occurs at progressively lower concentrations as the chain length increases. There also is a significant difference if the potassium salt is involved; i.e., the CMC value for thc Na oleate drops from 2.1×10^{-3} to 8.00×10^{-4} for potassium oleate.

Table 69.- CRITICAL MISCELLE CONCENTRATION FOR ANIONIC COLLECTORS

Surfactant	CMC. moles/litre
Na caprylate	3.5×10^{-1}
Na capriate	9.4×10^{-2}
Na laurate	2.6×10^{-2}
Na myristate	6.9×10^{-3}
Na palmitate	2.1×10^{-3}
Na stearate	1.8×10^{-3}
Na elaidate	1.4×10^{-3}
Na oleate	2.1×10^{-3}
Na dodecyl sulphate	8.2×10^{-3}
Na dodecyl sulphonate	9.8×10^{-3}
Na dodecyl benzene sulphonate	1.2×10^{-3}

Industrially important fatty acid collectors

The majority of the fatty acids used industrially are of vegetable origin, rather than animal fats. Tall oil fatty acids are by far the most consumed organic reagents in non-metallic ore flotation. Their main components are unsaturated oleic, linoleic and linolenic acids, and the saturated palmitic and stearic acids. All commercial products also contain a varying amount of rosin acids, which are frothers and normally do not act as selective collectors.

At ambient temperature, the effectiveness of these fatty acid collectors is a function of solubility, which favours the shorter chain alkyl groups. For example, at a pulp temperature of 24°C there is very low activity with palmitic acid, which has a melting point of 63°C; but if the pulp is conditioned with the collector at 70 to 75°C, recoveries rise to the 80 to 90% range. Note that it is the conditioning temperature that is important, as this is the point in the process at which collector adsorption on the mineral surface occurs, thus the flotation banks do not need to be heated if the collector concentration and conditioning pH is controlled correctly. Varying the conditioning temperature will affect recoveries more than concentrate grade, thus it is an important variable to study in the optimisation of a circuit.

As has been noted previously, interfering ions are a greater problem in non-sulphide flotation. The presence of calcium or iron in solution can result in activation of the gangue with collector complexes. Here, pH can be an important variable to correct these problems.

Sulphonate type reagents were first developed as anionic collectors for selective concentration of low grade iron ores or for the removal of iron contamination from glass sands, feldspars and other ceramic raw materials. They now are employed in the flotation of barite, celestite, fluorite, apatite, chromite, kyanite, mica, rhodochrosite, cassiterite, etc..

Sulphonates are water soluble or dispersible liquids or pastes. They are normally fed to the conditioner as a 5 to 30% water solution or dispersion. Most of them have frother properties, so mixtures of sulphonates are a possibility in obtaining grade and recovery without the addition of auxiliary frothers. Total dosage is generally in the 130 to 2,000 g/t range.

Cationic collector chemistry

The only cationic collectors used in metallic mineral processing are the amines. These reagents ionise by protonation. They are classified as primary, secondary, tertiary or quaternary amines, depending on whether they have one, two, three or four hydrocarbon radicals bonded to the nitrogen atom. Primary, secondary and tertiary amines are weak bases, whereas the quaternary amines are strong bases, and are completely ionised at all values of pH, while the ionisation of the other amines is pH dependent, so that the metallurgically active concentration is pH dependent. The next table lists a representative group of collector amines:

Table 70.- NITROGENOUS CATIONIC FLOTATION AGENTS

n-amylamine	$C_5H_{11}.NH_2$	Primary aliphatic amine
n-dodecylamine	$C_{12}H_{25}.NH_2$	Primary aliphatic amine
Di-n-amylamine	$(C_5H_{11})_2.NH$	Secondary aliphatic amine
Tri-n-amylamine	$(C_5H_{11})_3.N$	Tertiary aliphatic amine
Amylamine hydrochloride	$(C_5H_{11}.NH_3)Cl$	Salt of primary amine
Tetramethyl Ammonium chloride	$((CH_3)_4N)Cl$	Salt of a quaternary amine
Aniline	$C_6H_5.NH_2$	Primary aromatic amine
p-toluidine	$CH_3.C_6H_4.NH_2$	Primary aromatic amine
Benzylamine	$C_6H_5.CH_2.NH_2$	Primary aliphatic amine
Diphenylamine	$C_6H_5.NH.C_6H_5$	Secondary aromatic amine
Alkoxypropylamine	$R-O-(CH_2)_3NH_2$	Approx. mol. wt. 210
	$R = C_8 - C_{10}$	mixture
Technical coco	RNH_2	Approx. mol. wt. 203
Primary amine	$R = C_8 - C_{18}$,	55% C_{12}
Tallow amine	RNH_3Ac	Primary C_{18}
Acetate		R = approx, 96% C_{18}

Representative ionisation constant for some typical amines are:

Methylamine	4.4×10^{-4}	Trimethylamine	5.5×10^{-5}
Dimethylamine	5.2×10^{-4}	Dodecylamine	4.3×10^{-4}

Miscellization: is a characteristic of hydrocarbon chain surfactants. It is important in interpreting the molecular behaviour of flotation systems. Miscelles are aggregates of collector ions of colloidal size that form by van der Waal's bonding between hydrocarbon chains. They form at a critical miscelle concentration (CMC) because the hydrocarbon chain is non-ionic, causing a mutual incompatibility between water complexes and the hydrocarbon chains. These aggregates cannot be seen, but their presence is detected because of light scattering when a light beam is passed through the solution. The free energy decrease that the system experiences when this phenomena occurs is significant - 0.6 kcal per mol per CH_2 unit. For dodecylamine, therefore, the free energy decrease is 7.2 kcal/mole on formation of miscelles.

Table 71.- CATIONIC AMINE COLLECTOR STRUCTURES

Primary amines	$R-NH_2$
Secondary amines	$R(R')N-H$
Tertiary amines	$R(R')N-R''$
Quaternary ammonium salts	$RR'R''R''N^{+}\ Cl^{-}$
Diamines	$R-N^{+}H(CH_2)_xN^{+}H_2$
Triamines	$R-N^{+}H(CH_2)_x-CH(N^{+}H_2)-(CH_2)_xN^{+}H_2$
Pyridinium salts	$C_5H_5N^{+}H\ Cl^{-}$
Morpholinium salts	$R-NH(CH_2-CH_2)_2O^{+}\ Cl^{-}$
Sulphonium salts	$R-S^{+}(R')(R'')-Cl^{-}$

Flotation systems that employ amines as collectors include: silica flotation in cleaning phosphate rock; separation of KCl from other salts using primary aliphatic amines; cleaning beryllium minerals, using coco or tallow amines; upgrading spudomene in lithium mineral processing; etc.

The structures of typical cationic amine collectors are shown in Table 71.

Solubility: as the length of the hydrocarbon chain increases, the solubility of the amine is reduced. The effective ionic solubility, though, is pH dependent.

Table 72.- MOLECULAR SPECIES SOLUBILITY, pK_A, AND POLAR HEAD DIAMETER FOR CATIONIC COLLECTORS (from R.W. Smith, 1988)

Collector	pK_{a1} at 25°C	pK_{a2}	Solubility mol/litre	Diameter of polar grp Å
n-dodecylamine	10.63	-	2×10^{-5}	3.7
N-methyl-n-dodecylamine	11.0	-	1.2×10^{-5}	
N,N-dimethyl-n-dodecylamine	9.74	-	7.2×10^{-6}	
di-n-hexyl amine	11.01	-		
N-(n-dodecyl)-propyl-1,3-diamine	6.8	9.3	3.6×10^{-7}	
n-decylamine	10.63	-	5×10^{-4}	3.7
n-tetradecylamine	10.62	-	1×10^{-6}	3.7

Factors affecting selective flotation

Leja (1982) has pointed out that carboxylates, such as oleic acid, can chemisorb, as well as adsorb, dependent on the ionic salt content of the pulp (Fig. 57).

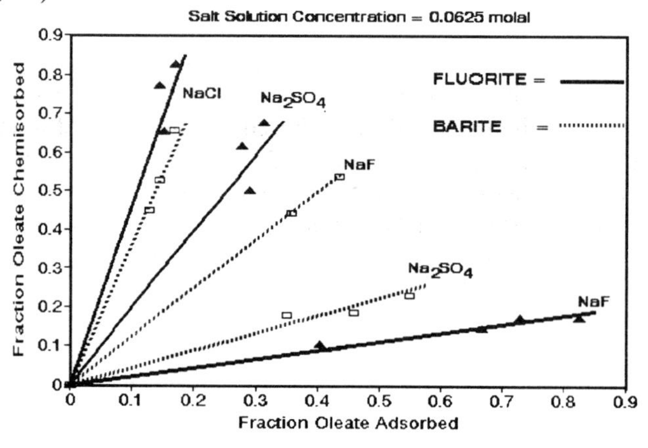

Fig. 57.- Effect of brine composition on the ratio of chemisorption

Note that sodium fluoride inhibits chemisorption of fluorite, and that the sulphate ion does the same for barite. These data are interesting in that they suggest the mechanism by which anions can influence collector selectivity, but it must be noted that there is not clear evidence whether the oleic acid salt is ionised or unionised on the mineral surface. Leja points out that co-adsorption of non-ionised carboxylic acid contributes to an increase in overall hydrophobicity of a particle, not so if the carboxylic acid is ionised.

As in the case of sulphide minerals, excessive amounts of collector will inhibit flotation of inorganics. The mechanism visualised is the formation of a second collector layer on the particle surface which is "upside-down", with the carboxylic anion exposed to the liquid phase, generating hydrophobicity. Addition of an oil will neutralise this extra layer and help flotation. This is probably one reason why fuel oil is a popular collector in copper roughers when molybdenite is an important by-product, as the primary xanthate collector can over-absorb on the molybdenite's naturally floatable surface layer and could well act as a depressant.

Carboxylic acids become more and more insoluble as the alkyl group lengthens. In addition, the acids and soaps are powerful frothers, even more so if the alkyl groups are unsaturated; therefore they normally can be used as collectors without the need to add a separate frother, an ability which is frequently a disadvantage. Because of their limited solubility, an increase in pulp temperature reduces the amount of collector required for a given recovery, and winter weather requires additional amounts of collector to maintain recovery.

Metallic oxide minerals mill practice

According to Sutherland and Wark (1955), in the early days of iron flotation oleic acid salts were known to be promoters for haematite, as Sulman and Edser (1924) patented their use of oleates or stearate soaps with sodium silicate to separate haematite from feldspars. Based on later work of Keck et al, they report that for 100 g/t oleic acid and 50g/t of terpineol as the frother, the optimum pH for recovery is a little above pH=7. As recovery is inhibited below pH=5.5, they conclude that oleic acid itself is not the collector. Calcium is a depressant at very low concentrations, as are metals such as ferric cations, and Pb^{2+}, Mn^{2+}, Sn^{2+}, Cu^{2+}, and Sn^{4+}, as these all form insoluble soaps with the fatty acid.

The SME Mineral Processing handbook reports, in some detail, current flotation experience at the Republic Mine in Michigan's Northern Peninsula, where they are processing 22,000 tpd of a specular haematite ore containing

36.5% Fe, and producing a concentrate with 65.4% Fe and 4.95% SiO_2, with a recovery of 46%. The operator's experience is that conditioning is crucial to minimising the amount of collector required. The collector is a distilled tall oil containing 91% oleic acid, 6% rosin acid, and 3% unsaponifiables. The amount of collector required is about 600 g/t of ore, ten times the amount reported by Sutherland and Wark, and laboratory work at the mine has shown that very intense conditioning will halve the amount of collector required for a given recovery. The flotation circuit treats 18,000 tpd, with a sand/slimes split using two conditioners, each consisting of a number of 4.5 x 4.5 x 5-ft cells equipped with 24" ship-type propellers driven at 290 rpm by motors drawing 12.5 hp. The pulp density is important, and is maintained at 65 to 70% solids, which gives a residence time of 5 to 8 minutes. Recovery is reduced in winter, so steam injection is used to maintain pulp temperatures at 70°F. Similarly the Hanna Mining Co.'s Groveland mine in Dickson County, Michigan is processing 15,000 tpd of an ore containing magnetite, haematite and iron silicates assaying 34.5% Fe, which they upgrade to 64.4% Fe and 6.3% SiO_2, using flotation, elutriation and magnetic concentration. Overall recovery is 42.2%. In their flotation section the deslimed elutriator underflow is conditioned at 66% solids in a bank of ten 60 cu ft conditioner cells equipped with ship-type propellers driven by 25 hp motors. Retention time is 10 minutes, and the conditioning pH is held at 5.8 to 6.0. The reagents used are:

Reagent	g/t	Point of addition
Petroleum sulphonate	400	1st conditioner
Fatty acid	150	1st conditioner
Fuel oil	200	1st conditioner
Sulphuric acid	500	2nd conditioner
Sodium silicate	150	1st rougher and 1st cleaner

Industrial minerals mill practice

The use of flotation to upgrade industrial minerals has become more popular as a consequence of declining reserves of high grade material. The products usually treated include barite, beryllium, borates, calcite (limestone), clay, graphite, gypsum, feldspars, fluorspar, glass sands, iodine, kaolin, kyanite, mica, pegmatites, phosphates, potash, quartz, rare earths, spudomene, talc, vermiculite, wollastonite, and others. The sources used for these summaries are the SME Mineral Processing Handbook (1985), Taggart (1945) and Redeker and Bentzen (1986).

Barite ($BaSO_4$) is found in combination with fluorite, celestite, quartz, galena, sphalerite, and many other minerals. Its primary use is in the formulation of drilling muds, for which the composition is important to obtain the maximum specific gravity. If used for chemical processes, the specification is usually +98%; while for drilling muds the specification is a specific gravity of 4.0 to 4.3.

Fine grained deposits, or the rejects from a gravity concentration operation, are frequently upgraded by flotation. The material is ground to liberation size (which can be minus 40 down to 325 mesh, depending on the ore). The coarser material is deslimed in cyclones, then floated using petroleum sulphonates, or mixtures of sulphonates. The drawback to these flotation reagents is that if the end product will be used in drilling muds, the collector must be destroyed to avoid frothing during drilling. To do this, the concentrate must be heated to 200 - 260°C.

Outside the U.S.A., barites are frequently associated with sulphide minerals and are recovered from the tailings of the plant. In this case the collector used can be oleic acid, fatty acid sulphonates or sodium laurate in an alkaline circuit. Barite is easily depressed by acidifying the flotation circuit with sulphuric acid. If quartz is present it can be depressed with the addition of silicates to the pulp.

Bauxite is a rock rather than a specific mineral. It consists of hydrated alumina minerals (gibbsite, boehmite, diaspore, etc.) contaminated with quartz, iron oxide minerals and titania. Calcite, siderite, mica, kaolinite, pyrite, and marcasite are also found as minor contaminants.

A case history of flotation is quoted by Norman Holme (1986): the ore's minus 20 mesh washer tails, which amounted to about 45% of the original plant feed, were discarded. The plus 20 mesh was ground to minus 35 mesh and deslimed. The sands were conditioned for 2 minutes at 60% solids with 3.3 kg/t of natural petroleum sulphonate, a similar quantity of fuel oil, and 1.1 kg/t of sulphuric acid. The pulp was then diluted to 14% solids and floated at a pH of 3.1. After 2 stages of cleaning, about 86% of the alumina in the deslimed feed was recovered in a concentrate representing 76.9% of the deslimed feed weight.

Beryllium is obtained commercially from beryl ($3BeO{\cdot}Al_2O_3{\cdot}6SiO_2$), bertrandite ($4BeO{\cdot}2SiO_2{\cdot}H_2O$), phenacite ($2BeO{\cdot}SiO_2$), chrysoberyl ($BeO{\cdot}Al_2O_3$), and barylite ($Be_2BaSi_2O_7$).

According to Paul Chamberlin (SME, page 29-7), beryl has not been treated successfully industrially, though a complicated laboratory processing scheme has been developed for the pegmatite type ore. The process starts with desliming, then adjusting the pulp pH to the 3 - 5 range with sulphuric acid and removing the mica by flotation with tallow amine acetate. Tailings from the mica float are dewatered, washed several times, and diluted to 25% solids. Hydrofluoric acid is added to maintain pH 3, and the silicates (beryl, feldspar etc.) are floated with coco or tallow amines away from the silica. The silicate concentrate is then scrubbed with calcium hypochlorite to remove reagents, the clean pulp thickened, washed, and diluted to 60 to 70% solids and conditioned with a fatty acid or petroleum sulphonate. The flotation is performed at pH 6, if the collector is a fatty acid, or reduced to 3 to 4 with H_2SO_4 when using a sulphonate. The final beryl concentrate must be cleaned several times to obtain a satisfactory grade. This step quite obviously might be improved using column flotation.

Borates: the four most important boron minerals are listed in Table 73. Colemanite is the only mineral that has been concentrated by flotation on a regular basis.

Table 73.- COMMERCIALLY IMPORTANT BORON MINERALS

Mineral	Formula	B_2O_3 %	sp. gr.	Water Solubility
Ulexite	$NaCaB_5O_9 \cdot 8H_2O$	42.9	1.96	nil
Colemanite	$Ca_2B_6O_{11} \cdot 5H_2O$	50.8	2.96	nil
Borax	$Na_2B_4)_7 \cdot 10H_2O$	36.5	1.72	soluble
Cornet	$Na_2B_4O_7 \cdot 4H_2O$	50.9	1.91	slowly soluble

Colemanite ores containing 20 to 30% B_2O_3 must be upgraded. For flotation concentration, they are ground to minus 20 mesh and deslimed with cyclones. High clay ores may require various stages of desliming. The pulp is washed, dewatered and conditioned at 20 to 40% solids with 500 to 1,000 g/t of a mixture of sulphonate collectors. When carbonates are a problem, starch, quebracho or dextrin can be added to depress them. No frother is required when floated at the natural pH of 8 to 9. The rougher concentrate is then normally cleaned to about 45% B_2O_3. Erratic frothing can usually be corrected by an improvement in the desliming step.

Table 74 shows the glass fiber market specification that marketable colemanite must meet in the U.S.A.. Flotation is also employed to separate borax from sylvite in the borax process. The collectors that are normally used are oleic and naphthenic acids, to which xylene, turpentine and kerosene may be added. The flotation can be activated with barium salts. Boric acid is naturally floatable. This property has been used for many years in the nitrate industry, which has recovered boric acid from the iodine plant acidification step when the world price and the ore was favourable. It also has been used to process colemanite by decomposing the ore with SO_2, to form boric acid. This is done by grinding the ore finely, suspending the solids in a saturated boric acid solution, and then adding the gas. The reacted pulp is passed on into flotation cells, where a small quantity of frother is added, and a boric acid concentrate collected. The concentrate may be further cleaned, or dissolved in a hot, saturated boric acid, from which pure acid is crystallised by cooling.

Table 74.- GLASS FIBER COLEMANITE SPECIFICATIONS

	Maximum %	Minimum %
B_2O_3	44.0	36.0
SiO_2	2.5	-
Fe_2O_3	0.5	-
Al_2O_3	1.0	-
CaO	32.51	31.0
MgO	1.0	-
Na_2O & K_2O	1.0	-
SrO	1.9	-
SO_3	1.5	-
H_2O (free)	1.0	-
As (ppm)	100	-
MnO (ppm)	<10	-
Cr_2O_3 (ppm)	<10	-
Ni (ppm)	<10	-
Co (ppm)	<10	-
Cu (ppm)	<10	-
V (ppm)	<10	-

Chromite has been obtained by flotation from chrome ores employing a mixture of natural petroleum sulphonates and crude tall oil with fuel oil as collectors. The process consists of crushing and grinding the ore to flotation size, followed by scrubbing and complete desliming. The sands are then conditioned at 60 to 80% solids with the collector mixture. The reagent dose normally used is 500 to 1,000 g/t of a fatty acid-fuel oil mixture, plus 100 to

250 g/t of sulphonate. In some cases the addition of sulphuric acid to reduce the pH to around 5 will markedly improve the selectivity. With some lode ores, the addition of sodium fluosilicate, at the rate of 1 to 2 kg/t, will do the same.

If the ores contain considerable amounts of carbonates, these must be floated off first using sodium silicate and fatty acids in an alkaline circuit, following which the feed is deslimed and floated as above. If excess magnetite is present, this should be removed by magnetic separation, as it will float with the chromite.

Cement rock, if of poor quality, can be upgraded by flotation. The following is a good testing case history, reported by Holmes (1986), so is quoted in extenso.

American Cyanamide studied a low grade limestone assaying 56% $CaCO_3$, 20% volcanic glass, 5% feldspars, and lesser amounts of quartz, opaline silica, clay, biotite, hornblende and zircon. Liberation was only achieved after grinding to 82% minus 200 mesh and was found to be impossible to deslime due to losses (25 to 45%) of soft calcium carbonate to the slimes. On flotation it was found that the use of a synthetic modified sulphonate produced a very tough, difficult to break, froth, so testing was performed with a 1:1 mixture of heavy fuel oil emulsified with water. The collector emulsion, plus a frother, was then stage added in increments of 50 g/t for a total of 300 g/t. In addition, tests were run to determine the effect of preconditioning the pulp with sodium silicate and soda ash. Stage concentrates were kept separate to permit calculation of cumulative grades and recoveries.

All test runs produced acceptable concentrate grades (78% $CaCO_3$) and recoveries after cleaning the rougher concentrate. The best results were obtained at 3 minutes of conditioning with 1,000 g/t sodium silicate and 500g/t soda ash (pH = 9.3), followed by flotation with 300 g/t of sulphonate promoter, 300 g/t fuel oil and 150 g/t frother, stage added to the six flotation stages, obtaining an overall $CaCO_3$ grade of 81.4% and recovery of 92.6%.

Feldspar: glass and ceramic grade feldspars are produced from alaskite (a uniform granite-like ore), or from more isotropic pegmatites.

The alaskite ores in Spruce Pine, North Carolina, according to Redeker and Bentzen (1986), have the chemical and mineralogical composition shown in Table 75. The run-of-mine is crushed and wet ground and then rid of its minus 200 mesh fraction by cycloning and classifying. The fines, though high in feldspars, are discarded as waste.

Table 75.- TYPICAL COMPOSITION OF SPRUCE PINE ALASKITE

	Wt %	Mineral	Wt. %
Na_2O	5.1	Soda Feldspar	42.9
K_2O	3.4	Potash Feldspar	14.7
CaO	0.9	Lime Feldspar	6.4
Al_2O_3	15.4	Quartz	28.0
SiO_2	74.4	Muscovite	7.5
Fe_2O_3	0.4	Iron-garnet minerals	0.5
Loss on ignition	0.4	Clay	very low

The coarse product is differentially floated in three stages to remove the undesirable iron-garnet minerals and quartz. The first stage consists of 3 to 5 minute conditioning of a concentrated pulp (55 - 60% solids) at a pH of 2.5 to 2.7 with 750 g/t H_2SO_4, 125 g/t of tallow amine acetate, 500 g/t of fuel oil, and 50 g/t of frother. The pulp is diluted to 20 - 30 % solids, which increases the pH to 3.0 - 3.3, and floated for 2 to 3 minutes in wooden acid-proof Denver DR-type flotation machines. In North Carolina these consist of 50 cubic foot cells in a trough with five flotation mechanisms. The mica concentrate is cleaned and screened, and plus 80 mesh sold to mica producers for further processing. In the second stage the iron-garnet impurities and the remaining mica is floated with anionic collectors. The tails from the first flotation are cycloned to 65 - 75% solids and conditioned for 5 minutes, at pH 2.2 to 2.5, with 200 g/t of sulphuric acid, 250 g/t of petroleum sulphonate and 25 g/t of frother, then again diluted to 20 - 30% solids, which increases pH to the 2.8 - 3.2 range, and floated for 5 minutes. The concentrate of iron-garnet minerals joins the waste slime stream.

The third stage consists of the cationic separation of feldspar from quartz in a hydrofluoric acid circuit. The tails from the second circuit are again dewatered to 50 - 60% solids, and conditioned for 3 minutes at pH 2.5 with 600 g/t of HF, 200 g/t of tallow amine acetate, 50 g/t of kerosene, and 25 g/t of frother. The dilution to 20 - 30% solids increases pH to the 3.0 - 3.5 range, and the product is floated for 5 minutes making a glass grade feldspar concentrate ready for shipment. A small portion is further upgraded by dry grinding in pebble mills and iron scavenged with high intensity magnetic separation, prior to selling it to the pottery industry. The quartz tails can be cleaned by floating off the impurities, sized and dried for sale as a high purity product for use as a fused quartz (see Fig. 58).

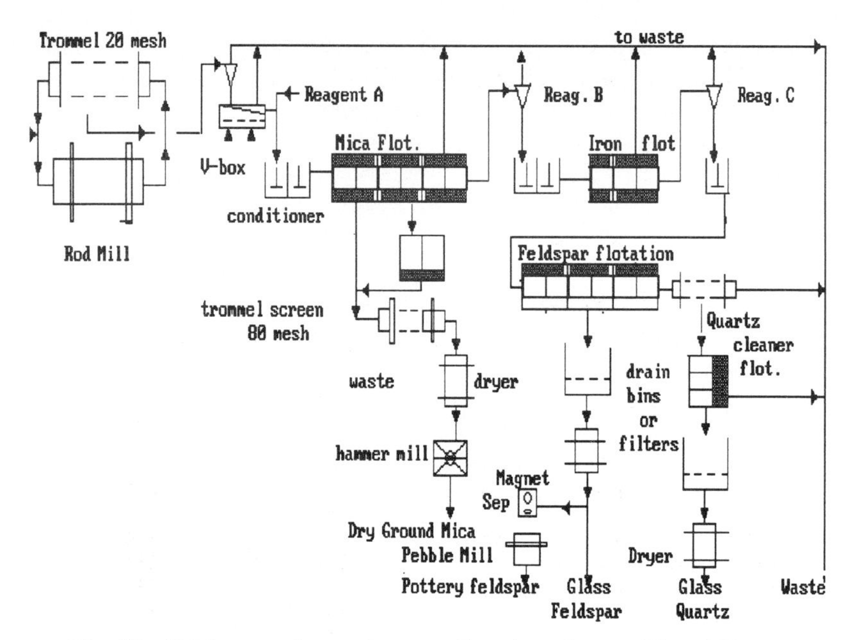

Fig. 58.- Feldspar, mica and quartz flotation, Spruce Pine, NC

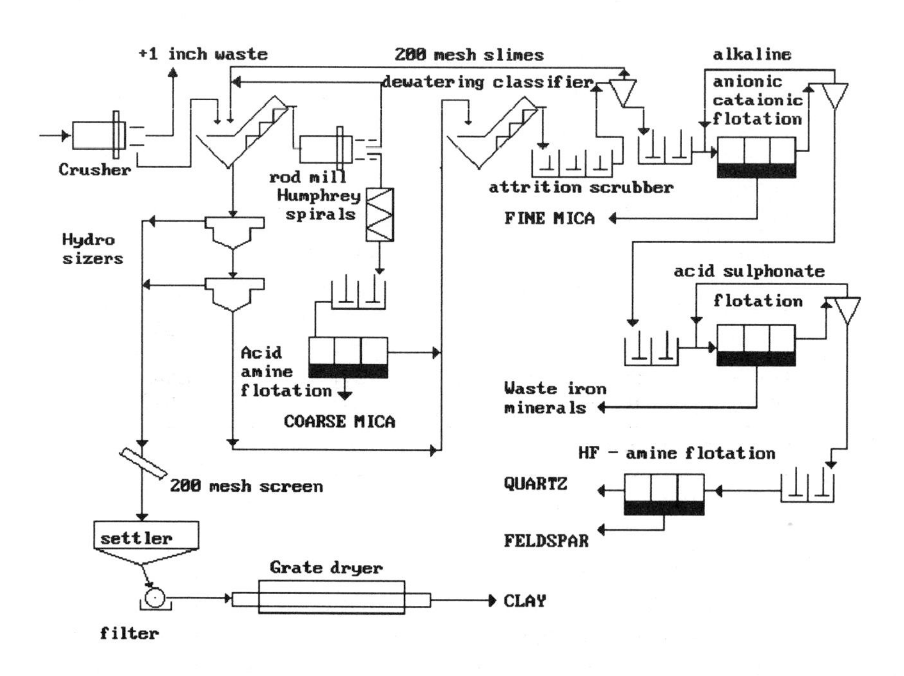

Fig. 59.- Weathered decomposed pegmatite processing, North Carolina

When the raw material is *weathered pegmatites*, the circuit used is similar, except that in the first stage, after wet grinding in a rod mill, a very coarse, fairly pure mica can be recovered with a trommel equipped with 1/8 " screens. The minus 1/8 inch material is passed to Humphrey spirals, where, because of the shape factor, the flat mica concentrates in the outer sides, where it is removed by splitters. The impure mica is cleaned by acid circuit flotation with oleylamine acetate as the collector (Fig. 59).

Table 76.- REAGENTS EMPLOYED IN THE FLOTATION OF FELDSPAR-MICA-QUARTZ-LITHIA IN 1985 (all values in thousands, except costs)

FLOTATION	TOTAL pounds	TOTAL value	Cost US$/t.ore
Collectors			
Unspecified amine	125	$117	0.483
Amine salt	287	252	0.333
Fatty and rosin acid soaps	1,000	350	1.500
Fatty acid derivative	251	107	0.877
Petroleum sulphonate	210	133	0.234
Fuel oil	886	96	0.668
Total	2,759	1,055	
Frothers			
Aliphatic alcohol	40	21	0.101
Pine oil	39	24	0.037
Polyglycol	8	1	0.060
Polyglycol ether	35	25	0.095
Petroleum based blends	48	34	0.334
Total	170	105	
Depressants			
Hydrofluoric acid	525	202	1.144
Lignin derivatives	35	1	0.300
Activator			
Hydrofluoric acid	599	271	0.813
pH regulators			
Sulphuric acid	1,743	56	1.543
Caustic soda	797	91	0.598
Flocculants			
Aluminium salts	351	10	0.140
Unspecified polymer	33	36	0.040
EFFLUENT TREATMENT			
pH regulators			
Caustic soda	719	14	3.420
Lime	1,299	55	2.421
Flocculants			
Unspecified polyacrylamide	5	6	0.040
Polyacrylate	55	52	0.253
Polyethyleneimines	11	9	0.800
Unspecified polymer	29	38	0.009

source:USBM

In 1985, according to the Minerals Yearbook, there were ten mills operating, with a daily capacity of 2.65 million tonnes/year of ore. Actual operations were at 71% of capacity. Concentrate produced amounted to 235,000 t of quartz, 550,000 t feldspar, 60,000 t mica, and an undisclosed amount of lithia. The reagents used are shown in Table 76.

Fluorspar: most fluorspars must be upgraded for marketing. Typically the ore is washed and sized in trommels. Minus 10 mesh is floated, while the coarser fraction usually is purified in a heavy media separation. The flotation feed is wet-ground to 100 mesh, and the pulp heated to 55 to 85°C and conditioned 5 to 10 minutes with 500 to 3,500 g/t of Na_2CO_3 and NH_4CO_3 to disperse the clays and slimes, plus 250 to 500 g/t of quebracho to depress carbonate minerals; at which point 250 to 1,000 g/t of fatty acid is added as the collector, and the pulp conditioned for an additional 5 - 10 minutes. Then the pulp is diluted to 25% solids, and the fluorspar is floated.

The concentrate then needs multiple cleanings prior to reaching specifications. The temperature is maintained during all the operation

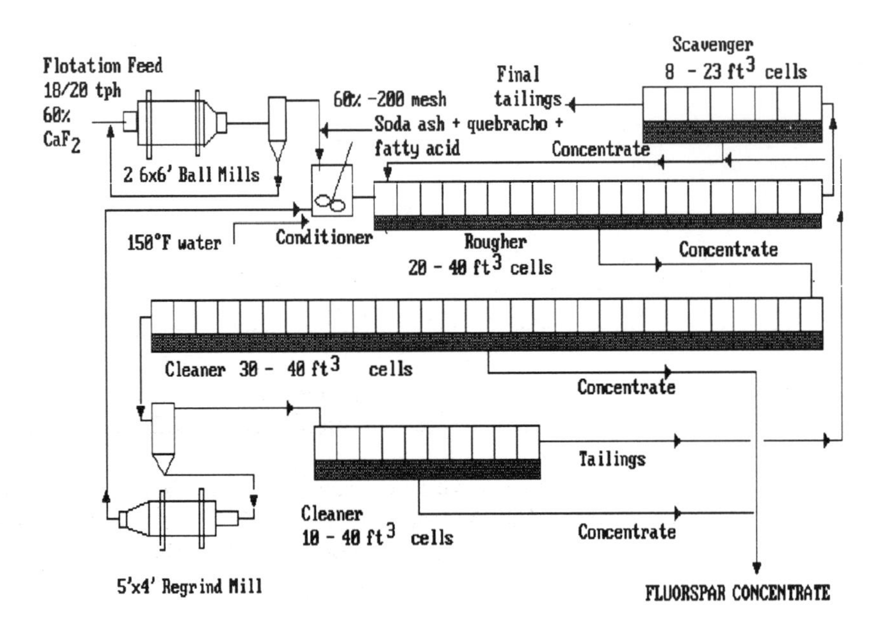

Fig. 60.- Fluorspar mill, Reynolds Mining Corp., Texas

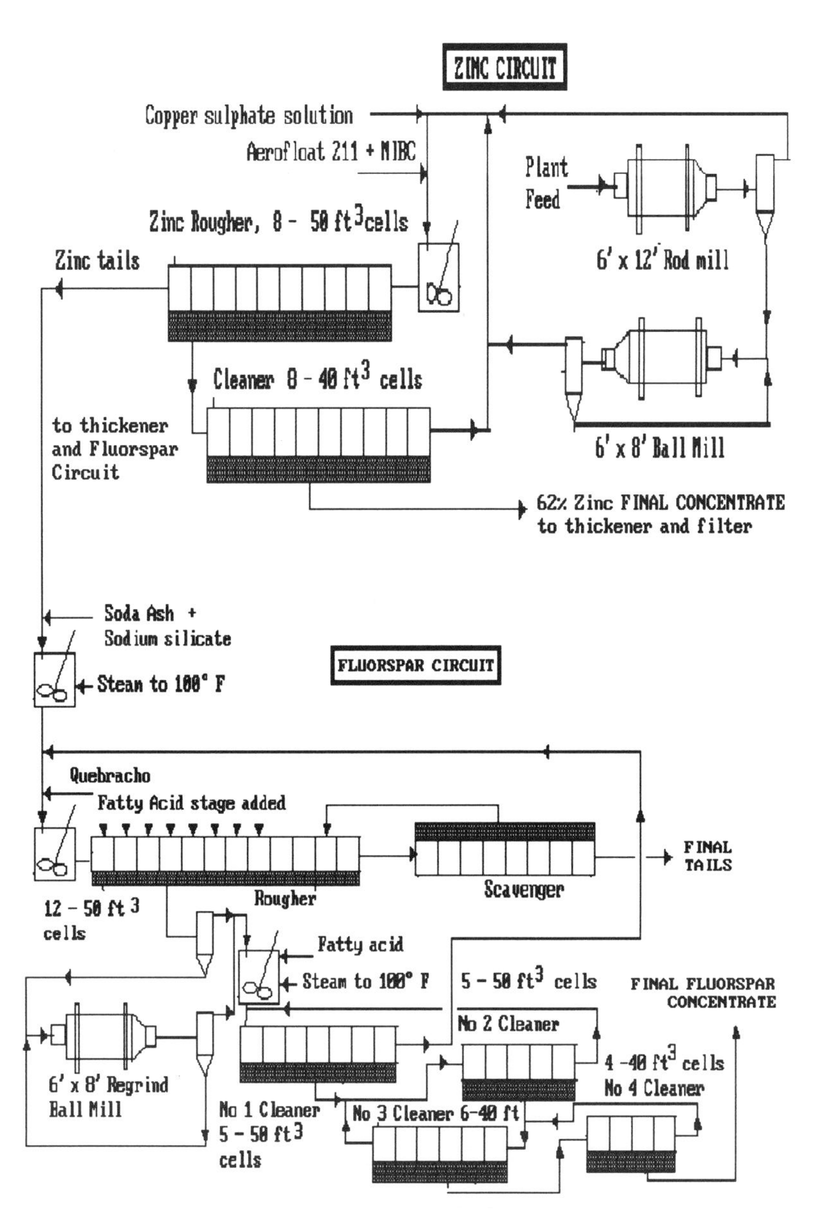

Fig. 61.- Mill circuit for a fluorspar-zinc heavy media concentrate (Reagents and conditions shown in Table 77)

through the injection of live steam. Laboratory tests indicate that a saponified fatty acid can be employed at a lower temperature if the gangue is siliceous rather than carbonaceous.

Table 77.- REAGENTS FOR FLUORSPAR, SULPHIDE AND BARITE FLOTATION

Reagent	Consumption g/t	Reagent	Consumption g/t
Lead-pyrite float		Zinc float	
Na ethyl xanthate	45	Na diisopropyl di- thiophosphate	150
MIBC frother	20	Copper sulphate	220
Zinc hydrosulphide	variable	MIBC frother	25
Fluorspar float		Barite float	
Soda ash	650 - 1500	Sodium cetyl sulphate	220
Sodium silicate	700	Stepanflote 24, Alcolac EC111, MIBC frother	45
Quebracho	700		
Starch	500		
Oleic acid collector	220		

Fluorspar is supplied in three grades: acid, ceramic and metallurgical; they must meet the specifications shown in Table 78.

Table 78.- FLUORSPAR GRADE SPECIFICATIONS

	Acid	Ceramic #1	Ceramic #2	Metallurgical
CaF_2	97	95 - 96	80 - 90+	60 (effective)
SiO_2	<1.5	<3.0	<3.0	<15
S	<0.1	-	-	<0.3
Fe_2O_3	-	<0.12	<0.12	-
Pb	trace	-	-	<0.5

% effective CaF_2 = %CaF_2 - 2.5 x % SiO_2

Glass sands: sand is ubiquitous, but rarely of a suitable quality even for use as aggregate in concrete, and certainly very rarely of the chemical quality required in glass making, quartz melts or as a flux in smelting. It is also a very large volume, low value, commodity. Normal processing requires washing and grading as a minimum.

Sands of a quality suitable for glass making are frequently obtained by purifying river or beach sand by reverse flotation. If the impurities include iron and heavy metals, gravity and magnetic methods to remove the impurities may be justified. When the impurities involved are carbonates and feldspars, these can be removed by flotation. Depending on the impurities present, glass sand flotation can be done either in acid or basic circuits. If the sand is to be floated in an acid circuit, which is usually the case when it is contaminated with high amounts of refractory heavy metals, it is normally attrition-scrubbed at 60 to 80% solids, followed by desliming. The flotation feed, also at over 70% solids, is conditioned with an acid at a pH of about 2 plus or minus 0.5, and with 250 to 1,000 g/t of petroleum sulphonate as the collector, depending on the amount of impurities present. At the end of the conditioning period 50 to 100 g/t of frother is added (mostly nowadays, a polyglycol ether frother is used), and fuel oil is added when slimes are excessive. The pulp is then diluted to 20 to 40% solids for flotation in conventional cells. The overhead contains the iron minerals, heavily stained quartz, and the heavy refractory minerals, and is usually sent to waste in most plants.

The alkaline flotation route is chosen when high carbonate content is the impurity, but again dependent on the impurities involved. The commonly used collector is a fatty acid. Conditioning is done at 65 to 75% solids, with the pH adjusted to about 9.5 plus or minus 0.5, using caustic soda or soda ash, and with about 500 g/t of fatty acid. At the end of the conditioning, a frother is added. Frothers used include pine oil, the water soluble glycols, or MIBC, with dosage a function of the frother and the sand quality - usually 25 to 50 g/t. Fuel oil can be used to help disperse the fatty acid and control the froth if the impurities tend to vary. Solubilizing the fatty acid with caustic has allowed conditioning at a lower pulp density than that used with raw fatty acid or fuel oil/fatty acid emulsions. Conditioning at too low a pulp density has been one of the major reasons for unselective quartz flotation with fatty acids, according to Redeker and Bentzen (1986). After conditioning, the pulp is diluted to 20 - 40% solids and floated. The overhead contains the carbonates, and is usually discarded.

If the sand is heavily contaminated, the quartz will be floated with a primary amine. High solids conditioning, at a neutral or slightly acid pH, is carried out with 50 - 200 g/t of the amine collector, followed by flotation.

The commercial sand specifications cover the size and the impurities contained. For containers and float/flat glass, the sand should be less 20 plus 150 mesh. Closer screening is required for fused and optical grades of glass. Chemical specifications are usually tolerances, rather than minimum values. Undesirable are colour producing metals, such as chromium, nickel, manganese, cobalt, and copper. Iron is less deleterious in small amounts for clear glasses, and actually desirable for amber and green glasses. Refractory heavy metals are always excluded.

Table 79.- SPECIFICATION FOR COMMERCIAL QUARTZ AND FELDSPAR

Composition and physical characteristics	Quartz Sands Foundry (1) (2)	Quartz Sands Glass (3) (4)	Feldspar Concentrate Glass	Feldspar Concentrate Ceramics
Total SiO_2 %	95 - 99	97 - 99	65 - 68	65 - 68
Quartz, %	90 - 96	96 - 98	NS	8.0 max
Al_2O_3 %	0.5	0.3	18.5 ± 0.6	18.0 min
K_2O + Na_2O_2, %	0.3	0.2	12.5 ± 1.0	13.0 min
Fe_2O_3 %	NS	0.3 - 0.03	0.1 max	0.08 max
LOI, %	0.2	0.5 - 0.2	0.5	0.4

In 1985, in the USA, there were eight mills upgrading sand by flotation. They processed 2.5 million tonnes of sand, and produced 2,163,000 t of clean sand; a concentration ratio of less than 1.2:1. Reagent use is shown in Table 80.

The following case history from Mathieu and Sirois (1984), describing a proposed Canadian operation for separating quartz from feldspars based on sea sands, shows the effect of conditioning times, collector dose and pH on the separation. Note that recoveries are highly sensitive to pH. The collectors used were petroleum sulphonate, tallow amine acetate, propane diamine dioleate, and a Hercules, Inc. experimental iron mineral collector identified as CX-62. Conditioning in all cases was done at 65% solids, and the flotations at 25% solids. The specification targets are detailed in Table 79.

Table 80.- REAGENTS EMPLOYED IN THE FLOTATION OF GLASS SAND IN 1985
(all values in thousands)

FLOTATION	TOTAL pounds	TOTAL value	Cost US$/t.ore
Collectors			
Diamine	91	$76	0.140
Fatty and rosin acid soaps	2,639	812	1.180
Petroleum sulphonate	143	35	0.420
Petroleum derivatives	596	70	0.885
Total	3,469	993	
Frothers			
Pine oil	4	2	0.007
Polyglycol	15	13	0.150
Petroleum based blends	3	3	8.400
Total	22	18	
Depressants			
Sodium silicate	20		0.100
Activator			
Sodium hydroxide	36	3	0.730
pH regulators			
Sulphuric acid	1,922	95	2.465
Caustic soda	1,885	173	1.425
Dispersants			
Sodium Silicate	344	24	3.440
EFFLUENT TREATMENT			
pH regulators			
Sulphuric acid	210	7	0.563
Flocculants			
Unspecified polymer	20	17	1.150

Source: USBM

The schematics show the laboratory lock tests employed and the flow sheet of the pilot plant. Tables 81 and 82 detail the experimental design and results.

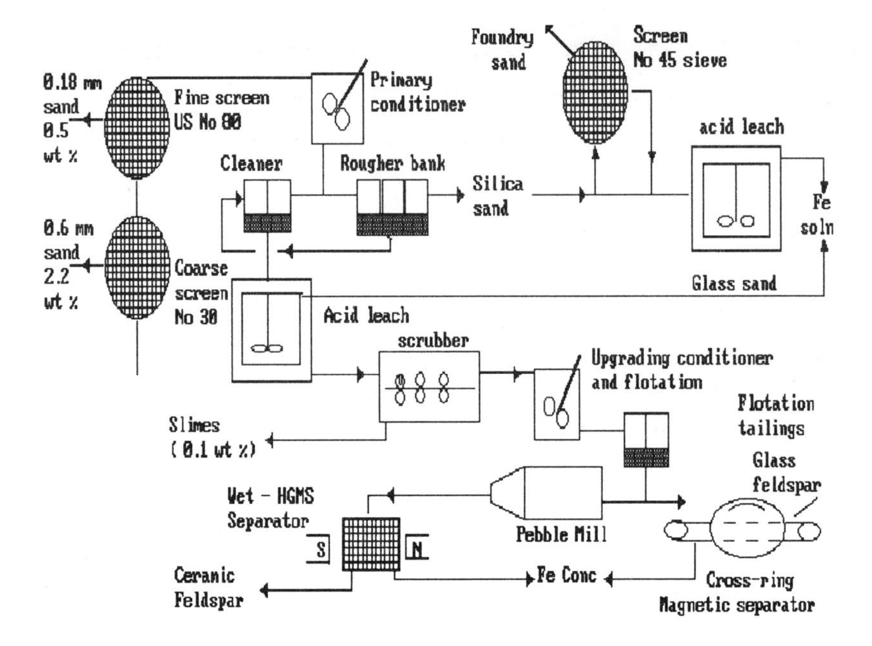

Fig. 62.- Pilot plant flowsheet for sand beneficiation

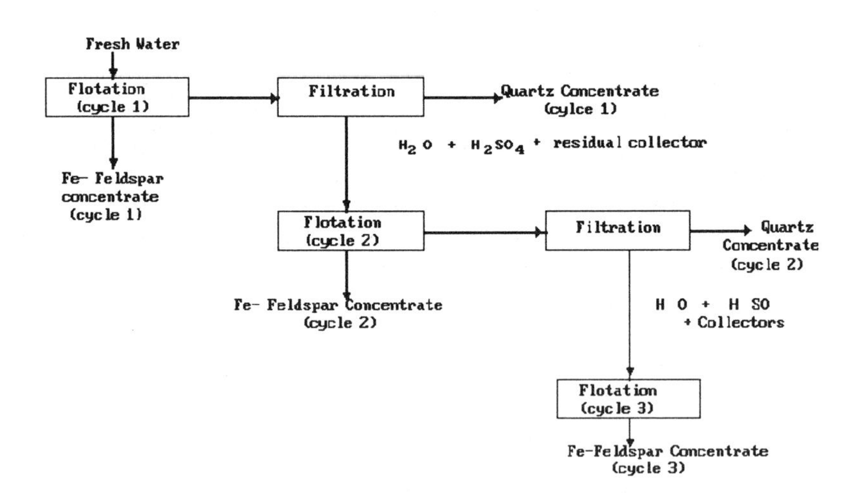

Fig. 63.- Lock cycle tests and pilot plant - Glass sands

Table 81.- EXPERIMENTAL DESIGN AND RESULTS - GLASS SANDS

Process: Feldspar-iron flotation: single stage versus two stage

Operation	Time min	Density % Solids	HF	Reagents kg/t H_2SO_4	Collector	pH
Test 1						
Fe conditioning	5	65	-	1.2	1.2 PS	2.1
Fe flotation	4	25	-	-	-	2.4
Feldspar conditioning	5	60	1.0	0.2	1.3 TAA	2.2
Feldspar flotation	4	24	-	-	-	2.5
Test 2						
Fe conditioning	5	65	-	-	0.5 CX	6.9
Fe flotation	3	25	-	-	-	6.8
Feldspar conditioning	5	60	1.0	1.1	0.5 PDD	2.0
Feldspar flotation	3	24	-	-	-	2.5
Test 3					PS + PDD	
Fe conditioning	5	65	-	1.3	0.3 + 0.45	2.9
Fe flotation	3	25	-	-	-	2.4

Reagents and Conditions

PS = petroleum sulphonate; TAA = tallow amine acetate
PDD = propane diamine dioleate; CX = Hercules CX-62 iron collector

Test	Products	Wt%	Analysis, % Feldspar	Fe_2O_3	Quartz	Distribution Feldspar	Fe_2O_3	Quartz
1	Fe Conc	1.6	10.4	4.90	79.7	1.7	34.0	1.4
	Feldspar	16.0	55.2	0.49	42.7	89.1	34.0	7.7
	Quartz	82.4	1.1	0.089	98.2	9.2	32.0	90.9
	Feed (calc)	100	9.9	0.23	89.0	100.0	100.0	100.0
2	Fe Conc	1.0	13.1	5.55	78.1	1.3	27.3	0.9
	Feldspar	16.4	55.1	0.43	43.0	88.2	34.6	7.9
	Quartz	82.6	1.3	0.094	98.1	10.5	38.1	91.2
	Feed (calc)	100	10.2	0.20	88.9	100.0	100.0	100.0
3	Fe/feldspar	17.4	54.0	0.81	43.8	90.1	64.2	8.6
	Feldspar	82.6	1.2	0.096	98.1	9.9	35.8	91.4
	Feed (calc)	100	10.1	0.22	100.0	100.0	100.0	100.0

Graphite: graphite ores are classified as flake, vein and amorphous. The most valuable, flake graphite, is found in metamorphosed feldspars. To process it, it must be carefully ground to liberate, but not degrade, the coarse flakes. It is naturally hydrophobic, so was historically the first ore ever concentrated by flotation (the Bessel brothers in the 1880s). Despite the natural floatability, the separation of feldspar, quartz, mica and marble gangue is normally improved by the addition of a small amount of kerosene, and floated with pine oil or MIBC as the frother. Selectivity is improved by adjusting the pH with caustic soda to the 7.5 to 8.5 range. Cleaning with regrinding usually is ineffective because the graphite will smear over the surface of the gangue on grinding, and make it floatable. Because of this, and the natural interleaving of mica with the flake graphite, it is not possible to produce grades of over 95% by flotation. Super purity graphite is made by chemically leaching out the impurities from the concentrate.

Table 82.- EXPERIMENTAL DESIGN - GLASS SANDS FLOTATION - EFFECT OF VARIABLES ON FELDSPAR-IRON FLOTATION FROM QUARTZ

Constants	Variables		Results				
			Quartz Conc		Feldspar-iron concentrate		
					Feldspar		Fe_2O_3
			SiO_2 %	Recov %	Content %	Recov %	floated %
I		2.0	97.6	96.5	73.0	85.4	53.4
Collectors		2.1	97.8	95.5	68.1	86.9	56.9
PS, 250 g/t	pH	2.2	97.9	94.6	64.7	87.6	54.0
PDD, 500 g/t		2.3	98.0	92.7	58.3	88.6	59.0
Conditioning time, 5 min		2.4	98.1	88.7	50.2	90.8	65.7
II	Condi-						
PS, 150 g/t	tion	3	98.3	89.1	50.6	92.0	60.0
PDD, 400 g/t	min-	5	98.1	95.3	66.9	89.6	-
pH 2.2	utes	8	98.0	96.5	68.9	87.2	-
		10	98.4	93.1	57.3	91.2	54.0
III	Col-	0.45	97.4	97.3	76.4	86.7	55.2
	lect-	0.55	98.1	94.9	67.1	89.5	57.2
	or	0.65	98.2	93.0	59.3	90.0	62.4
Conditioning	dose	0.75	98.2	90.1	52.5	90.6	64.1
time, 5 min	kg/t	0.85	98.3	87.7	46.6	91.7	68.6
	with	0.75	98.1	91.5	55.9	90.1	59.6
	re-	0.65	98.2	90.7	53.9	90.6	63.2
pH 2.2	cycl-	0.55	98.0	91.0	54.9	90.6	60.3
	ing	0.45	98.2	90.5	53.4	91.0	60.8
		0.45	98.0	90.9	54.5	89.6	59.9

Gypsum rarely has been concentrated by flotation industrially. In the laboratory it is easily floatable using sodium oleyl sulphate, fatty acid sulphonates, sodium taurate, and primary and secondary amines. Alum is reported to help selectivity. If gypsum is to be depressed, it can be done with sulphuric acid, gelatin, or tannic acid, depending on the circuit employed.

Iodine is produced in Chile from the strong sodium nitrate mother liquor, which contains over 500 g/l of salts. The iodine is cut with SO_2 in a packed absorption tower, followed by a reactor, where the effluent of the packed tower is mixed with fresh mother liquor. The reaction between the iodide from the SO_2 tower with the iodate in the original mother liquor produces metallic iodine. The effluent from the reactor is a dilute iodine pulp which is fed to standard small Denver rubber lined cells, without the use of reagents. The concentrate is washed with water and the pulp melted under pressure to purify the iodine. The only operating problems encountered are when the nitrate ores are high in boron minerals which also react with the SO_2 and are converted to boric acid. Boric acid will float without reagents, and contaminate the iodine. When the ores treated are very high in borax, the effluent from the tower can be sent to flotation to recover boric acid, prior to entering the cutting tank where the solid iodine is precipitated.

Kaolin: of the clay minerals, kaolinite, which is used as a filler for paper, as well as in the manufacture of porcelain, is the main component that requires extensive treatment to meet selling specifications. Clays are normally washed, suspended in water, and "fractionated" into two sized fractions by means of centrifuges or hydrocyclones. The fine white material is used for paper filling. This fraction is frequently contaminated with 2 or 3 % of anatase (TiO_2), which gives the product a yellow tinge which cannot be bleached. To remove the anatase, a process known as ultra-flotation is employed, where the anatase is removed by "piggy backing" on a second mineral, which adsorbs it. The second mineral is usually ground limestone. The mixture is floated with the usual limestone reagents (soda ash, fatty acid, and fuel oil), and the resultant limestone concentrate removes all the anatase. Kaolinite clays with a brightness of over 90 are produced by this method.

Kyanite is an aluminium silicate used primarily in the production of steel furnace refractories. To upgrade the mineral, it is wet ground in rod or ball mills to 45 to 75 microns (200 to 325 mesh), deslimed, thickened, conditioned and floated. To remove the pyrite, which is sold as a colouring agent in the glass industry, a xanthate rougher and cleaner flotation is employed. Conditioning is done at a neutral pH with 50 g/t of ethyl xanthate, and then 25 g/t of pine oil is added as the frother.

A second flotation step is used to remove the small amount of residual clay, mica and pyrophyllite contained in the kyanite. The pulp is conditioned with 50 to 100 g/t of cocoamine acetate, at a neutral pH, and the contaminants floated off.

The final step in the purification process is a bulk flotation of the kyanite. To do this the pulp is thickened to 70% solids and given intense conditioning with 1 to 1.5 kg/t of petroleum sulphonate at a pH of 2.3 to 2.6 (sulphuric acid). The pulp is diluted to around 30% solids and the kyanite is floated and given 2 or 3 stages of cleaning, adding more sulphuric acid to keep the pH below 2.6. The concentrate is conditioned with sodium silicate to remove the flotation agents, dewatered and dried. Dry magnetic separation of the kyanite removes iron minerals and iron stained material which the petroleum sulphonate also activates. Overall kyanite recovery is 80 to 85% of the feed.

Phosphates: most of the phosphate used for fertiliser production is obtained from phosphate rock which is found as unconsolidated or consolidated sedimentary surface deposits composed of coarse calcium phosphate pebbles mixed with fine grains, all intermixed with layers of sand and clay. Many deposits contain slimes and contaminant iron phosphates, alumina and other foreign matter. Hard rock deposits generally are of a sufficiently high grade that they do not need to be concentrated for sale.

As is usual with industrial minerals after mining, which is frequently by dredging, or hydraulically, the run of mine is first screened to eliminate inert coarse material, then washed to get rid of as much slimes and clay as possible, and crushed and ground to liberate the mineral. The usual mineral in Florida is collophanite, a fluorapatite carbonate with substitutions in the lattice, which contains 31 to 36% P_2O_5.

The phosphate rock is separated from the diluting quartz by selective flotation with fatty acids, derived from tall oil, a by-product of the paper industry. To properly disperse the fatty acid it can be presaponified with caustic or ammonia, or this mix is added in high per cent solids to the conditioning step. Fuel oil is also added to reduce the amount of high cost fatty acid required to have good recoveries. The primary phosphate flotation is not sufficiently selective to eliminate all the stained quartz and slime contaminated gangue. The cleaner steps are more complicated than for other minerals because of surface coating and activation of the gangue, therefore it is necessary to strip off the collector from the primary flotation by attrition with sulphuric or fluosilicic acid and float off the quartz in cleaner, using an amine collector.

Reagent usage in the primary flotation is 400 to 1,000 g/t of fatty acid, about a tenth of that amount of caustic or ammonia to obtain pH of 8 - 9 in the conditioner, plus 500 to 2,000 g/t of fuel oil. Conditioning time is 5 to 8 minutes, and flotation 3 to 6 minutes at 20 - 30% solids. The de-oiling step requires sufficient sulphuric or fluosilicic acid to wash at pH 3- 4. The cleaner quartz flotation requires 50 - 100 g/t of amine collector, plus 25 to 150 g/t of kerosene at the head of the cells, and 3 to 6 minutes flotation time at 25 to 30% solids.

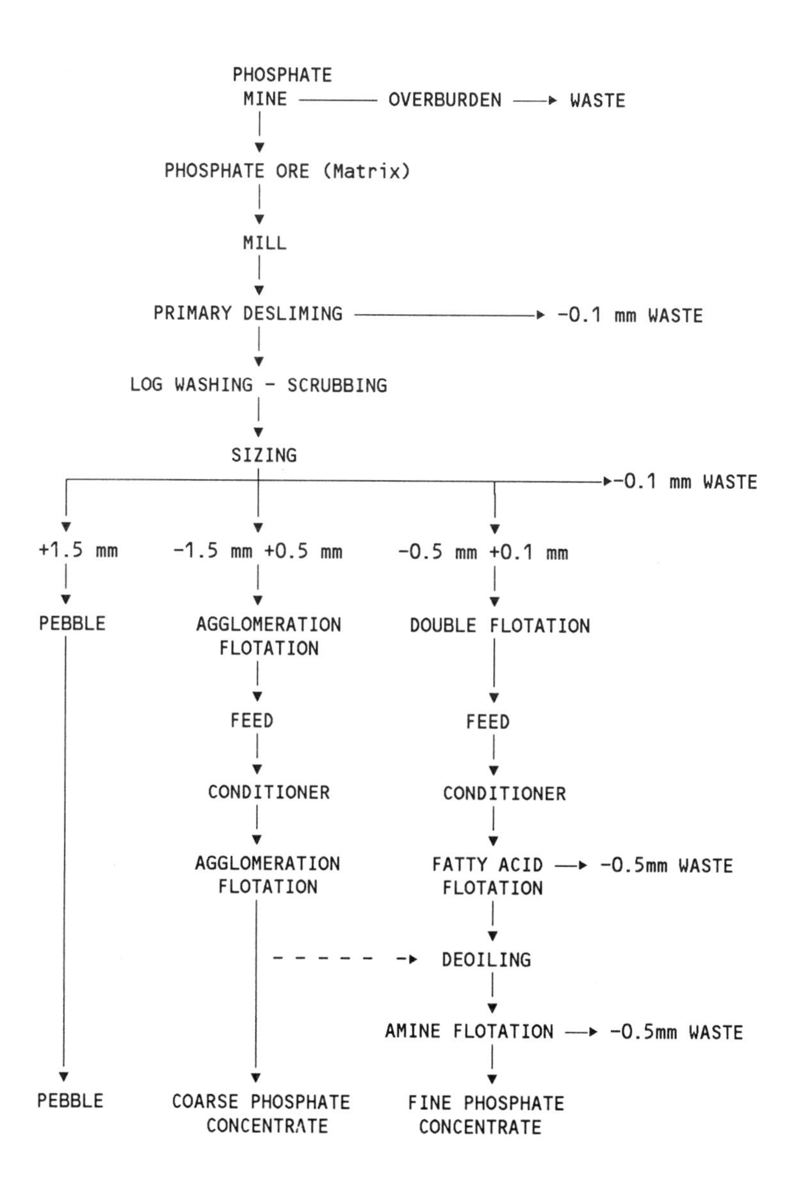

Fig. 64.- Flow diagram for a phosphate rock plant

Table 83.- REAGENTS EMPLOYED IN THE FLOTATION OF PHOSPHATE ROCK
(all values in thousands)

	TOTAL		Cost
	pounds	value	US$/t.ore
Collectors			
Secondary amine	4,847	$1,600	0.223
Unspecified amine	4,800	2,880	1.200
Amine derivative	6,885	2,444	0.221
Quaternary ammonium salt	1,644	553	0.319
Fatty and rosin acid soaps	151,447	23,015	1.419
Fatty acid derivative	10,770	1,268	0.121
Petroleum sulphonate	110	50	0.300
Petroleum derivatives	676	68	0.074
Fuel oil	130,943	12,521	1.268
Kerosene	8,863	1,025	0.170
Total	275,410	47,950	
Frothers			
Polyglycol ether	82	41	0.019
Unspecified polyol	13	11	0.400
Petroleum based blends	50	32	0.200
Total	145	84	
Depressants			
Sodium silicate	1,170	139	0.190
Starches, cellulose	374	52	0.350
pH regulators			
Sulphuric acid	134,866	3,398	1.320
Caustic soda	5,649	455	0.526
Ammonia	28,362	2,626	0.294
Flocculants			
Anionic polyacrylamide	19	17	0.003
Dispersants			
Sodium silicate	126	19	0.009

During 1985, in the USA, there were 22 flotation plants in operation at an average of 75.4% of capacity. They treated 109,715,455 t of phosphate rock grading 9.94% P_2O_3, using 3,480 gallons of water, 12.3 kW-hrs of power, and consuming 0.54 lbs of grinding balls per tonne of ore processed. Phosphate produced was 24,620 tonnes of 31.21% P_2O_3. Reagent consumption is shown in Table 83.

Potash: the most common potash ore is sylvite, which consists of a mixture of potassium and sodium chloride, usually assaying about 23% KCl, 72% NaCl and 4% others, which includes clays, iron oxide and silica. The separation of salt from KCl by flotation is relatively straightforward. In the early process the NaCl, activated by lead or bismuth ions, was floated off with anionic organic reagents and fatty acids, the KCl reporting to the tails. Concentrates containing over 60% K_2O could be obtained, but at a cost of rather high reagent consumption.

Most current plants are doing the opposite, floating off the KCl (sylvite) and leaving the halites in the tails. The run-of-mine ore is crushed and classified dry, then wet ground in a saturated NaCl and KCl solution, down to liberation size (usually minus 4 plus 65 mesh). In the conditioning step, clay depression reagents such as starch are added at a dosage dependent on the ore grade, then the pulp is conditioned with the flotation reagents at 50 to 75% solids. The normal collectors are primary aliphatic amines of about C_6 to C_{24} carbon chain lengths, used at a rate of 250 to 1000 g/t. Polyglycol ether frothers are common, with fuel oil added to control the froth. Flotation is then carried out at 20 to 35% solids, for 3 or 4 minutes, after which the concentrate is cleaned. It is washed in a saturated KCl solution, and/or water, and dried.

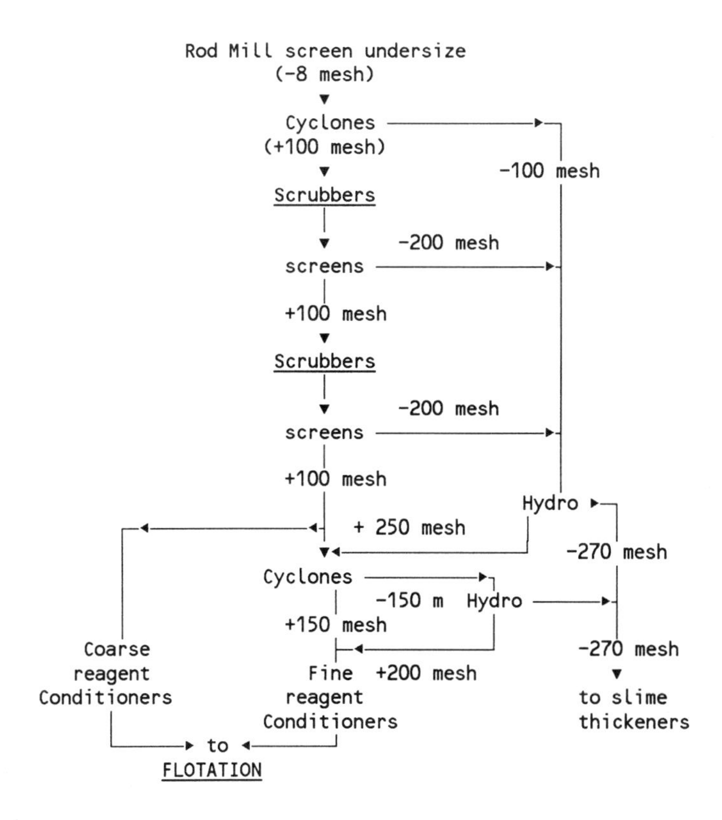

Fig. 65.- Potash scrubbing-desliming-reagent conditioning system

Most important in amine flotation of KCl is good scrubbing, desliming and depression of the clays. The pulp is vigorously scrubbed in agitated tanks or tumblers, to remove clay particles from the crystal surfaces, and to disperse the clay in the brine. Ores with over 3% slimes are scrubbed in stages, with each stage followed by a screening to remove the slime-laden brine. The screen oversize is repulped with clear brine at each stage. An optimum power input must be determined and pulp densities are critical in effective scrubbing, as dilute pulps reduce particle contact. The presence of carnallite produces problems as, when it dissolves, it adds magnesium chloride to the brine, and causes partial crystallisation of KCl and NaCl. If carnallite content fluctuates when the low cycle occurs, the chlorides tend to crystallise on the original crystal surfaces, while during the higher carnallite content period the recrystallised chlorides tend to remain in the slurry as extremely fine crystals that will be lost in the desliming steps. The slime removal circuit is complicated, as can be seen from Fig. 65, which shows a typical procedure.

Potash collectors, according to the SME Handbook, are primary amines produced from beef tallow, which is a mixture of palmityl, stearyl, and oleyl amine, with alkyl groups containing 16, 18 and unsaturated 18 carbon-chain atoms. Hydrogenation converts the oleyl to stearyl amines, saturating nearly all the groups, which increases hydrophobicity and surface adhesion. Another quality on the market is distilled tallow amines. The final collector is produced by converting the amines to acetates or hydrochlorides, which in the flotation plant is neutralised to 90 - 97% with caustic.

Slime depressants in common use include starches, guar gum, dextrines, and synthetic polymers, such as polyglycols and polyacrylamides. High molecular weight polyglycols are effective, but affect frothing. Most amines in commercial use contain small amounts of low molecular weight amines, such as capryl, lauryl and myristyl, which provide frothing. MIBC, in potash flotation, can be used when there is over-frothing, to provide more selectivity. Typical reagents used (according to the SME Handbook), and their consumption, are shown in Table 84.

Table 84.- REAGENTS USED IN POTASH FLOTATION

Dispersants, silicates, etc.	50 - 100
Slime depressants, starches	450 - 1000
Others	70 - 120
Collectors, amines	50 - 150

There were five potash flotation plants in the USA in 1985, operating at 70% of capacity. Ore treated had 11.3% K_2O and amounted to 10,090,900 tonnes, from which 1,517,270 tonnes of 59.15% potash was produced. The flotation reagents consumed are shown in Table 85.

Table 85.- REAGENTS EMPLOYED IN THE FLOTATION OF POTASH IN 1985
(all values in thousands)

	TOTAL pounds	TOTAL value	Cost US$/t.ore
Collectors			
Primary amine	675	$454	0.138
Quaternary ammonium salt	230	83	0.069
Fatty and rosin acid soaps	990	198	0.350
Petroleum derivatives	26	10	0.060
Fuel oil	1,953	213	0.425
Total	3,874	958	
Frothers			
Aliphatic alcohol	525	294	0.054
Polyglycol ether	1	1	0.002
Petroleum based blends	246	39	0.112
Total	772	334	
Depressants			
Gums and dextrans	1,732	332	0.208
pH regulators			
Hydrochloric acid	156	6	0.059
Flocculants			
Anionic polyacrylamide	46	42	0.008
Unspec. polyacrylamide	17	31	0.009
Unspecified polymer	424	215	0.192

Scheelite: flotation concentration of this tungsten mineral is typically with a high grade fatty acid collector, though at several plants the fatty acid is supplemented with synthetic modified sulphonates. Holmes (1986) reports improved results with the use of 100 g/t of sulphonate collector to replace the normal 400 g/t of oleic acid. The sulphonate collector was emulsified with No. 6 fuel oil at a 1:1 ratio in water. A grind of 75% minus 200 mesh was followed by a bulk sulphide float in which the pulp was conditioned at 30% solids with 2,000 g/t of sodium silicate followed by a two stage addition of 100 g/t of emulsion per stage, as well as 7.5-15 g/t of glycol frother. The pH remained at 9.7-10.0 throughout roughing and cleaning. The results are shown in Table 86.

Table 86.- SCHEELITE FLOTATION RESULTS

Product	Weight %	Assay WO_3 %	% Recovery WO_3
Feed	100.00	0.54	100.00
Sulphide conc.	4.66	0.41	3.52
WO_3 concentrate	1.88	25.24	87.58
Middlings	6.92	0.32	4.10
Tails	86.54	0.03	4.80

The competing concentrate recovered with 400 g/t of oleic acid was only 7 to 11% WO_3, at the same recovery level. Holmes notes that grinds as fine as 95% minus 200 mesh may be necessary to provide adequate liberation of the scheelite. Sodium silicate additions of 500 to 2,000 g/t to the rougher and each cleaning stage may be necessary for gangue depression and slime dispersal. If carbonate gangue is present, quebracho (10 - 50 g/t) may be needed to depress the carbonate in the cleaners, though care must be taken not to over-dose as quebracho affects sulphonate collection power. If the process water is hard, soda ash should be used to reach the operating pH of 8.5 to 10.5.

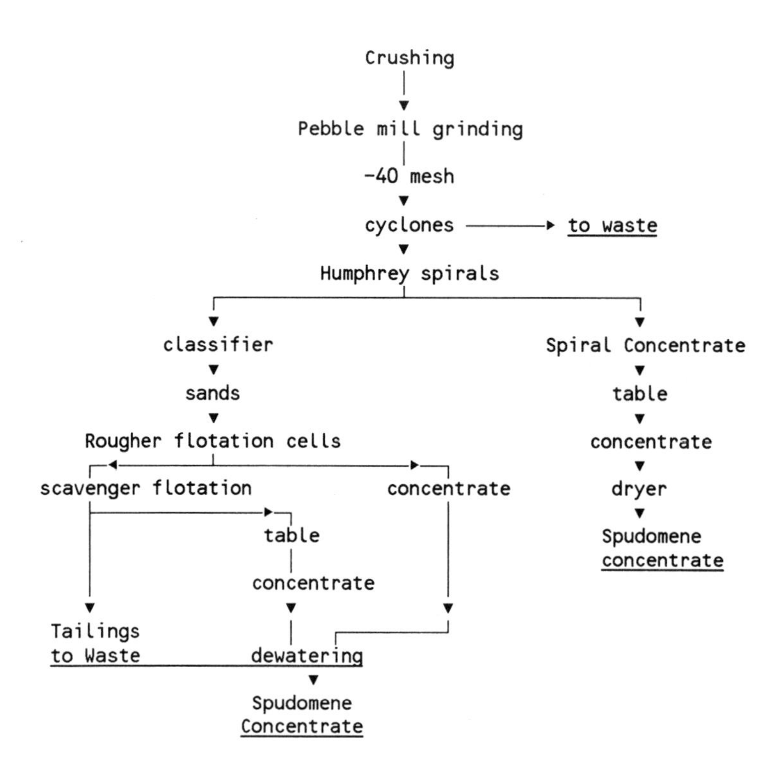

Fig. 66.- Foote spudomene flotation circuit

Spudomene and co-products (Pegmatite ores): lithium is obtained from its ores only in the U.S.A., otherwise the lithia concentrate is marketed directly for use in the glass industry by operations in Australia, Canada, and Africa. There were two spodumene producing operations in America: Foote Mineral, at King's Mountain, North Carolina, utilising a combination of gravity separation and flotation to make a coarse spudomene concentrate from spirals, and a fine concentrate from flotation (see Fig. 66 for a simplified schematic); and Lithium Corporation of America, at Hill City, South Dakota, recovering spudomene, mica and a feldspatic sand. A typical mineral assay of the Kings Mountain (Fig. 67) spudomene ore is summarised in Table 87.

Table 87.- MINERALOGICAL ANALYSIS OF KINGS MOUNTAIN SPUDOMENE ORE

Mineral	Wt. %
Spudomene	15 - 25
Potash feldspar	12 - 15
Soda feldspar	28 - 33
Quartz	25 - 35
Muscovite and others	5 - 15

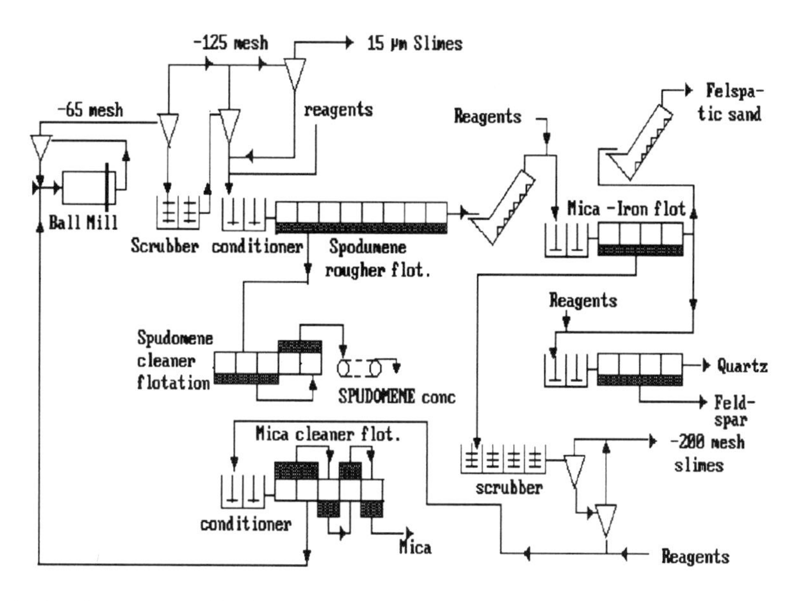

Fig. 67.- Spudomene and feldspar-quartz-mica flotation, Kings Mt, NC

In both processes the ore is wet-ground to flotation size and about 250 g/t of NaOH added to the ball or pebble mill feed. The ball mills operate on a closed circuit with cyclones delivering a -65 mesh product to the process. The pulp is then diluted and recycloned to remove the -400 mesh fines.

Table 88.- FELDSPAR, QUARTZ, SPUDOMENE, MICA AND CLAY MINERALS RECOVERED AND SOLD FROM NORTH CAROLINA PEGMATITES IN 1977

	Glass Spar		Pottery Spar			Feld spatic	Spudomene			Quartz Sand		Clay		Ground Mica	
Trade name	F-20	K-40	NC-4	K-200	Foote Spar	sand	Glass Chem.	Ceramic	Low Iron	Glass	K-30	Hallosite	KM-25	P-80 dry	Oil Well
						Chemical Analysis, weight %									
SiO_2	68.00	67.10	68.15	67.10	68.70	79.20	63	64.12	64.80	92-98	99.03	47.90	53.8	44-48	44-48
Al_2O_3	19.00	18.30	18.88	18.30	19.27	12.10	24.7	26.50	26.25	0.5-3.0	0.23	36.91	32.10	31-38	31-38
Fe_2O_3	0.07	0.07	0.07	0.07	0.06	0.06	1.7	0.9 max	0.10	0.1-0.15	0.20	0.30	1.40	5-max	5-max
CaO	1.85	0.36	1.60	0.36	tr	0.52	tr	tr	tr	-	tr	na	tr	-	-
MgO	tr	tr	tr	tr	tr	tr	tr	tr	tr	-	tr	na	tr	-	-
K_2O	3.75	10.10	4.50	10.10	3.69	2.62	0.50	0.27	0.14	0.5-2.0	0.23	1.05	2.20	9-max	9-max
Na_2O	7.15	3.80	6.70	3.80	7.91	4.80	0.30	0.32	0.35	-	0.08	0.25	0.13	0.5-2.5	0.5-2.5
Li_2O	-	-	-	-	0.08	0.35	5.6-6.5	7.22	6.83	-	-	-	0.08	-	-
Ign loss	0.13	0.26	0.10	0.26	0.25	0.35	0.30	0.50	na	wet	0.11	13.50	10.30	4-5.5	4-5.5
						Approximate Mineral weight %									
Potash spar	22	60	26	60	22	16	3	2	1	2	-	3	5		
Soda spar	61	32	57	32	67	41	3	3	3	4	-	2	1	5	5
Lime spar	9	2	8	2	-	2	-	-	-	-	-	-	-		
Quartz	8	5	9	6	11	41	9	5	6	90-95	+9 9	2	10	5	5
Spudomene	-	-	-	-	-	-	85	90	90	-	-	-	-	-	-
Clay Minrls	-	-	-	-	-	-	-	-	-	-	-	90	80	10	10
Muscovite	-	-	-	-	-	-	-	-	-	-	-	3	4	80	80
Tyler mesh															
+20	0.1	-			0.1		-		-	-	-	-	-	-	-
-48	42.0	-			na		1.5		-	-	43.2	-	-	-	-
+100	85.5	56.15			35.1		20.0		0.5	-	82.1	-	2.2	1-2	10-70
+200	97.2	90.05			75.1		50.0		90.0	100	97.5	tr	14.4	13-18	70-90
+325	-	-			-		-		-		-	1.0	30.1	78	10-30

After a second thickening, scrubbing and recycloning to remove fines, the main product is thickened. The -400 fines are deslimed and sized in hydrocyclones to a 15 micron cut. The -15 goes to waste and the +15 is sent to conditioning. Conditioning, with 750 g/t of tall oil fatty acid (having 5 - 7% rosin acid content), plus a small quantity of polyglycol frother, is done at 55% solids. This pulp is then diluted and floated at 30% solids, and cleaned and recleaned several times.

Flotation tailings are dewatered and conditioned for iron mineral and mica removal by flotation with petroleum sulphonate in an acid circuit. The low iron feldspatic sand is sold to glass makers. Flotation cleaned white mica is also produced from the previous flotation's concentrate. Typical specifications (Redeker and Bentzen (1986)) are shown in Table 88.

Talc is found in sedimentary rocks altered by regional or contact metamorphism, and in ultrafine igneous rocks which have been serpentinised. Associated minerals include serpentine, quartz, calcite, dolomite, magnetite, and in some cases nickel and iron sulphides.

Talc ores are ground to 65 - 200 mesh for flotation. Although in a pure form talc has natural flotation properties, its ore requires differential flotation, so it is usual to add a frother and fuel oil to the flotation cells in staged small quantities. If only carbonate gangue is present, small additions of amine collectors can increase talc recovery. If the ore contains quartz, mica, etc., amines are to be avoided. When nickel and iron sulphides are present, these can be depressed with citric acid. If graphite is present, the talc can be depressed by raising the pH to over 12, and graphite floated.

Magnetic separation can upgrade the brightness of talc, and is necessary if nickel and chromium is a contaminant.

Vermiculite is an interesting curiosity because it at times is upgraded by cationic table flotation - a turn of the century process.

Wollastonite is a calcium metasilicate with a density of 2.8 - 2.9, used as a filler in paint and tiles. To produce a white fibrous material suitable for the market, it is upgraded by:

1.- removal of heavy minerals by wet or dry magnetic separation
2.- removal of calcite by anionic flotation
3.- removal of quartz and silicates by cationic flotation

Chapter 13

FLOTATION TESTING

Planning a test

The simplest studies of the floatability of an ore involve at least 25 clearly identifiable variables. A full-scale study of the processability of a mineral will require consideration of at least 100 variables. General knowledge of the interaction between these variables is sparse, therefore care in formalising the experimental design is essential, rather than just desirable. A flotation recovery optimisation experiment, such as one in which the only change studied is the collector dosage, is a waste of time, because the variable studied is not independent of other variables held constant, so the effect on mineral recovery detected can be dangerously deceptive. The key problem in designing all good flotation experiments is to see that the important interactions between the variables are reliably identified.

Tests to evaluate mineral dressing schemes are expensive. The number of independent variables involved is vast. Table 89 lists typical flotation affecting variables such as depressant, modifier, collector, etc. In fact, each designation hides a group, rather than an individual variable, so each item may involve 5 to 10 controllable conditions; therefore, what looks like a relatively short list, is in fact well over a hundred independent variables. Handling this great number of potential variables requires careful experimental design to reduce the number of experiments required for unambiguous results, so careful design is essential to detect all the interactions; if this is not done, the data generated can calculate out "not statistically significant" because of the noise in the system from the random variations in the ore sample. Thus, the importance of designing the experiments to maximise interaction information cannot be over emphasised.

The simplest interaction between variables, frequently overlooked in testing, is the one between the frother and the collector. One reason is that it is practically impossible to obtain reliable scale-up of data on the effect of a frother change from any type of laboratory testing machine, or testing procedure. The result of this, in mines with years of operation, is that the frother and collector doses are not effectively rechecked when the ore gradually changes as the mining operations go deeper; so today many of these mills are using an excess of collector and frother in their roughers, and this excess is not only costly, but results in an undetected lower than optimum recovery.

Table 89.-FLOTATION TESTING VARIABLES

Optimisation of Reagent Suite - uniform mineral	Development of a dressing scheme for a new mine
mineral dissemination in the ore	number of samples needed to
crushing and grinding conditions	cover variability of ore body:
size distribution (liberation)	ore hardness, abrasiveness,
age (time in contact with air)	and grinding index
pulp percent solids	natural pH of deposit
reagent addition point	ground water content
conditioning time	mining methods
flotation machine design	applicable concentration methods:
air rate	gravity, leaching,
impeller velocity	flotation, or combination
residence time of the pulp	determination of the optimum
pulp temperature	treating circuit
air flow rate	optimum liberation sequence,
mill altitude (oxygen availability)	i.e. regrinds and locations
collector type	within the circuit
collector quantity	process water choices
frother type	by-product recovery possibilities
frother quantity	waste disposal
pulp pH	plus all the flotation variables
modifier type (lime, soda ash, etc.)	already listed
modifier quantity	
activator type	
activator quantity	
depressant type	
depressant quantity	
water hardness	
heavy metal cations in solution	
operator skill	

In considering the type of work that a flotation laboratory should be capable of handling, if it is to be an adjunct to a working mine, it is important to note that in an existing mill the easiest to change variables, which directly affect mill through-put, and therefore economic performance, are: the reagents; the pulp dilution; and/or an increase or decrease of the rate of feed to the grinding circuit, which necessarily results in coarser or finer grinding. Generally, after a mill is in operation, the pressure on the metallurgist is to increase tonnage without sacrificing recovery. He therefore is normally routinely monitoring the replacement of the existing reagent suite, or the step, if this is incorporated in the mill circuit. There is somewhat less freedom in changing the type of reagents used, and in circuit changes such as sand/slimes separations.

Experimental design

In planning an experimental programme, the following five steps should be involved:

1.- a clear definition of the objectives of the experimental programme and of the hypothesis that will be used to analyse the results;
2.- an experimental design that will quantify the hypothesis and provide the data to reach the stated objectives;
3.- the performance of the experimental plan and the collection of the data;
4.- the analysis of the data obtained, and
5.- the formulation of the conclusions and recommendations.

Usually these five steps are a good framework for the organisation of the final report on the technical study.

In planning the experiment, it is good practice to write out the main points of the mechanistic hypothesis used as the basis for the design, and all the identifiable variables. The variables that will be tested should be specifically listed, indicating which will be varied and which will be kept constant. Those that cannot be controlled should be clearly identified. At this stage, a priori information on which of the variables are dependent should be noted, as well as any possible interactions. Written lists of variables are helpful when evaluating results during the experimental programme, as too great a familiarity with the mill operations result in not spotting the effects of secondary or uncontrolled and unmeasured variables. These lists are also important to determine the confidence level of the results that will be required to arrive at a cogent management decision, and to confirm, by measurement, that the testing programme has met these criteria. Although these are the ideal factors to fix the experimental design, in practice the experiment will have to be trimmed to conform to the limited research funds available.

A semi-empirical method of designing a flotation experiment can be best described using a specific example. A copper mill, after an initial period with recoveries of around 90%, experiences a change to a more refractory ore, and recovery drops to 83%. The metallurgist is asked to determine if the recovery can be improved by changing the frother and collector dose, or if new reagents are required.

As a first step, most mill metallurgists' response to such an assignment will be to run a series of standard flotations at a constant frother dose, varying the collector amount. If the ore sample has been sent to an outside laboratory, on the other hand, the first step will be to determine a standard grind and flotation time for the test series. This is done by running a series of flotation tests to determine the ball mill grind that results in the same rougher recovery as that obtained at the plant. This exploratory series of tests frequently gives the service lab float technician a better feel for the ore than the in-house metallurgist. In both cases, the tests will be run for a standardised grinding time, with the length of the batch flotation time equal to, or some arbitrary fraction of, the residence time in the plant rougher.

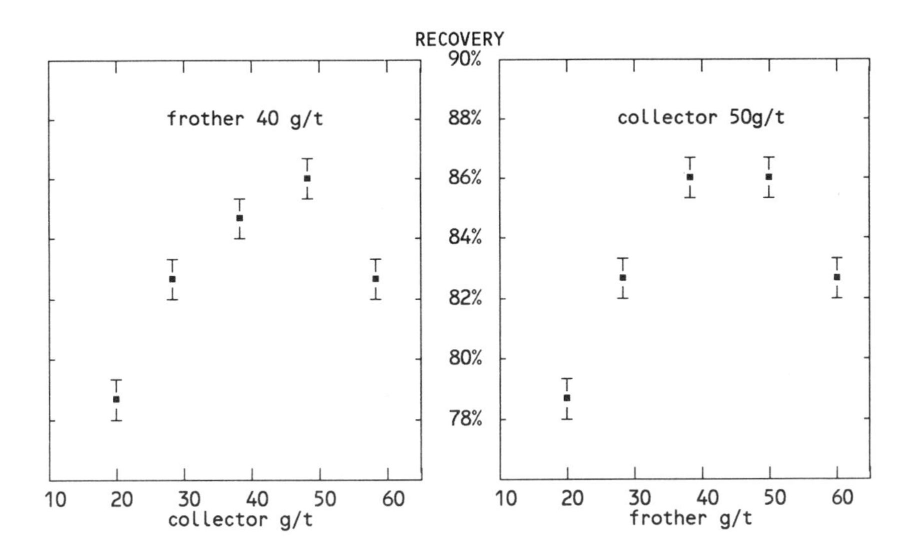

Fig. 68.- Results of 2 kg laboratory flotation tests, single variable, South American porphyry

To visually analyse the results for interactions, the data can be plotted as grams per tonne frother versus g/t collector, with the recovery as a response. Fig. 68 illustrates this stage in testing. It is based on scattered data taken from a series of experiments carried out at a large, privately owned copper mine in Chile, which had operated for a number of years with a collector feed of 50 - 65 g/t, and around 40 g/t of a polyglycol ether frother. It was shown, after more than a year of laboratory work, that, without changing the reagents, a considerably improved recovery could be obtained at collector feed rates of under 20 g/t and a frother dose of about 60% of this value.

As can be seen, this initial data gave no indication of the results that were eventually obtained in the long testing programme; i.e., that the optimum recovery at this mill is attained at collector and frother doses of 20 g/t of collector and about 12 g/t of frother. On the contrary, the preliminary experiment suggests that the mill is operating at close to optimum recovery, with 60 g/t of collector, and that, if any change should be made, it would be to mill test an increase in the frother dose to 50g/t.

As we know that invariably there is an interaction between the frother and the collector in this type of test, the more meaningful way to plot these two sets of data is in terms of the recovery as the main response. This is shown as the first graph of Fig. 69. Visual examination is still ambiguous, and if there is any trend visible it would support the idea of mill testing a slight increase in the frother addition. Instead, if a full factorial had been done initially, the results would have been as shown in the second graph of Fig. 69 (the data shown is taken from scattered results which do not complete the factorial). At a cost of about 30 flotation experiments, as compared to the 12 runs of Fig. 68, the conclusion that equal, or better recoveries could be obtained at less than half the reagent consumption would have been indicated a year earlier.

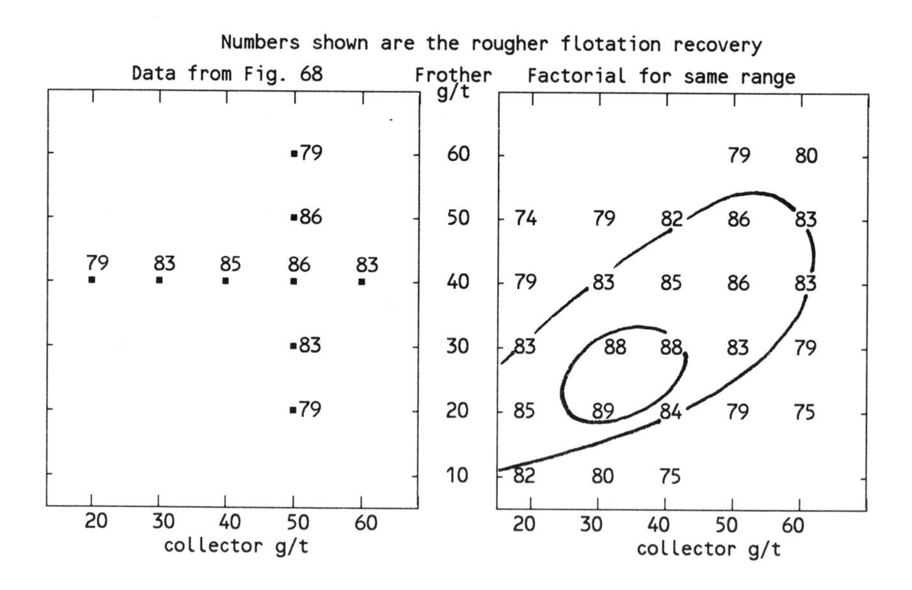

Fig. 69.- Results of Fig. 68 replotted compared to a two-variable factorial design covering the same range

Although this two variable factorial design defines the interactions adequately, it does require more experiments than a more sophisticated design. These are some general rules on the design of experiments which can provide more efficient information:

1.- *symmetry* in the experimental layout simplifies calculation considerably, as well as helping in the visualisation of the responses;
2.- the *number of levels* used in the experimental design fixes the shape of the response curve that can be detected:

a) *two* levels detects trends only;
b) *three* levels can detect the existence of a response curve, but not its shape;
c) more than three levels progressively refine the shape of the response curve.

Examining the factorial in Fig. 69, it is obvious that the experiment can be simplified by using a repetitive unit cell based on an experimental group of the four corners of a square. This reduces the factorial to a two variable two level experiment, which can only detect linear responses; adding the centre point converts it to a three level design yielding considerably more information. This type of design, the corners and centre of a square, is a popular layout for two variable designs, as is the corners of a cube plus a centre point, in three variable problems, where tendencies can be seen clearly by visual inspection.

In Fig. 70, the square with a centre point design is applied to the example. The initial design consists of the square with a replicated centre point. The existing plant conditions correspond to the left upper corner of the first square (shaded). The unit difference in frother and collector dose is chosen to yield a change in recovery of at least twice the experimental error. The next group is picked, after the flotation results are in hand, based on: (a) that the next experimental group has at least one common point with the prior group (doubles as a check of the experimental error of the next set); (b) that the next centre point lies on an axial or diagonal line, starting from the prior centre point, following the overall slope of the four corner points. Thus, with five experiments, one generates a semi-factorial 2 variable 3 level design that is easy to analyse, relatively quick to detect a maxima in a response surface, but ambiguous if there is more than one maxima in the surface because only three levels are studied. It is desirable that all the points be replicated; this is not exorbitant in cost because the ten tests required are a convenient one shift flotation laboratory load for a float technician and aide. A very important point in the design is that all the flotations of each unit cell must be in a random order to avoid unconscious bias of the operator.

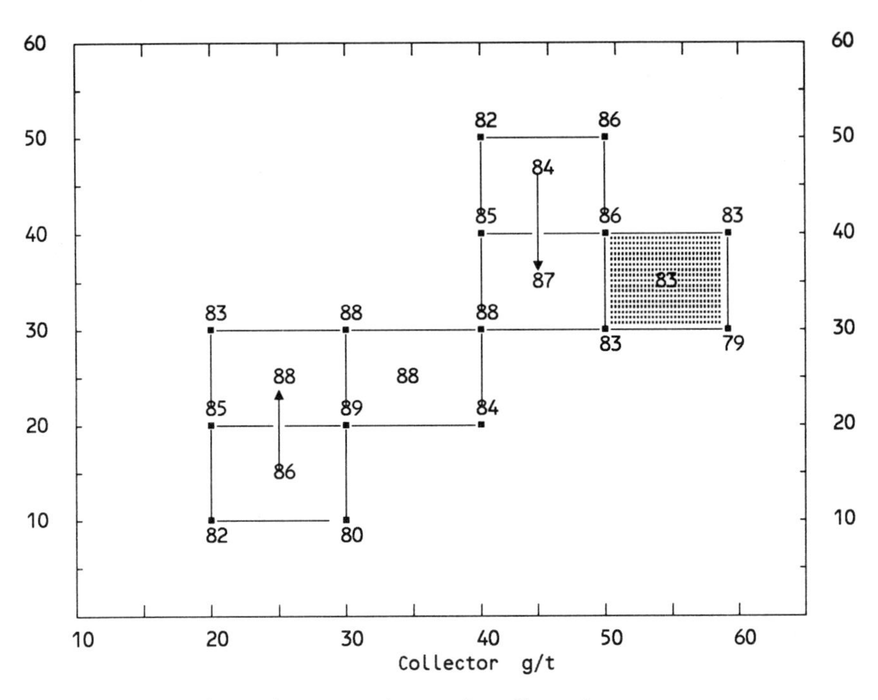

Fig. 70.- Experimental design based on a square pattern search path

If one is designing a three variable experiment, for example the effect on recovery of changes in the frother and collector dose in combination with changes in the system pH, it is obvious that the above procedure can be converted into an empirical design based on a body centred cube layout. The minimum number of experiments involved in this case are the 8 tests at the corner conditions plus a replicated centre, i.e. again 10 flotations which can be done in a single shift. If it is desired to replicate all runs, the unit experiment takes 2 operating days (or shifts). This number of tests can be conveniently carried out in one shift, while a complete 3 x 3 factorial involves 27 flotations, plus replications, which would take nearly a week's work, and the results from tripling the number of experiments will provide only marginally more complete a picture of the system reaction.

Fig. 71 shows a convenient form of recording the experimental design on paper, which can also be used to plot the recovery response. The x and y-axis can represent the frother and collector dose, respectively, and the z-axis the pH. The recoveries recorded are from the prior experimental series. This is a good point to indicate that pH is not normally a continuous function of collector and frother dose, as it will show a recovery discontinuity at the natural alkalinity of the ore (pH between 7 and 8.5), therefore a cube which straddles this region will give ambiguous results.

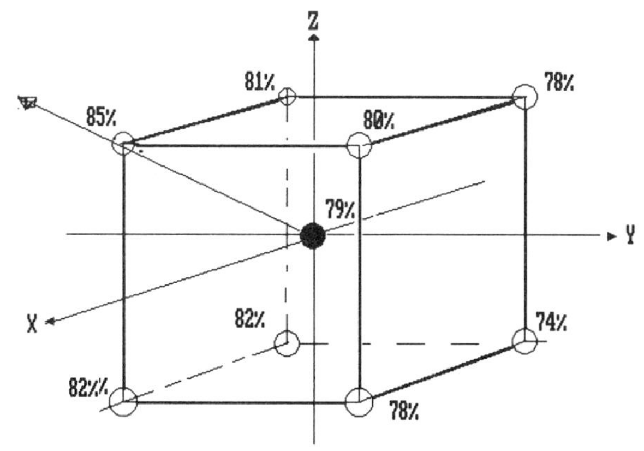

Data: Run #	x Frother g/t	y collector g/t	z pH level g/t lime	Response Recovery percent
1	15	35	150	82%
2	15	85	150	74%
3	15	85	350	78%
4	15	35	350	82%
5	30	60	250	79%
6	45	35	150	82%
7	45	85	150	78%
8	45	85	350	80%
9	45	35	350	85%

Fig. 71.- Cubic experiment for a truncated 3 level by 3 variable test module

As can be seen, this simple experimental design method can give a surprising amount of information from a very few experiments, particularly when one considers the very high error estimated for the laboratory recovery measurements; but if, say, the effect of varying the amount of a sulphidising agent must also be studied (i.e. if one must extend the experiment to four variables), experimental design for visual evaluation becomes much more difficult. The obvious choice of repeating the cubic experiment at three different levels of sulphhydrate is not easy to interpret because of the possible inflections in the response surface.

As mentioned in Chapter 6, Lekki and Laskowski (1971) compared batch flotation and Hallimond tube tests by measuring the copper recovery surface for the chalcocite-ethyl xanthate - alpha terpineol. Their results are shown in Fig. 72. An obvious conclusion from these data is that Hallimond tube collector data cannot be scaled up, but that experiments employing 1 to 3,000 ml laboratory batch flotation machines can be sensitive enough to detect the existence of a maxima, and that the operating conditions and dosages yield recovery and grade results that approximate those obtained with plant tests in which both the collector and the frother dosage is varied over a broad range.

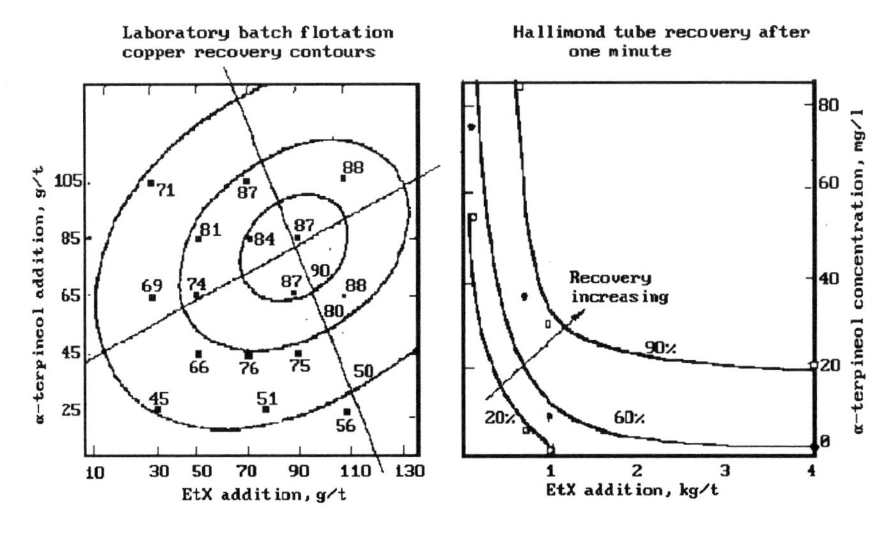

numbers by the data points = recovery %.

Fig. 72.- Effect of varying frother and collector dose on copper recovery contours for chalcocite - batch tests vs Hallimond tube

Based on this preliminary work they later (1975) published the only detailed results of a two level, two variable flotation factorial, including its mathematical analysis for the interaction between potassium ethyl xanthate and three different types of frothers: one with collector properties, alpha terpineol; another inert, diacetone alcohol; and a frother with distinct surface active properties, octanol. The responses measured were recovery and rougher concentrate grade.

The results of the recovery and grade measurement for alpha terpineol and diacetone alcohol are shown in Fig. 73, and for octanol in Fig.74. Note that for octanol the recovery assays were not completed because the concentrate grades obtained with this strongly surface active frother were not of

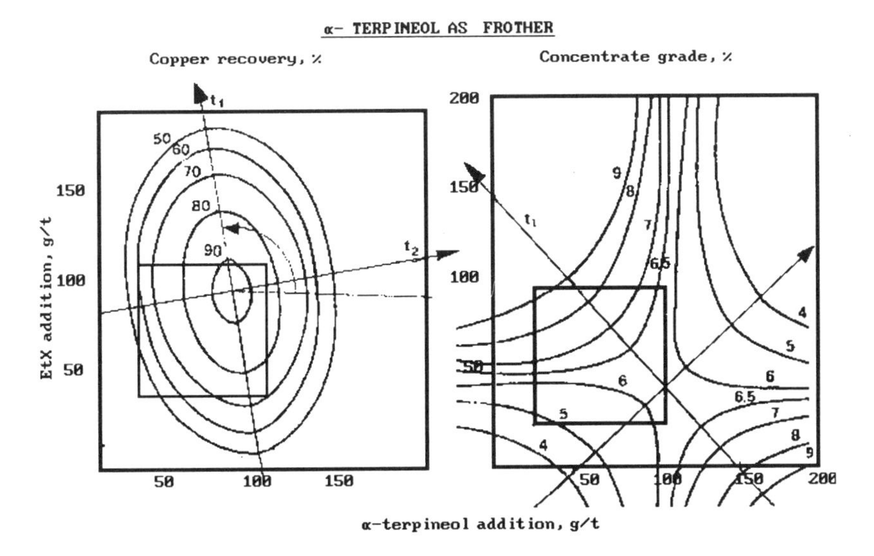

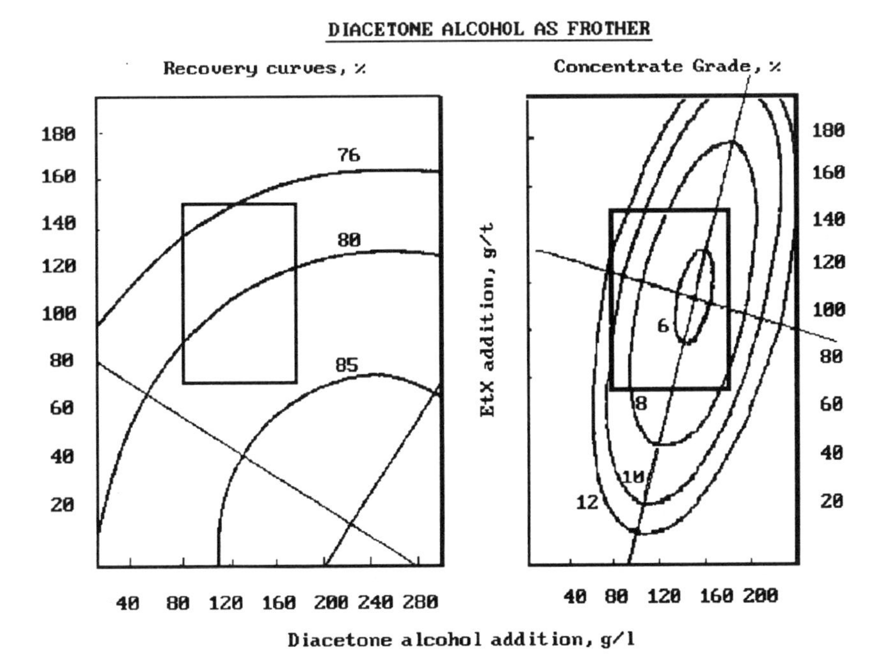

Fig. 73.- Cu recovery and concentrate grade as a function of the interaction between the frother type and the collector (EtX), at varying dosage

Table 90.- INFLUENCE OF ETX AND α-TERPINEOL CONSUMPTION ON GRADE OF CONCENTRATE (ß), AND COPPER RECOVERY (R).

$w = g_o + g_1x + g_2y + g_3x^2 + g_4y^2 + g_5xy$			
Second order fit		w = ß	w = R
Regression coefficients	g_o =	2.17236	-40.9696
x - dosis of αT, g/t	g_1 =	0.041448	1.941365
y - dosis of EtX, g/t	g_2 =	0.0680189	1.069910
	g_3 =	-0.000026	-0.010184
	g_4 =	0.0000326	-0.004522
	g_5 =	-0.0006911	-0.002463
Sum of squares removed by regression		15.6739	4442.935
Remainder of sum of squares		4.643579	134.502
Estimate of the population variance		0.4643579	13.4502
Value of F_{calc}		6.75079	66.0649
F value for 10.5 degrees of freedom $F_{0.01}$		5.64	5.64
Canonical equation	u_1 =	-0.00035	-0.004266
$z = u_1t_1^2 + u_2t_2^2$	u_2 =	-0.00034	-0.014404
Point of maximum response	f_o =	6.08756	91.41737
equal to the canonical	x_o =	103.34	83.7665
coordinates	y_o =	52.14	95.4777
	ϕ =	132°34'	101°45'

IDEM - for DIACETONE ALCOHOL			
$w = g_o + g_1x + g_2y + g_3x^2 + g_4y^2 + g_5xy$			
Second order fit		w = ß	w = R
Regression coefficients	g_o =	60.79912	37.33319
x - dosis of αT, g/t	g_1 =	0.2234259	-0.30250
y - dosis of EtX, g/t	g_2 =	0.1104235	-0.15490
	g_3 =	-0.0004817	0.000935
	g_4 =	-0.0000326	-0.004522
	g_5 =	-0.0006911	-0.002463
Sum of squares removed by regression		126.338	127.4156
Remainder of sum of squares		23.888179	30.3678
Estimate of the population variance		1.137532	1.4318
Value of F_{calc}		22.21278	17.7979
F value for 10.5 degrees of freedom $F_{0.01}$		4.04	4.04
Canonical equation	u_1 =	-0.00011877	-0.00090
$z = u_1t_1^2 + u_2t_2^2$	u_2 =	-0.00062041	0.00052
Point of maximum response	f_o =	86.77	5.78097
equal to the canonical	x_o =	222.24	148.53
coordinates	y_o =	20.745	117.32
	ϕ =	121°43'	13°37'

Table 91.- INFLUENCE OF ETX AND OCTANOL CONSUMPTION ON GRADE OF CONCENTRATE (ß).

$w = g_o + g_1x + g_2y + g_3x^2 + g_4y^2 + g_5xy$	
Second order fit	w = ß
Regression coefficients	g_o = 10.78667
x - dosis of αT, g/t	g_1 = -0.9606
y - dosis of EtX, g/t	g_2 = 0.43835
	g_3 = 0.0033466
	g_4 = -0.0021295
	g_5 = 0.00024
Sum of squares removed by regression	15.697
Remainder of sum of squares	11.794879
Estimate of the population variance	0.655272
Value of F_{calc}	4.791
F value for 18.5 degrees of freedom $F_{0.01}$	4.25
Canonical equation	u_1 = 0.003349
$z = u_1t_1^2 + u_2t_2^2$	u_2 = -0.002132
Point of maximum response	f_o =
equal to the canonical	x_o = 139.5435
coordinates	y_o = 110.786
	ϕ = 1°43'
response contours shown in	Fig. 73 and 74

commercial interest. The calculated results for the three frothers are shown in Tables 90 and 91.

The authors note that the ellipsoidal recovery contours, with a plateau at 80 g EtX and 80 g a-terpineol per ton, are similar to those conditions found for the Polish ore from which the chalcocite was obtained, and that the angle of slope of the axes of the ellipsoid indicates the high interaction between the two reagents. They also say:

> "The relation between the grade of concentrate and the consumption of EtX and a-terpineol is shown in Fig. 72. A singular point is situated within the experimental range and enables to postulate the hypothesis of a different mechanism of cooperation between the collector and the frother at small dosages. An increase in the dosage of both gives a remarkable increase in the copper content of the concentrate. Under conditions of excessive concentrations of both reagents, however, the grade of the concentrate decreases and the flotation begins to be less selective. It is to be pointed out that in flotation practice the grade of

> concentrate is the main quality coefficient of the process and its decrease usually leads to the addition of new doses of reagents."

These data and their analysis is a more elegant presentation of the experience obtained at El Teniente, i.e., that nearly all established mills drift into an overdosage of flotation reagents.

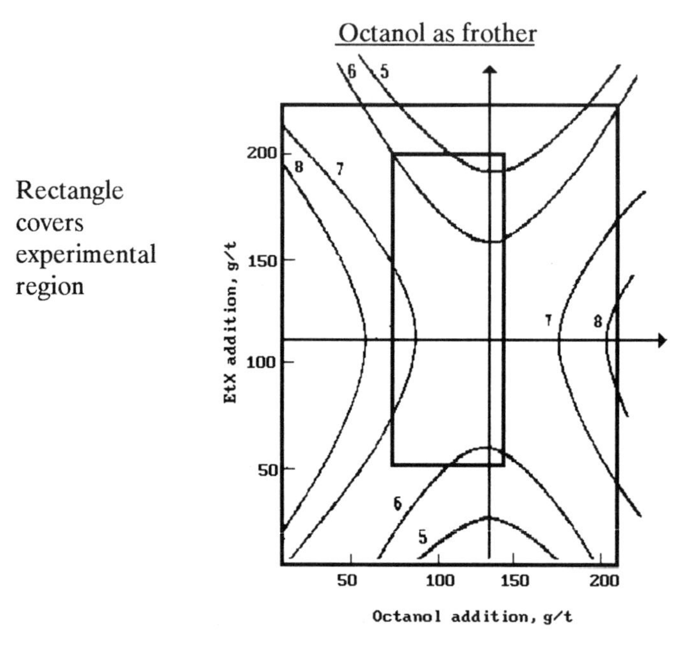

Fig. 74.- The interaction of octanol and EtX on copper recovery and concentrate grade

Lekki and Laskowski's work is important in supporting this last conclusion, as it is the result of sophisticated academic experimental work, which is only rarely duplicated in operating concentrators. Reagent suppliers were among the first to recognise the importance of trying to control the shape of the ellipsoidal recovery response surface, as a flat ellipsoidal plateau recovery response to changes in the frother and collector additions helps stable operation and customer satisfaction with a particular reagent suite.

Box-Wilson semi-factorial experimental design: When studying a three variable ore flotation problem, say the effect of collector dose, frother dose and pH on the recovery as a response, we noted above that, in a random order and with suitable replication to estimate experimental error for all possible combinations of $C_1F_1Ph_1$, we will have to perform experiments $C_1F_1Ph_2$, $C_1F_1Ph_3$, ... , $C_3F_3Ph_3$, or 27 tests (3^3). The results of these tests can be interpreted using regression analysis and an analysis of variance. If the number of variables or levels that must be tested are greater than three, the number of experiments required by a full factorial become exorbitant. For example, a two variable experiment at 5 levels will require 32 tests, while three variables studied at 5 levels require $3^5 = 243$ tests for a complete factorial. For these, five-level fractional factorials, or Box-Wilson experiments, are an effective alternative.

The simplest example of a Box-Wilson experiment is a two variable example, where the recovery response in a flotation is to be studied as a function of collector dose and frother dose, with other factors kept constant. To achieve this statistically, experiments are carried out and their results reduced to a regression function of the form:

$$y = b_0 + b_1x_1 + b_2x_2 + b_{11}x_2^2 + b_{22}x_1^2 + b_{12}x_1x_2$$

The general model (response function) includes three types of terms in addition to the constant b_0 :

1.- Linear terms in each of the variables $x_1, x_2, \ldots, x_p$
2.- Squared terms in each of the variables $x_12, x_22, \ldots, x_p2$
3.- Cross-product or first order interaction terms for each paired combination $x_1x_2, x_1x_3, \ldots, x_{p-1}x_p$

For the three-variable problem the response function would be

$$y = b_0 + b_1x_1 + b_2x_2 + b_3x_3 + b_{11}x_2^2 + b_{22}x_1^2 + b_{33}x_3^2 + b_{12}x_1x_2 + b_{13}x_1x_3 + b_{23}x_2x_3$$

The number of terms for p variables is therefore (p - 1)(p + 2)/2.

Limiting the interactions to first order is justified in most industrial experiments because the experimental accuracy does not require the higher order interactions for adequate description of the response surface.

Box-Wilson rotatable experimental design provides an easy way to calculate a 5 level system which gives as much information as a full factorial, but with a minimum of experiments. For a two variable experiment, the coordinates of each experiment are located at a central point surrounded by two squares rotated 90 degrees from each other. The arrangement is illustrated in Fig. 75, where the factor k equals half the range, and the range is the difference between the extreme experimental points, which in turn are determined as the approximate square root of p, where p is the number of variables to be studied. If the sides of the squares are designated by 2a, the radius of the circle (sphere, etc.,) in which all the outer points lie is given by r = ka, and the maximum range of each variable is 2ka. If the centre point is replicated, this two-variable design requires 10 experiments. For a three-variable experiment k = 1.73, and, if we assume that a = 1, the coordinates or the combination of levels for the three-variable design will be as shown in Table 92, and Fig. 76. The total number of experiments required are 15 plus replicates.

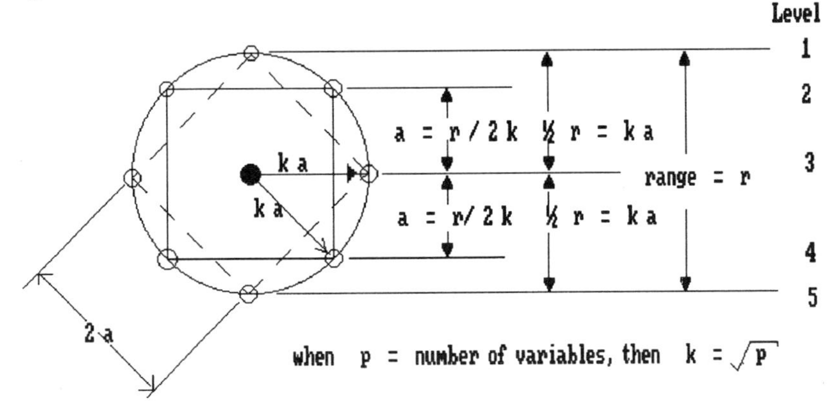

Fig. 75.- Box-Wilson rotatable two-variable five-level design

These design principles are applicable to more than three variables (at 5 levels), as they always include three types of combinations: the axial, factorial, and centre points. Thus axial points include each variable at its extreme levels, with the others at their centre point level. The factorial points include all combinations of intermediate levels (indicated by the plus and minus ones in Table 92). A centre point is a single test at the average level of each variable. To estimate experimental error, the centre points are usually repeated three to five times during the experiment. Designs for any number of variables can thus be laid out using these principles.

When testing has been completed, a regression analysis must be carried out to determine the coefficients of the response function. For convenience, this is done keeping the independent variables in coded form.

After the mathematical model has been determined, an analysis of variance should be carried out to evaluate its statistical significance. Having determined a model which is significantly representative of the system, an analysis can be made of the shape of the response function. The analysis can be contour plotting, or more sophisticated non-graphical analysis such as ridge analysis which characterises the function's behaviour in p dimensions.

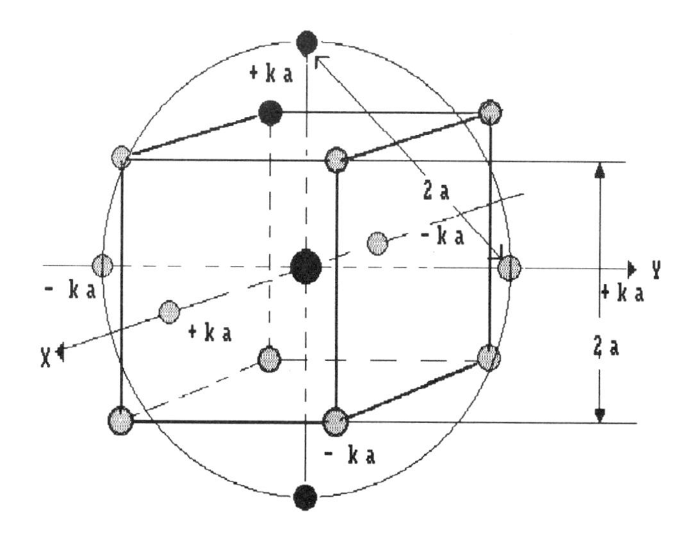

Fig. 76.- Box-Wilson rotatable three-variable five-level design

Most physical situations can be approximated by a quadratic function over a reasonable range of the variables. This is the reason why the Box-Wilson experimental design methods are industrially very popular.

Table 92.- THREE-VARIABLE FIVE-LEVEL BOX-WILSON DESIGN LAYOUT

Axial points			Factorial Points			Centre point		
$X = x_1$	$Y + x_2$	$Z = x_3$	$X = x_1$	$Y + x_2$	$Z = x_3$	$X=x_1$	$Y + x_2$	$Z=x_3$
1.73	0	0	1	1	1	0	0	0
-1.73	0	0	1	1	-1			
0	1.73	0	1	-1	1			
0	-1.73	0	1	-1	-1			
0	0	1.73	-1	1	1			
0	0	-1.73	-1	-1	-1			
			-1	-1	1			
			-1	1	-1			

Other on-line search strategies: The simplest technique - varying one variable at a time - usually consists of holding constant all manipulative variables, while a search is made, variable by variable, for changes that result in a rise in the desired response. It is assumed that when the response starts to decrease, a maximum has been passed; then that variable is held constant at the new value, and a search started with another variable.

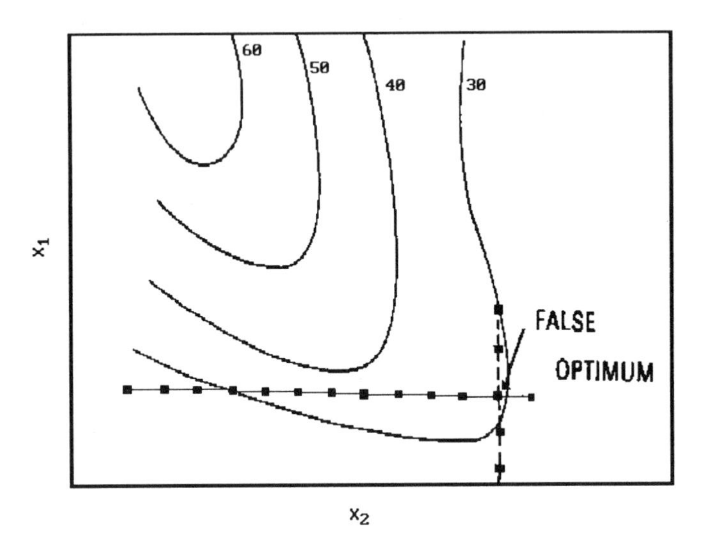

Fig. 77.- False optimum in a one variable at a time search experiment

Mular (1976) points out, using the response contours shown in Fig. 77 as an example, that it is impossible to avoid "false maxima" using this procedure. He believes that the steepest ascent method is sounder, as it is a steady-state strategy based on orthogonal designs. The procedure should be obvious from his example in Fig. 78. Here the calculated interactions from each orthogonal experimental group are used to lay out a series of individual runs designed to find a maxima. A new orthogonal group is then run with the centre point at the maxima. This sequence is repeated until the true maxima is found for the system. The procedure can be generalised to more than 3 variables. A cautious version of the steepest ascent method is EVOP, developed by Box (1969). A variant, shown in Fig. 79, also taken from Mular, is SSDEVOP: Simplex Self Directing Evolutionary Operations. Spendley et al (1962) minimises the number of runs performed, and sacrifices statistical inference, to speed up the experimental process. The experimental design requires n + 1 runs for 'n' manipulable variables. Mular describes the process thus:

> "To start a program, levels are selected for the starting simplex design and runs at each vertex completed. ... The vertex associated with the lowest measured response in a simplex is replaced by its mirror image found through the

face opposite the vertex to be reflected. Thus, there is a continuous movement away from the low results in directions most likely to produce higher results. The move is always to an adjacent simplex. The response at the new vertex is measured and the lowest result in the new simplex is dropped in favor of another mirror-image vertex and so on. The procedure is repeated until an optimum is approached. In the region of the optimum, successive simplexes will tend to hover about the optimum, moving around but seldom away from it."

Mular (1976, 1989) describes these experimental designs in more detail, including the curve fitting mathematics, and the mathematical evaluation of the statistical significance of the results.

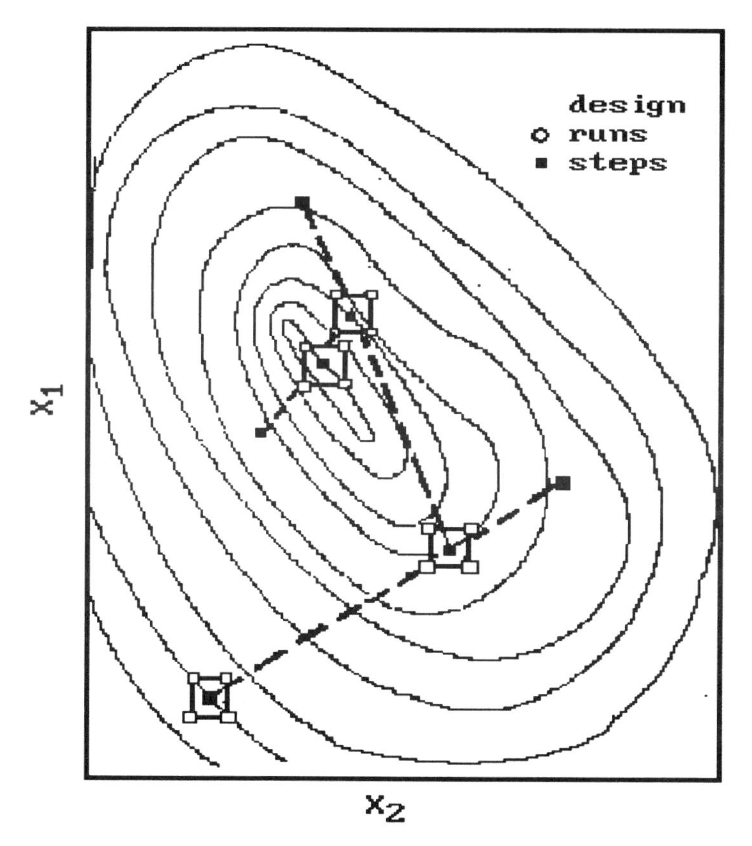

Fig. 78.- Steepest ascent experimental procedure result

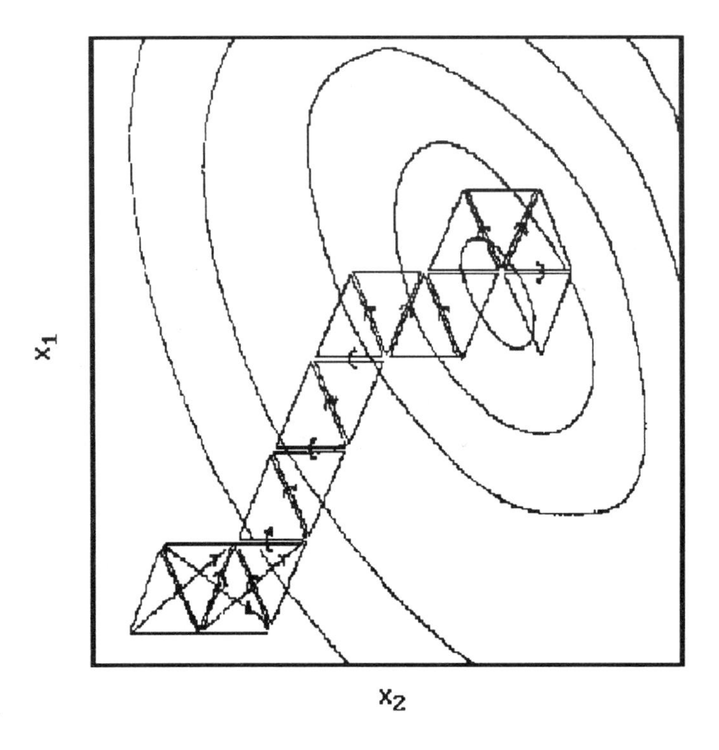

Fig. 79.- SSDEVOP: climbing a response surface, from Mular (1976)

Testing procedures

The type, size and investment in a mineral testing laboratory, and the specification of the equipment and staffing required, varies according to the testing objectives, and differs greatly if the work to be done is:

1.- concentrator performance monitoring and optimisation,
2.- supplier technical service - reagent selection or formulation,
3.- processability verification of a new ore,
4.- industrial development of new reagents, or
5.- the development of a mineral beneficiation process for a new mine.

The standard unit operations which must be provided in a flotation laboratory include:

sampling and sample preparation services;
crushing, grinding and grindability determination facilities;
microscopy and mineralogical evaluation facilities and skills;

a fully equipped flotation laboratory;
data processing and experimental design resources.

Services that can be obtained from independent laboratories, if promptly available, include:
chemical analytical services;
mineral surface physical and chemical characterisation;
equipment and skills; and
chemical synthesis capabilities.

Facilities that general purpose mineral testing laboratories usually have available are:
equipment to perform gravitational, electrostatic or magnetomotive separation testing;
classic gold processing facilities, including amalgamation, cyanide leaching, zinc precipitation, carbon in pulp testing, etc.;
a leach laboratory capable of simulating agitation, percolation and heap leach processes.

If the tasks envisioned include the development of the full mineral processing scheme for a new mine, including its process design and construction, all the above services must be available, and they must be supplemented with equipment to handle various thousands of tonnes of ore samples, and a complete line of crushing, grinding, screening and processing equipment of such a size and design as to inspire confidence in the accuracy of the scale-up data required for the detailed design of the mine.

The only purpose-designed facility capable of quantifying from scratch the design data for a new mineral, that the author is familiar with, is the CIMM in Chile. This mineral research and development complex was originally funded from Brussels in 1969 with a 15 million dollar grant. Today, after accumulating considerable equipment purchased to solve specific client problems, its facilities probably could not be duplicated for under 50 million dollars.

The existing installations of the "Centro de Investigacion Minera y Metalurgica" cover 20 or 30 acres of a hillside with three main multistory buildings. At the highest point, accessible to large tonnage trucks, there is a full scale multistage crushing plant capable of handling various hundreds of tonnes of ore per day, with a flexible screening circuit for the secondary and tertiary cone and short head crusher stages. The grinding equipment includes ball, rod and autogenous mills. Flotation facilities range from single cell bench and floor units to a pilot unit capable of processing about a tonne per hour of mineral, with cell banks that can be arranged so as to include scavengers and cleaners. There are two concrete sloped floor tanks to receive tailings, which can be used to pilot test ores with water reclaim, to evaluate the effect of circulating flotation reagents. Test flotation columns are available, as well as equipment to test different types of bubblers for flotation columns. The leach pilot facilities have percolation tanks with a comparable hourly ore treatment capacity, while the gold section includes amalgamation drums. Obviously the support facilities complement the larger scale equipment, with extensive analytical labs including all the standard testing equipment, supplemented with electron beam analysis and scanning electron microscopes.

All types of mineralogical studies are also performed at CIMM. The Institute is essentially self financing, as it provides routine analytical service to the mineral industry, as well as arbitral analysis for ore purchasing contracts, and charges for all pilot plant work. International and domestic mining companies have used its pilot plant and design services for essentially all new mines built in Chile in the past 15 years. In addition, contract mineral testing work has been done for mines in most of the South American countries. As a service to the smaller, locally owned mines, CIMM operates satellite analytical labs located within each of Chile's mining districts, to provide assay services. The organisation could serve as a model for other nations with large mineral wealth who need to develop indigenous mineral processing technology.

Equipping the flotation laboratory

In addition to such general facilities as ore reception, quartering and crushing areas, plus microscopes and other mineral identification and assay instrumentation, the minimum equipment required in a flotation testing laboratory is:

1.- Sample preparation, weighing and storage facilities, which should include a plastic bag heat-sealing machine and a freezer.

2.- One or more stainless steel ball mills driven by a variable speed rubber roller set. The mill is frequently home-made and consists of a stainless steel tube, about 6 to 12 inches in diameter by 10 to 18 inches in length, equipped with a quick opening, gasketed, end cover. The drive mechanism should be activated by an electric motor equipped with a reliable timer.

3.- A standard laboratory flotation machine, fitted with square glass, or transparent plastic, cells capable of handling samples of 500 gm to 1000 gm or more of ore . The bench area should have power, air and water outlets, a large diameter timer and, optionally, a distilled water supply.

4.- Adequate vacuum resources, as concentrate and tails collected during the test are normally dewatered in a large Buchner funnel.

5.- A good sample drying oven. In cool climates the drier can be a steam heated metal table, otherwise a home-made vented cabinet with slide-out shelves will do the job adequately without overheating the tailing and concentrate samples.

Flotation Cells: many different types of laboratory flotation machines have been used during the past thirty or forty years, but today only three or four types are in common use. All are mechanically agitated types with variable speed impellers. Some are sub-aeration types, with limited capacity to control the air flow to the agitation zone, and others require an outside source of low pressure air to generate the froth. Most laboratories purchase commercial machines with interchangeable glass or stainless steel test cells, sized for flotation tests of 500 gm, 1,000 gm or 2,000 gm ore samples; i.e., effective cell volumes are about 1.5, 3 and 6 litres, as the usual test starts with a pulp density of around 30 to 35% solids. In many cases the machines have interchangeable impellers so that cell scale-up geometry can be maintained reasonably constant for different types of industrial flotation cells. In most cases, for both sub-aeration and blown flotation machines, the airflow can be measured with a rotameter. Generally, rotor speed is adjusted to a higher rate than that used industrially because of the poorer agitation in the small cells, which can result in coarser particles in the pulp settling out. Because of the need to monitor this condition, glass and transparent plastic cells are preferred over stainless steel cells in most laboratories.

Attempting to make the flotation test independent of the skills of the operator, many laboratories use metal cells and equip them with mechanical paddles, and in some cases automatic pulp level controls. There are no good statistical studies on whether there is any benefit to mechanical removal of the froth. Klimpel, at Dow, will recommend this arrangement because he also is depending heavily on analysing his results on reagent performance in terms of the rates of flotation. As can be seen from Fig. 80, the US ASTM standards group for the coal industry also try to improve accuracy by specifying a mechanical paddle in their standard, while the Europeans, Figs. 81 and 82, propose a hydrodynamics modification to the cell (deflector block) which reduces froth skimming to a purely rhythmic operation, and possibly will not affect scale up.

Experience at laboratories specialising in metal sulphide ore flotation is firmly against the use of rotating paddles and automatic froth level control because it very significantly reduces the amount of information obtained from each flotation. A non-transparent (metal) flotation cell and mechanical paddling may be justified in two-component systems - mineral and gangue. Presumably there are only three rate curves to be measured in this case: mineral, gangue and water; and the three are linearly related by the material balance. In a typical copper porphyry one encounters at least 4 major target sulphide minerals for recovery (chalcopyrite, chalcocite, another copper sulphide, and molybdenite, silver and gold), and a surface active multicomponent gangue. Average rate of flotation curves for the gangue and non-gangue do not provide sufficient information for an intelligent choice of an optimum collector-frother reagent suite, and can be shown to have led to erroneous comparative conclusions in collector substitution studies. A good flotation technician, by observation of the colour of the froth, can approximate differential rate of flotation for the sulphide minerals, and can recommend, and check at laboratory scale, whether stage-added reagents would improve recovery, and determine the approximate point down the rougher flotation bank where the reagent addition should occur. Computer analysis of the rate data for individual minerals is technically feasible, but unlikely to inspire confidence in its results because of system noise.

Instrumentation: Absolutely essential is the availability of a rugged electronic pH meter with a probe that can be kept in the froth or dipped in when necessary. Also essential is a 10 or 12 inch diameter laboratory second-timer to be set up in full view of the operator. Otherwise the laboratory should be well equipped with accurate weighing machines, and at least one analytical balance to calibrate reagent feeders. Redox and in-line infrared or atomic absorption equipment is a luxury that is justified if a large amount of basic research is routinely included in the duties of the float lab. The other item that should not be skimped on is the type and quality of screen size measuring equipment.

Water: is a controversial item in laboratory testing. Undoubtedly the availability of deionised water for use in preliminary flotation testing trials is desirable, but usually, unless the community has unusually hard water, tap water is adequate for routine work. If the client has a water problem, and particularly if reclaimed water, containing residual flotation and flocculation reagents, is, or will be, the main source of process water, then storage facilities for large quantities of process water sourced from the normal mine supply should be made available at the laboratory.

Reagent Additions: water soluble reagents, such as xanthates, should be made up daily, and are generally fed to the float cells as relatively dilute solutions (decomposition rates for most xanthates are about 1% per day and the decomposition products can be depressants). Inorganic salts also are

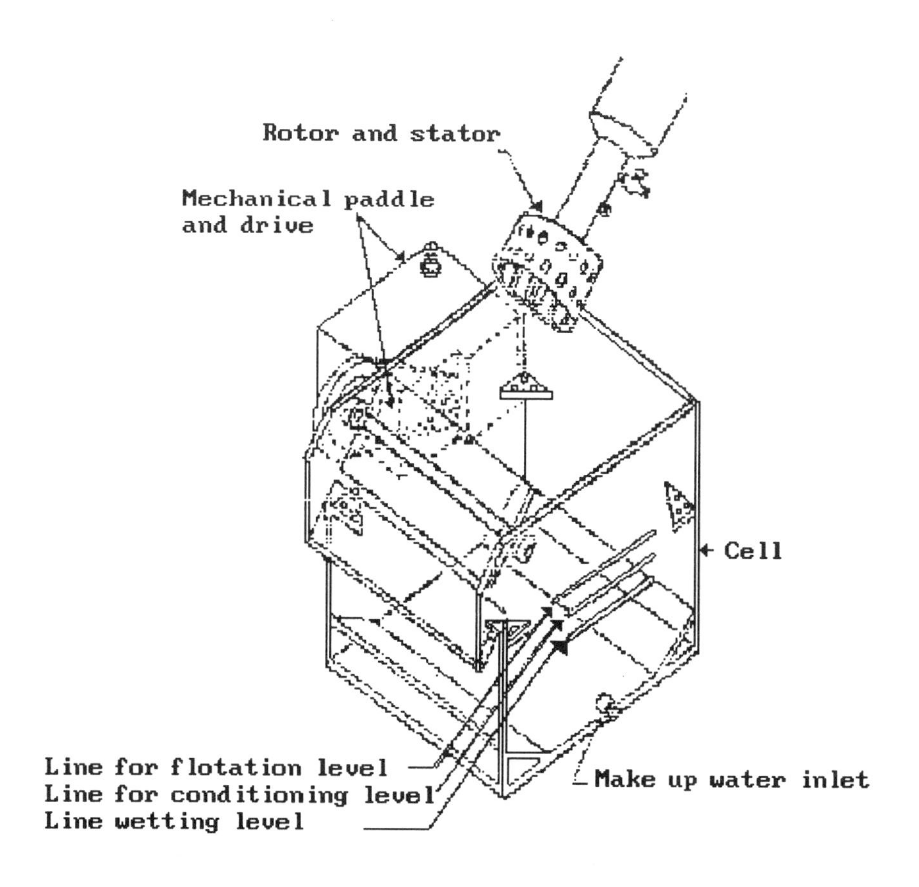

Fig. 80.- Standard 5.5 litre (2 kg) flotation cell from ASTM coal standards proposal, equipped with a mechanical paddle

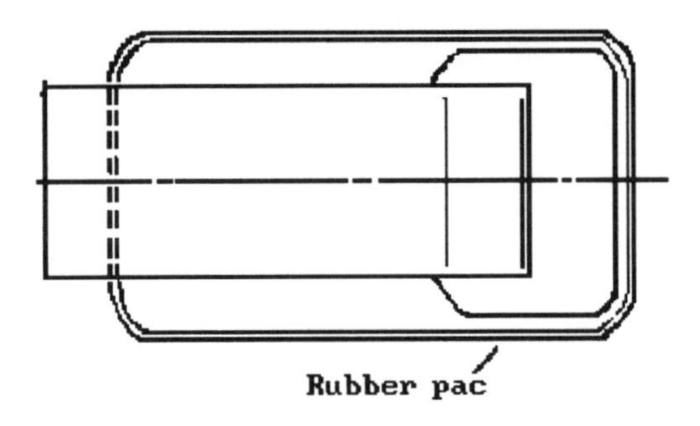

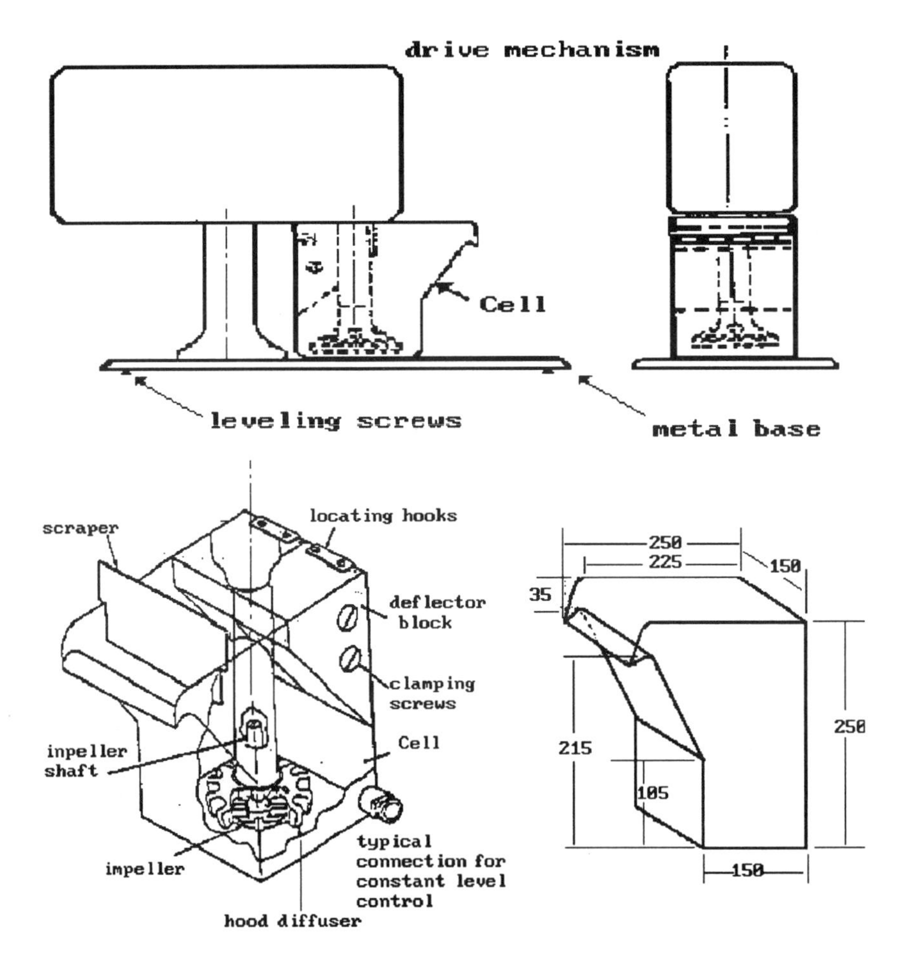

Fig. 81.- Flotation cell proposed by ISO committee on coal testing
Note that a deflector block replaces the mechanical paddles

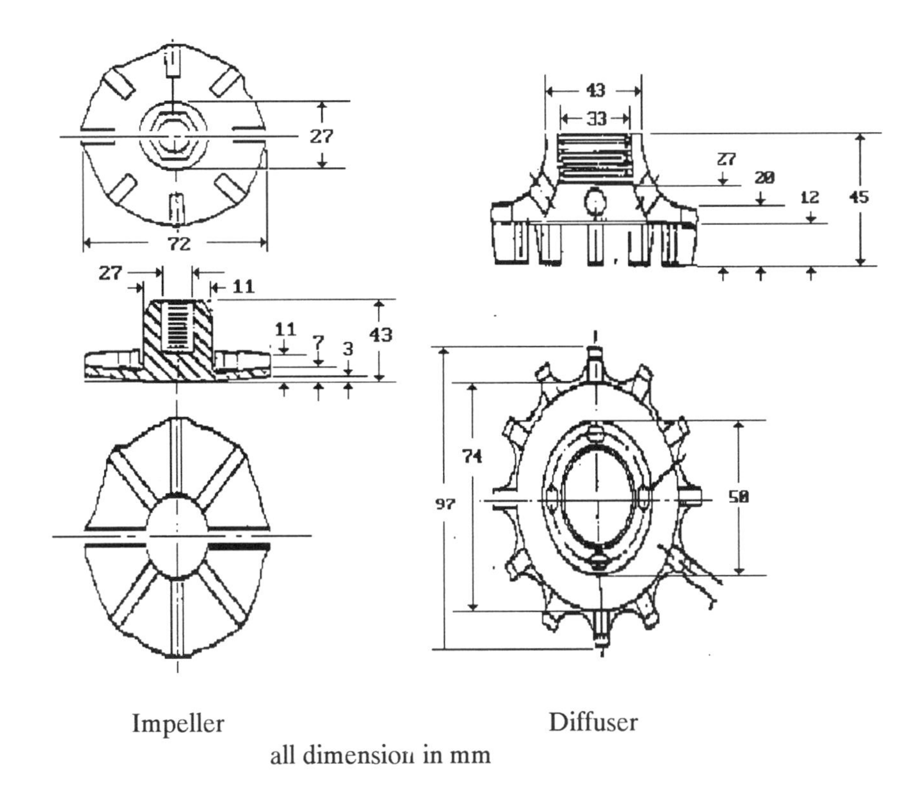

Fig. 82.- Flotation cell impeller and diffuser dimension for coal flotation machine, Fig. 81

normally handled as dilute water solution. Lime is handled as is usual in the mill, as a standard suspension.

Some texts recommend that non-miscible organic collectors should be dissolved in an alcohol, or another organic solvent, to improve the accuracy of feed in the lab. This very definitely is not recommended, as all the alcohols and many of the organic solvents have frother properties, and frother-collector interactions are known to very significantly change the flotation characteristics of the reagents. The phenomena does not necessarily scale up from the laboratory results. As an example of the drastic effects of organic additives, the El Teniente mine in Chile, for 15 years, used a mixture of lead-containing gasoline, plus MIBC, with their traditional xanthogen formate collector. This operating change was made because it was found to reduce the collector consumption in half, without affecting copper recovery, and in addition increased molybdenum recovery in the mill. But ore changes reverted this effect, so Teniente is back to using a considerably reduced amount of straight diethyl xanthogen formate.

Insoluble liquid organic collectors, such as the alkyl thionocarbamates, xanthogen formates or the xanthic alkyl esters, can be added to the flotation pulp using a hypodermic or any other method which produces calibrated drops. In labs where a number of stable oily collectors are routinely used, it has been found convenient to manufacture custom eye droppers by fusing and stretching small diameter Pyrex tubes. With a little practice the diameter of the stretched tube can be adjusted so that the weight of the drop produced by the different collectors is the same. The labelled eye droppers equipped with standard rubber bulbs are then kept in test tube racks. The fact that the same number of drops gives the same dosage for the different collectors reduces experimental errors during testing.

The point of addition for each of the collectors depends on usual industrial practice, but for most of the oily reagents half to three quarters of the dosage is usually added to the ball mill prior to starting to grind, and/or to a conditioning stage. The use of a conditioning stage and reagent addition at this point also conforms to normal mill preference, unless it is a variable under study. It is simulated in the laboratory by initial operation of the flotation cell with the air cut off.

Selection of a float technician should be combined with his training. In the copper metallurgical labs that the writer is familiar with, the normal procedure has been to obtain a large sample of a well disseminated copper ore, train the new technician in the standard batch flotation process used in the lab for a day or so, and then subject the candidate for the flotation test technician job to at least a hundred duplicate 500 gm or kilo flotations. A competent technician, with a helper, can do at least 10 complete flotation tests per day, so the selection cum training process must budget at least a couple of weeks' time. The candidate is then considered ready to take on the job when his tests per day meet the laboratory norm and his last 20 or 30 flotation test tails consistently check out to within 0.015% copper between flotations, with a reasonably constant concentrate grade. Usually if the float man does not have the knack, it will show up after the results of the first couple of days of testing. This tails assay figure is based on the experience of a good flotation laboratory, and means that in recovery terms an accuracy of better than one unit of recovery is not attainable, with a reasonable number of tests.

The data obtained from manual testing, by a good float man, will give considerably more information than an automatic machine, or even one only dependent on mechanical paddling, because, for example, the way the frother has to be added to maintain a stable froth and the time changes in froth colour can give important clues to the rate of flotation of the minor minerals which can be very important in evaluating scale-up.

Comparison of the performance of commercial laboratory flotation machines

A recent paper from South Africa, O'Connor et al (1987), provides data on the comparative performance of Denver, Wemco, Agitair and Leeds Autofloat machines (with square and round cells). The comparative testing was done with a blended gold slime residue obtained from an old dump near Johannesburg. The slimes contained 2.6 wt% pyrite and screened 75% -200 mesh (75 micron). The standard flotation conditions and the dimensions of the machines and cells used are shown in Table 93.

Table 93.- FLOTATION CELL DIMENSIONS AND STANDARD FLOTATION TEST CONDITIONS

Collector	90 g/t Na mercaptobenzothiazole
Frother	17 g/t Dowfroth 250
Aeration rate	6 l/min (STP)
Solids concentration	1250 rpm
Froth height	20 mm
Nominal cell volume	3 litre

Machine	Leeds	Agitair
Impeller diameter, mm	80.0	90.0
Peripheral speed m/min	314	353
Square Cells		
dimensions, mm top	200 x 140 x 170	155 x 158 x 145
bottom	140 x 60 x 120	155 x 158 x 145
Working volume, cm^3	4,000	3,550
Cell surface area, cm^2	280	245
Froth crowding ratio surf/vol.	0.070	0.069
Round Cells		

Machine	Denver	Wemco
Impeller diameter, mm	72.0	50.0
Peripheral speed m/min	282	196
Square Cells		
dimensions, mm top	190 x 150 x 140	
bottom	160 x 90 x 65	
Working volume, cm^3	3,590	3,350
Cell surface area, cm^2	285	305
Froth crowding ratio surf/vol.	0.079	0.091
Round Cells		
dimensions, mm diameter		197

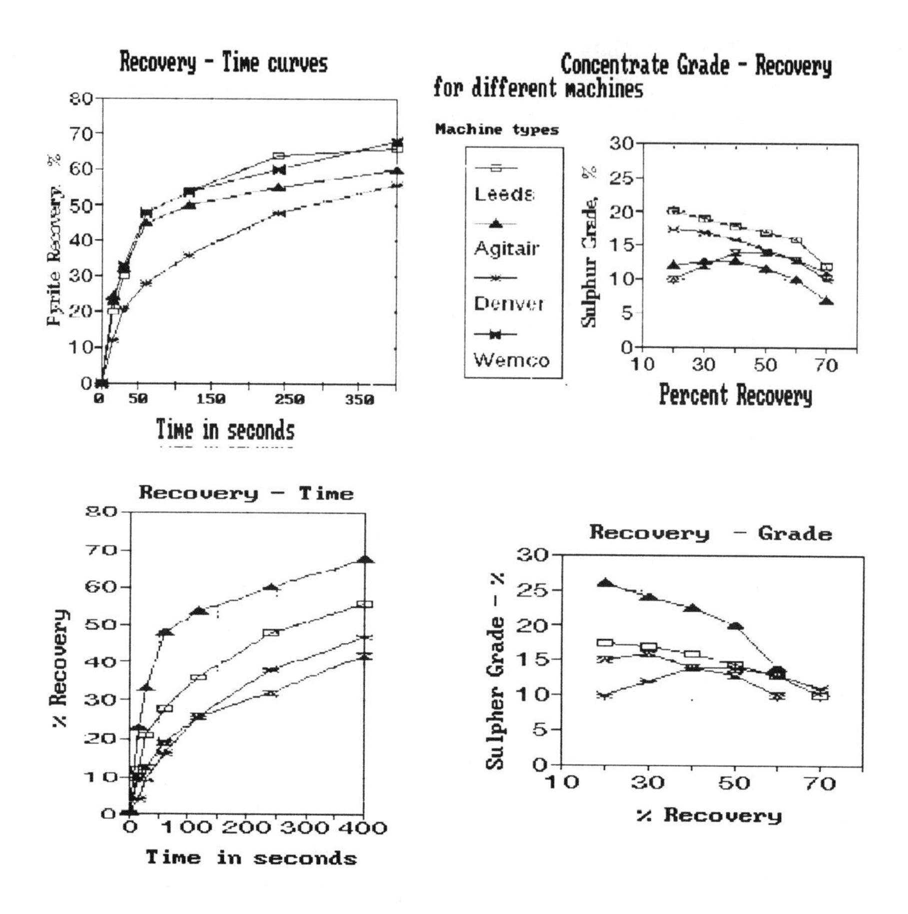

Fig. 83.- Comparison of the performance of commercial laboratory flotation machines

The results of time-recovery profiles and machine selectivity, expressed in terms of the grade recovery curves, are shown in Fig. 83. The authors chose to reflect the reliability of the machines by a comparison of the standard deviations of the individual points in a time recovery curve for the Leeds and the Denver machine. These data are in Table 94.

The authors' conclusions from these data are:

> "When the results of laboratory flotation studies are compared, the type of cell mechanism and the shape of the cells must be considered. The Leeds cell has been found to yield the most reproducible results over the entire flotation

time. This cell also performed best with respect to grades and recoveries. The use of a round cell caused slower flotation with poorer recoveries and higher grades."

It is unfortunate that the authors did not include a Hallimond tube in the devices studied, as by far the largest number of theories on the mechanism of flotation have been based on data obtained with Hallimond tubes and bubble contact angle measurements. The comparison was also only between different makes and designs of bench scale testing equipment, and not a verification of scale-up capability. As the purpose of a laboratory test is to simulate an industrial installation, the machine with the "best" kinetics and recovery in the laboratory might or might not provide the poorest scale-up data.

The data are useful as a guide to the selection of a machine for a specific experiment, as most flotation laboratories own various different types of machines, not because of confusion on which are the best, but because for different ores it is necessary to pick a machine that most closely follows the responses of the specific full-scale unit under study. This, and the ease of getting into a testing rhythm, is the key to the empirical preference of good flotation technicians between different machines.

As there is an incomplete description of how the different machines were operated, the validity of the standard deviation data in Table 94 cannot be evaluated. It is possible that it is more an evaluation of the float man, or men, who performed the tests, rather than of the inherent properties of the two machines.

Table 94.- REPRODUCIBILITY OF TIME-RECOVERY CURVES FOR LEEDS AUTOFLOAT AND DENVER FLOTATION MACHINES

Sulphur recovery, % (standard deviation in parenthesis) versus Time, seconds	15	30	60	120	240	480
Denver	-	12.7	24.6	35.9	47.8	58.6
		(1.81)	(1.98)	(2.24)	(2.63)	(2.54)
Leeds	59.1	74.8	86.8	91.2	92.0	92.2
	(0.62)	(0.65)	(0.99)	(1.18)	(1.31)	(1.31)

Sampling and sample preparation

Normally the most difficult part of doing reliable ore testing is to obtain a durable representative sample of the ore to be tested. The most readable, reliable, practical and comprehensive treatise on this topic is contained in the first 71 pages of section 19 (written by H.A. Behre and M.D. Hassialis) of Taggart's Handbook of Mineral Dressing. The SME Mineral Processing Handbook, Section 30, updates the descriptions of recommended mechanical sampling equipment, and is a good complement to the Taggart information.

Sample Size: the quantity required to provide a representative sample of a particular section of an ore body is nearly always underestimated by the client, so care to communicate the procedure for obtaining the test sample, and its minimum size, must be clearly specified by the test lab. Luckily the statistically required sample size usually is greater than the amount needed for the routine analysis and processability tests.

Taggart's recommendation on minimum representative sample size is approximated by the following relationship between sample size (m) and largest particle diameter (d):

$$m = kd^n$$

where the empirical parameters, based on Taggart's tables, are defined in Table 95 for sample sizes in pounds, and minimum particle diameters in inches:

Table 95.- PARAMETERS TO DETERMINE SAMPLE SIZE FOR DIFFERENT MINERALS

No	k	n	Ore Type
1	30,000	2	Gold
2	3,000	2	Silver
3	1,000	2	Low grade uniform distribution
4	9	1.5	Base metal, high grade
5	1,100	2.13	Low grade, highly variable
6	3,500	1.8	Average grade, variable

These values are plotted for the numbered curves on log paper in Fig. 84, expressed as sample sizes in kilograms, or tonnes, and particle sizes in millimetres.

The theoretical basis for this minimum sample size definition assumes that an infinitely large, inherently uniform, sample source is available with only short scale anisotropy. In practice, these conditions do not describe a mineral body, rather they are those of a specific shipment of ore, and only if the ore is of a uniform size and the property to be measured is the main component assay. Even under these very ideal conditions, the minimum size sample will only be representative if the sample gathering procedure is random and

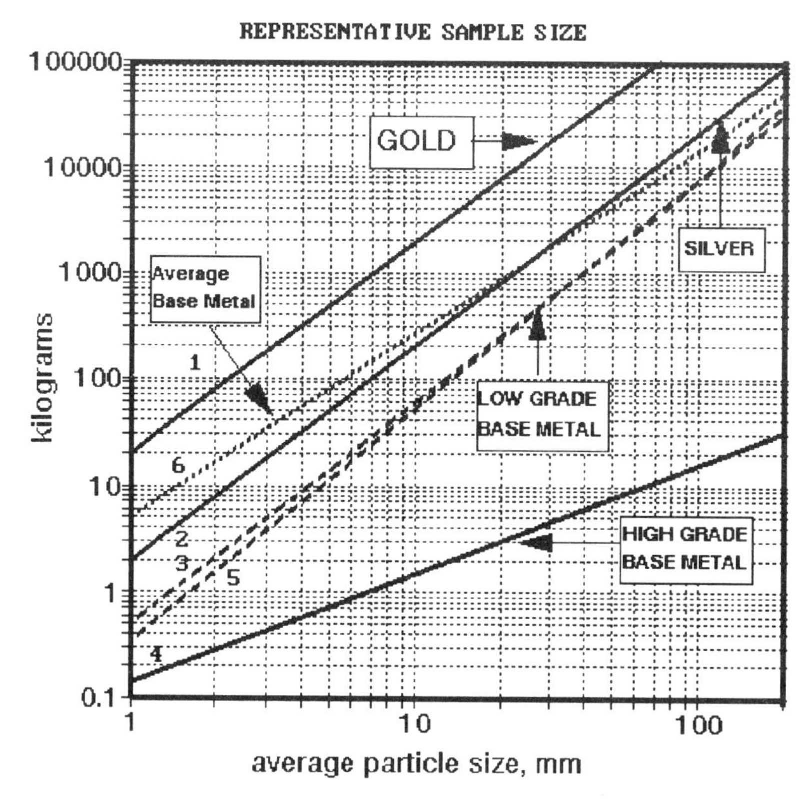

Fig. 84.- Minimum representative sample weight as a function of particle size

systematically unbiased.

To provide reliable samples for valuable ores where the error involves large sums of money, such as in the purchase of gold ores, Taggart describes sampling plants whose construction costs would be comparable to the investment required for a full size crushing plant. The designs include elaborate continuous sampling devices located at the different size mineral transport points, just to provide reliably representative samples for assaying. He quotes an amusing study made in the thirties which shows that, despite these precautions, gold assays favoured the seller or purchaser, depending on which party did the sampling.

If it is not really possible to obtain completely reliable samples of homogenised streams from a crusher circuit, obviously there is no practical basis for assuming that an ore-body can be sampled in any way that could be considered representative of the processability to be expected during a mine's lifetime. Not even a non-technical business planner underestimates this point when the ore is sourced from a vein deposit, or has a weathered cap, with secondary enrichment overlaying a primary ore. But even in the case of a giant low grade porphyry, it is easy to overlook the potential unreliability of samples, because even large disseminated low grade ore bodies are far from mineralogically uniform, usually having a barren central pipe and a non-uniform radial distribution of co-products, such as gold, silver, molybdenum and arsenic.

The pattern used to collect samples for reserve calculations of an ore body are determined by the need for geologic information on the formation, and are carefully evaluated three dimensionally, using arbitrary cut-off grades, to obtain a meaningful reserve figure for investment decisions.

In the preliminary evaluation of the processability of a particular ore, a similar approach must be adopted in obtaining and evaluating mineral samples for processability evaluation. Thus, the selection of samples to test for treatability should be done in cooperation with the geologic team, because changes in mineralogy will affect the efficiency of the recovery process and should be taken into account in planning the experimental programme that will be the basis for the concentrator design. Furthermore, the mining staff should also be included at an early date in planning the metallurgical test programmes as, particularly in large mines, it is routine practice to blend ores from different mine fronts to optimise processability. So certainly there should be fluid and systematic communications between the geologists, the mining planners, and the metallurgists. Careful attention to these details are very important during the sample planning process as drill cores, which are generally adequate to provide reliable data for reserve calculations, have only a very limited value in metallurgical testing, while adits, whose costs are high, are of little value in confirming reserves, of more value to the mine planners to give an idea of geologic integrity and extraction problems, but essential to the metallurgical study as these are the usual source of the larger ore samples needed to test processability. A poorly selected adit location can therefore result in costly design mistakes in a concentrator, while satisfying the information needs of the geologists and miners.

Although the most representative assay samples from an ore body are obtained if a very large sample is crushed to a small uniform particle size and then quartered systematically, finely crushed material is not ideal for an extensive flotation study, as the sample should have the minimum of prior crushing to assure that the material tested, on any one day, has as close as possible the same freshly generated surface history as the ore that will be fed to the head end of the flotation cells in the future concentrator. Therefore, for metallurgical purposes, it is important to choose a primary sample particle size that will not bias the results because of loss of high grade fines during quartering and transport, as well as to provide a minimum of old mineral surfaces. To keep stable samples, service labs usually will keep their ore samples in an inert atmosphere and at as low a temperature as possible, particularly when a sulphide ore is involved.

At a lab that the writer was associated with which dealt with repetitive work on optimising reagents for a particular mine, it was possible to maintain a large ore sample unchanged, in terms of its flotation characteristics, for a year or more. Samples, on arriving at the lab, were crushed to half inch size, quartered into convenient equal size samples, and stored in sealed plastic bags at freezer temperatures. At the same time a number of attempts to obtain untarnished samples from a mine in a humid location (Bouganville), where transit time was a number of weeks, was never successful. Nickel ores also seemed particularly prone to alteration, and had a very limited storage life.

O'Connor et al (1990), whose study is an excellent model of an experimental design and evaluation, provide evidence of the durability of pyrite and arsenopyrite in an 18 month study done on an East Transvaal gold deposit. They point out

> "The reproducibility of the flotation procedure has been shown to be excellent. In all tests the effect of the parameter being investigated was compared to that of a current standard run. This led to the discovery of the importance of aging on arsenopyrite recovery. ...
> The generally improved separation efficiencies using aged ore led to an investigation of methods to artificially age the 'fresh' ore. Initially the fresh ore was leached in order to see whether the aging process could be reversed ..... (this) improved pyrite and arsenopyrite recovery and hence reduced selectivity. The leached ore was then heated to 42°C in an oven for 13 days. ... this treatment did not simulate the aging process .. (and).. caused a significant reduction to pyrite recovery."

Sampling working concentrators

Sampling ore for flotation testing in an existing mine is easier because of the number of points where mineral streams have automatic sampling devices already installed for analytical purposes. In choosing where to obtain a flotation sample, the source should always be the coarsest available continuous stream of dry ore. Because of the aging phenomena, the source point chosen should also minimise the time required to obtain a composite that is representative of a particular period of operation or a particular region in the mining area. There usually is enough available information from the operating sampling system to make such a rational sampling plan possible. If the investigation required is the processability of an ore that will come from a future mine operation, the sample selection process and the storage precautions become the same as those required in a service lab, i.e., a larger than necessary sample of run of mine ore, which is then reduced in volume by the use of the quartering procedures recommended by Taggart to obtain a representative sample of the primary sample. This material should then be crushed to no less than plus half inch, and stored in nitrogen purged hermetic plastic bags in a freezer. A household freezer is the cheapest, and usually is quite adequate to keep samples for a year or more.

Sampling an operating mill for immediate flotation testing is usually easiest, and most representative, if periodic samples are cut by an automatic sampler from the belt feeding dry ore to the rod or ball mills or from the belt leaving the secondary or tertiary cone crushers, as long as the circuit does not have by-passes (fines from a grizzly) or recycles of oversize. The standard period chosen to make a composite will depend on the frequency of fluctuation in the ore characteristics and the objectives of the test. For very special cases a pulp sample from the ball mill discharge may be adequate, though normally it is to be avoided as the aging of the mineral surface under normal pulp conditions is impossible to stop and significantly affects floatability. It is not unusual that if the metallurgical problem to be solved requires testing pulps, then the laboratory flotation equipment must be set up in the mill building, close to the final ball mill or conditioner tank discharge. Freshly fractured mineral surfaces, in contact with water, will change their characteristics in seconds, then undergo a second, less rapid, aging, which is also uncontrollable, so results from pulp tests must be carefully evaluated.

Because the amount of information on the mill design and operating conditions, as well as the amount of metallurgical and mineralogical background information on the ore body, is vastly different in each case, these factors must be taken into account in fixing the minimum sample size that will be representative and adequate not only to complete the planned test programme, but also to provide a reserve duplicate sample to confirm results if third parties, such as banks, need convincing.

Summarising, the testing objectives in a flotation laboratory generally are to determine:

1.- the general processability of a particular ore with conventional reagents,
2.- the amenability of the ore to being processed in an existing concentrator, and the operating conditions required at start-up,
3.- the reagents and conditions required for optimum metallurgical results when an ore's characteristics have changed, or are expected to change as the mining fronts advance,
4.- optimisation of an existing operation, or processing circuit, and/or the determination of the relative merits of a set of new reagents as compared to existing reagents in use in a specific concentrator.

To meet these objectives, representative samples of the test ores or the various ores that will be processed in future, must be obtained, and properly protected from alteration. The minimum sample size required will rarely be less than 5 to 10 tonnes, if extensive flotation testing is required. In the case of a major investment in a gold mine, where finely disseminated ore is the norm, representative samples for plant design usually range in the 200 to 1,000 tonnes of run of mine ore.

If the ore to be treated has been stockpiled, sourced either from low grade overburden or waste, or, in the case of precious metals, an alluvial gravel, there are statistical sampling models, such as those in Taggart, or Gy, which relate the size of the representative sample with the particle size of the disseminated metal and the average size of the blasted ore, to the size of the sample required. A representative sample of a gravel may be orders of magnitude smaller than if the ore is in the form of 10 inch lumps.

Preliminary sample treatment and characterisation: The evaluation and assays that samples are submitted to, prior to flotation testing, depend on the tester's familiarity with the particular mining operation involved, and the scope of the study to be undertaken. In all cases the sample should be checked for size uniformity and the existence of extreme fines. If fines are prevalent, they should be assayed independently to check if their metal content is substantially different in composition to the bulk of the ore. If so, the sampling method used in the mine or by the geologists should be specifically reviewed to assure that the sample has not lost fines and is really representative of the ore body to be tested.

If the bulk samples received at the laboratory are wet, and this is not representative of the deposit, they should be rejected and replaced with fresh dry samples. If the ore-body or the waste heap is naturally wet the samples should not be dried, as long as the test work can be carried out promptly and the samples can be kept sealed to maintain the moisture content constant during the testing period. Otherwise a test flotation should be done with a representative sample which has been split into two portions, and one portion dried by the most convenient low temperature process available. If the flotation test indicates that dried samples do not duplicate the tests performed on the wet samples, the only alternative that might permit delayed testing of the wet ore is freezing quartered and sealed ore batches. Whenever naturally wet ore is to be tested, additional replicating standards should be incorporated in the test programme to monitor mineral surface aging during the test period.

On receipt, a normal sample should be screened and the plus half inch material crushed separately, usually in a laboratory jaw crusher. The under half inch fraction can then be screened to remove the minus quarter inch and the oversize reduced to quarter inch with crushing rolls. The whole sample should then be homogenised and split into about 10 kg portions using standardised equipment. A rotary splitter (such as a spinning riffle) is convenient for this step. The 10 kg samples are then sealed in plastic bags, and, if the storage time will be long, placed in freezers.

Unknown samples from an exploration programme or from someone else's property should be carefully sampled and a representative common sample submitted to a reasonably complete chemical analysis; the exact elements that should be included will depend on the nature of the ore. In addition, physical properties such as density, hardness, abrasiveness, grinding index etc., should be determined. Microscopic mineralogical examination should include a number of random samples to evaluate crystallographic uniformity of the ore. The report should quantify the minerals identified and give an estimation of dissemination. This is also the stage at which some idea of the grind that will give good liberation can be estimated by judicious choice of metallographic samples.

Otherwise the extent and the type of primary non-metallurgical data needed will depend on the objectives of the test planned, and therefore how elaborate an experimental design will be required to accomplish the purposes of the test at minimum cost. For most copper properties, the head assay should include total copper and total non-sulphide copper, as a significant oxidised mineral content would require exploring more complex flow schemes and retreatment loops, as well as the possibility of sulphidising prior to flotation.

Testing routines

The procedure used in a flotation testing series must take into account the known surface instability of metallic sulphide ores and the possible aging effects that can occur with all minerals; thus, it must follow a very strict sequence and rhythm. Unless the concentrator will have an unusual location, such as at extreme altitude or latitude, the tests are all run at the ambient temperature of the laboratory. It is recommended, though, that a pulp temperature record be kept for each test, and duplicates be included morning and afternoon, as a long test series may involve a systematic rise in the pulp temperature during the day, and produce a spurious trend in the recoveries.

1.- A day's supply of ore should be ground to minus 10 mesh using double rolls, then homogenised and weighed into accurate sample sizes for each flotation. Usual rougher flotation standards are 500, 1,000 or 2,000 gram lots.

2.- The clean mill is opened and placed on its end, and the ball charge added. This charge normally consists of around 10 kg of ball bearings in half to one and half inch size. Some operators like to use a mixed charge to mimic the wear in the full scale mill. Whether this helps the size distribution of the ground ore to be closer to the plant material is a matter of opinion. In any case, whatever procedure is used, it should be adhered to strictly. This is then followed by the ore sample and enough water to make up a 50 to 70% solids charge. At this time usually a standard amount of lime is also added for pH control, and about half the planned collector charge, if the collector is a non-soluble oil. The motor timer is then set to the previously determined standard grind, which usually will be in the 3 to 10 minute range.

3.- For the next stages the use of a prominently visible large second timer is essential as, for reliable flotation results, the need for extreme consistency in the time between the different operations cannot be over-emphasised. Usually the tests are carried out by the flotation technician and a helper. The start of the grind should have been coordinated with the flotation test, so that when the mill has stopped, a clean flotation cell is available. The transfer of the ground charge to the cell is usually carried out using a coarse mesh screen-equipped funnel of a slightly larger diameter than the grinding mill, which fits over the flotation cell. The mill is turned over onto the funnel, which retains the ball charge, and the remains of the ore charge adhering to the mill walls and balls is washed out with the water that will be used in the flotation test. Usually the mill pulp densities are in the 25 to 40% solids range. In the flotation tests the pulp density varies during the time of the test and should usually be set somewhat higher than that used in the mill at the start of the batch flotation.

4.- The cell is placed on the machine stand and the agitator lowered into position. The time between the end of the grinding cycle and the start up of the flotation machine agitator should be kept as uniform as possible during flotation tests in each series. To be able to do this, each step of the transfer procedure should be such that it is carried out in the same time at the start and the end of the day when the operators are tired, so it should be consistent but unhurried.

5.- The exact sequence followed after the start of the agitator will depend on the ore and the experimental design. Usually there is a short conditioning time which is used to adjust pH and add the frother and any other of the reagents required for the particular test being run. The reagent addition time should be consistent and recorded, as should the time of each later addition of reagents. Flotation machine settings, such as agitator speed and air rates, should be standardised and recorded. Power consumption measurements during the course of the flotation can be of value but usually are not essential.

6.- After the start of aeration, the consistency of the rhythm and depth of scraping, plus the pattern in which the froth from the rear of the cell is recovered, is the secret of accurate reproducible flotation results. The procedure is certainly not the same for every operator, but always must be consistent. If there is significant change in pulp density, the air flow rate valve is adjusted periodically to standardise the air supply with the rotameter. Depending on the rate of flotation of the different minerals contained in the ore being tested, the timing of the scraping can consist of a sequence such as a stroke every 15 seconds for the first 2 to 4 minutes, followed by 30 second intervals for a similar period, and continuing at 60 second intervals thereafter. The pulp and froth levels are kept constant by washing down the agitator stem and the cell sides with flotation water from a clip-controlled rubber tube fed from an elevated tank. This washdown also must be at fixed time intervals, say every 60 seconds. Frother is added as needed during the whole period of the test. A common problem in testing is over-frothing. This preferably is foreseen and controlled by limiting the amount of frother used in each addition, or in some cases by varying the time in which stage addition of the collector is done.

7.- Normally the concentrate is collected for equal time periods, separating each batch. The collection process ordinarily covers shorter periods at the start of flotation. During all the flotation period the colour of concentrate coming over the lip should be noted, as this can give clues to the rate of flotation of the different species.

8.- The length of time that a given ore will be floated will depend on the plant design and the kinetics of the system. Usually tests run from ten to twenty minutes each, but can be over an hour if the limiting recovery for a number of the mineral species is being studied.

9.- Generally, concentrate samples are collected on circles of filter-paper and dewatered on a large Buchner funnel by vacuum filtration, and the full sample kept for assay or blending to obtain a common concentrate assay sample. The pulp remaining in the cell is also treated in the same way. All samples are dried in a low temperature tray oven or on a steam table, and carefully weighed. Printed data sheets are a must, as they help keep the data collection standard between different operators and serve to remind the operator when certain of the data must be recovered at fixed time intervals.

With a skilled two-man team, about ten flotation tests per day can be completed. In laying out the experimental programme it is important to keep this sort of limitation in mind as the unit - grinding, flotation, and sample collection - cannot be broken up.

Tests in which concentrate cleaning has to be checked, or where the effect of primary reagent effect on by-product separation efficiency such as moly circuits is involved, usually require starting with 2 kg ore samples and testing the cleaning step in smaller, 100 to 200 gram test cells.

Test procedures used by a reagent supplier and a mill

The following describes the flotation procedure used at the Dow Chemical Company when it was active in the collector market; it is summarised from a paper by David J. Collins, presented July 1970, at the 6th Annual Intermountain Minerals Conference, Vail, Colorado:

Ore samples received from the mines, which range in size from 75 to 500 lbs., are first crushed in a 4 by 6 inch laboratory jaw crusher which discharges on to a 12 by 30 inch low head vibratory screen equipped with a 20 mesh deck. Oversize is then further crushed in an 8 by 5 inch roll crusher which also discharges on to the vibrating screen. Undersize is collected in large trays placed under the assembly. Dust control is achieved using vacuum collectors located on top of the screen and crushers.

After crushing, the samples are mixed for 30 minutes in a 2 cu. ft. twin shell V blender, and then riffled down to provide suitably sized charges for splitting into 500 g samples.

Following splitting, the samples are weighed to 500 g, and stored in heat sealable plastic bags. .. For those ores which oxidise rapidly, the plastic bags can be purged with nitrogen, sealed, and then stored in a freezer. The bags are easily opened and empty cleanly.

All samples are wet ground in a 8 by 8 inch 304 type stainless steel ball mill using a 304 type stainless steel ball charge. The ball charge is made up as follows:

Ball size	No. of balls
1.25"	14
1.00	41
0.75	99
0.50	178

total ball weight = 7.4 kg

The mill is rotated at 70 rpm using standard laboratory rolls. Following grinding, the mill is dumped into an aluminium container which has an aluminium rod screen on the bottom. The balls are easily washed using a spray type nozzle and, due to the lightness of the assembly, the balls can be agitated with a minimum of effort. The slurry is caught in a funnel and discharged directly into the flotation cell. Speed is emphasised here: the whole operation, discharging and washing, takes from 30 to 60 seconds. Using two identical cells, this means that the next batch of slurry is ready for flotation immediately after the previous flotation test has been completed. Water generally employed is deionised water which is supplied to all washing hoses. In addition, the laboratory is equipped with water storage and pumping facilities which can be tied into the equipment water supply system, and enables tests to be conveniently carried out using special water solutions.

Flotation Testing: The flotation cell that Dow employed had been completely modified to substitute for the usual square glass unit. Their cell was constructed from a 4 1/4 qt. 304 type stainless steel beaker with the conventional 'quiescent zone' and lip. They felt it desirable to incorporate the beneficial properties of a circular cross-sectional cell, and thus eliminate dead zones which would allow material to settle and cause variation in flotation results. The full cell width lip was retained in the modification. Stainless steel punched plate baffles were installed on the standpipe immediately above the agitator. These reduced excessive swirling, particularly of the froth column. The cell was also fitted with a bracket, mounted on the side of the unit, to which was attached a 10 rpm paddle used for froth removal, as they thought that this device would greatly eliminate human error or bias in a flotation test. Experience in other laboratories indicates that this unbiased operation provides less reliable and complete data than a good operator who can visually identify the different minerals in the ore and their relative rate of recovery.

The machine was supplied with supercharged air. The rate of flow was monitored using a glass rotameter, and maintained at a constant rate relative to the cell surface area of 0.038 $l/cm^2/min.$ (or about 7.7 l/min.). Water for initial make up was added via a pipe which fed through the standpipe of the machine. Pulp levels were measured at the beginning of each test and this level was kept the same from test to test. Impeller speed was monitored, and usually maintained constant at 1750 rpm.

The water, from a graduated plastic squeeze bottle, was used to wash down the sides of the cell and standpipe during flotation, and was added at the same rate as the volume of froth removed. The volume of froth removed was monitored using a graduated receptor pan. It was felt that maintaining a constant pulp level, and hence a constant froth column height, greatly enhanced experimental reproducibility.

The speed with which tests could be effected naturally depended on the conditioning and flotation times. As an example of what could be done, Collins indicates that, with relatively short conditioning (grind 4 minutes, condition 1 minute, float 4 minutes), two men, one handling grinding and sample filtration, and the other running the flotation, carried out up to 66 flotation tests in one day.

Product handling: Tailing samples were filtered in a laboratory pressure filter which had a quick release 1/2 inch inside diameter air valve on the lid, through which slurry was added with the filter lid secured. This prevented spillage of material if the filter shell was accidentally jostled. The filter shell was made of 304 type stainless steel, and was lighter than the commercial shell, it was non-rusting, thus non-contaminating to the samples, and released more cleanly from solids than the conventional filters. The filter was operated at 85 psi air pressure, which in general gave rapid filtration. Those samples which filtered slowly were flocculated. Concentrate was then recovered with a 6 inch diameter conventional vacuum filter.

This test sequence has been described extensively because it conforms generally to good operating practice in competent flotation laboratories, though the modified cell and the use of a mechanical paddle is not recommended except for very special studies. As noted, in batch flotation tests, the float technician can obtain important information on the relative rate of flotation of the different minerals when he hand strokes the froth out of the cell. This added information probably compensates for any improved uniformity of results which may or may not be obtained by using mechanically driven paddles.

At the El Teniente laboratory, the current procedure followed with in-house flotation tests are to grind a one kilogram sample of ore crushed to 100% minus 10 mesh in a 7 by 8 3/4 inch ball mill (Fig. 85a and b) loaded with 127 1" ball bearings. The grind target is 25% +100 mesh, which is met by grinding somewhat over 10 minutes. A sub-aeration machine is used, without a metered air feed. The ball mill charge is dumped into a standard glass cell using a perforated screen to retain the balls. These are washed with mill process water, and this wash is made up to give a final flotation charge with 33% solids. Flotation is carried out for a fixed time, usually less than 10 minutes. The froth is stroked off once every 7 seconds, the stroke starting at the rear of the cell and swinging around the spindle. Make up water is added after each stroke, from a squeeze bottle, the amount regulated by a level marked on the impeller housing. Both the concentrate and the tails are dewatered by vacuum filtration in a porcelain Buchner funnel mounted on an Erlenmeyer flask. A flocculant is added to the tails to reduce filtration time. All samples are dried in an electric tray oven.

David Collins discusses the equipment used at Dow Chemical Company, but not how the conditions at which tests should be carried out are chosen. We should add that the preliminary screening of the processability of an ore, and most mill optimisation studies as well as reagent replacement tests, are done simulating the rougher stage, without attempting to determine the final concentrate grade attainable. Batch flotation tests to determine optimum conditions on subsequent separations, such as cleaners, scavengers, etc., and the separation of more complex ores, such as moly/copper concentrates, or lead, zinc, silver operations, require so called locked cycle tests, which are very difficult to carry out reproducibly.

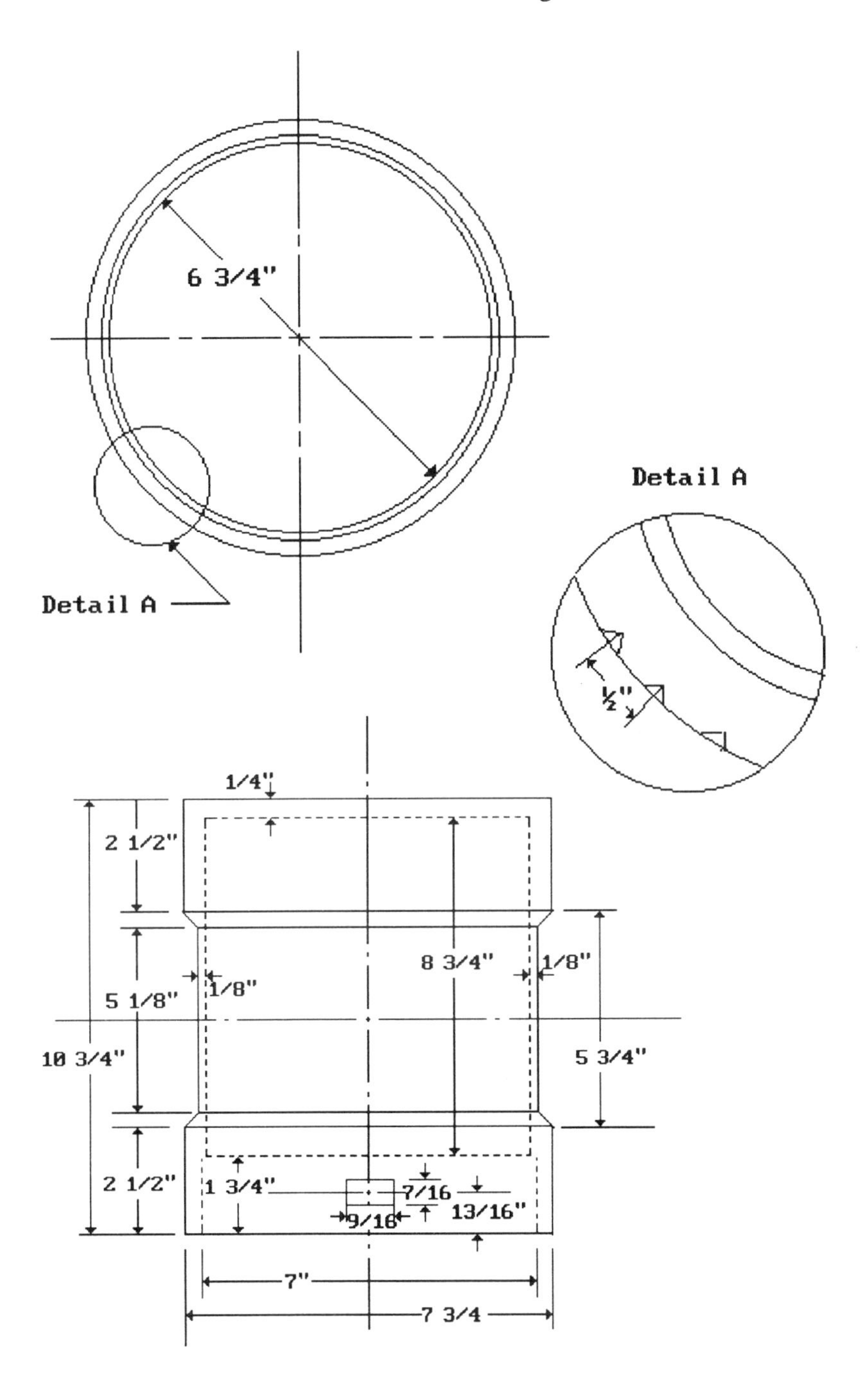

Fig. 85a.- Standard 2 kg ball mill used in Chile (body)

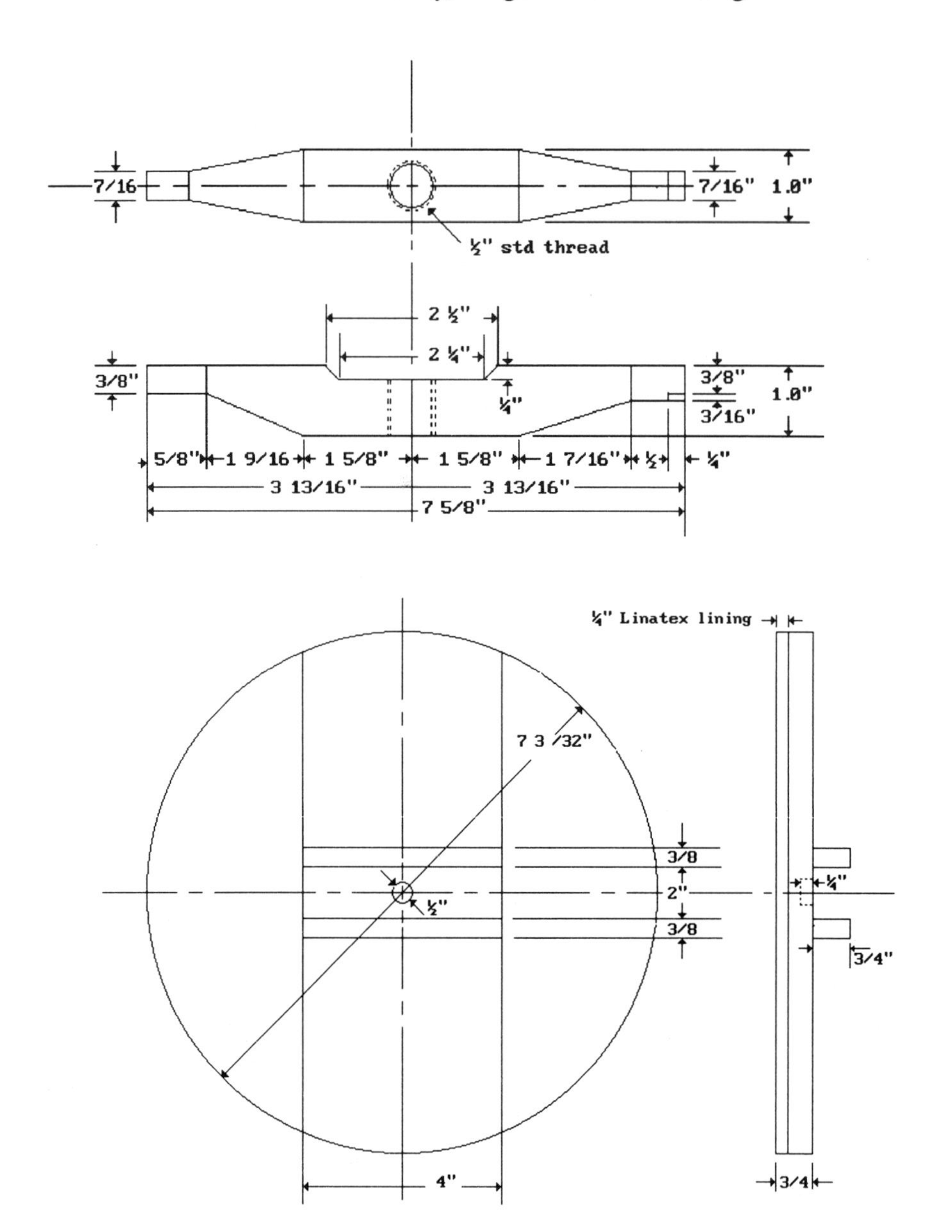

Fig. 85b.- Standard 2 kg ball mill used in Chile (cover and retainer)

Recommendations on setting standard test conditions

The procedure to be followed in running a batch flotation test with a new ore must start out with setting up appropriate grinding time and flotation test conditions. If the test involves an existing concentrator, the first step is setting the standard conditions which will be used as a comparison in the evaluation of results. These are arrived at empirically because the continuous industrial process will be approximated by a batch laboratory test. To do this, the exact rougher flotation conditions must be obtained from the customer. The information to be obtained from the customer (or mill superintendent if the lab is in-house) must include:

1.- representative head, tails and rougher concentrate grades for the type of standard ore which will be used during the test programme, including any mineralogical information which is available;

2.- the standard mill and grinding section operating conditions which resulted in the average tail, head and concentrate grades reported have to be obtained in detail. This information should include all reagents employed, dosage and addition points, as also pulp pH wherever it has been measured, plus any information on observed changes in the natural pH of the ore pulp;

3.- an annotated flow sheet, showing recycle streams, which will allow the calculation of all the solid and liquid mass balances, pulp densities, average pulp residence times in each of the operating stages, starting with the different stages of wet grinding through the conditioners, rougher flotation and scavengers. This is a good document to use to record any unusual temperature conditions;

4.- the operating data should include all pertinent mineral assays done during normal plant performance monitoring, and screen analysis of the feed, cyclone recirculation, rougher pulp feed, tails and rougher concentrate if available. Also the operators should be asked for their opinion of the rate of flotation of the different minerals in the ore as, from the colour of the concentrate overflowing the flotation cells along the length of the rougher, a good observant operator has some idea of how fast the different mineral species float.

The scope and nature of the data required should be set up a priori. Ideally the ore sample which will be used for the flotation testing should be collected concurrently with the operating data detailed above, in such a way that at the end of the data collection period a representative, stable, uniform sample is made available for the flotation test programme. Experience indicates that the sample size collected should be at least twice as much as the maximum amount that the experimental design calls for. As a tonne of ore can be packed in about three 55 gallon drums, which can be handled in a standard pickup truck, it is better to have an excess of sample available from the start of the test.

If the ores to be tested are the feed to an existing mill, the grinding conditions for the sample can be set as a given percentage retained by a 100 mesh screen, or better, they can be set up by trial and error to match the flotation results obtained under the mill's current operating conditions. To do the latter, a standard test flotation time is picked that is some ratio of the mill rougher residence time modified from previous experience with a similar ore. The pH, conditioning time, reagent feed quantities and conditions, including location (mill or flotation cells), and the pulp densities used by the mill, are generally duplicated as closely as possible. Usually, though, frother addition sequence and dosage is adjusted arbitrarily to provide an "adequate" froth. The amount of frother used in each test is of course recorded. Based on this standard laboratory flotation sequence, a series of ore samples are ground in the laboratory ball mill for different times and floated. From the tailings assays obtained, and the behaviour of the rougher concentrate grade, the standard grinding time and the standard flotation conditions are fixed. If the mineral recovery results of the series do not approximate the mill conditions, the grinding media load, or size distribution, or details of flotation conditions may need to be varied, and another standardisation series run. Usually the rougher recovery and concentrate grade in the lab is found to be similar to the plant, but the mineral particle size distribution and the frother dosage rarely coincide with the plant averages.

Using this procedure to set the grinding and laboratory flotation machine variables, such as air rates, agitator speed, sample times, froth dipping rate, etc., an experienced flotation technician can scale up his collector dosage and structure conclusions quite reliably. In all cases, though, the flotation results of changing reagents, dosages, pH, or grind must be confirmed by duplicating the experiments on a single flotation row (in older plants with smaller cells) or in a pilot flotation cell row fed directly from one of the concentrator ball mills. It is very rare that the effect of laboratory frother changes made in batch bench level flotation test will correlate with the plant. Therefore nearly all frother optimisation is done on a pilot row, followed by a full scale plant trial. As a consequence, laboratory collector testing is usually done in the lab with the standard mill frother, and if one finds that a proposed new collector requires a frother change to operate satisfactorily, scale up to the mill will require optimisation of the frother in a full scale mill test. Because of the danger of recovery losses, mill superintendents will be reluctant to do this without a very significant potential recovery improvement.

If the ore to be tested is completely new (a case that is routine for contract concentrators which purchase ore from small miners), or when a process development project for a new mine is involved, the grinding conditions for the test flotation are picked to approximate the size distribution obtained or obtainable in the full scale plant. Matching the size distribution of mill feed exactly is impossible, as industrial ball mills are invariably operated in a closed circuit with the oversize material from a cyclone or rake classifier returning to the mill feed. Under some circumstances, the use of a laboratory scale rod mill could help to change the particle size distribution of the laboratory grind to better approximate a full scale ball mill. Under no circumstances, though, should dry grinding in a disk pulveriser be considered, as the fines produced by pulverisers and the iron contamination make the resultant flotation results very erratic.

Laboratory ball mills: Collins describes the Dow Chemical ball mill as a commercial 8 x 8 inch laboratory unit. In Chile the standard mill is about 7 inches in diameter by 8.8 inches long, much lighter than an 8 x 8 inch mill, and easily made in a normally equipped machine shop from standard thin wall stainless steel pipe. The use of rubber lined stainless steel for the walls does not seem to reduce the iron contamination as compared to plant mills, as long as the balls used are standard steel ball bearings. CODELCO's sophistication of adding a Linatex rubber lining makes for a quieter laboratory, but otherwise probably has a negligible effect on the tests. The drive mechanism usually is a table equipped with two rubber covered rolls driven by a variable speed electric motor. These tables do double duty in the metallurgical labs and are used in agitation leach experiments or to prepare and homogenise reagent solution by rolling the glass bottles. The rolls can also be used to drive ceramic grinding mills if iron contamination is one of the variables important to the flotation.

At El Teniente the ball charge is 127 1" ball bearings, weighing approximately 10 kg. Ore is crushed to 100% minus 10 mesh and the normal ore charge to the mill is 1,000 g with 500 cc of plant process water (67.7% solids). The laboratory grinds flotation feed to 25% +100 mesh, which for current ores requires about 10 minute grinds.

One reason that flotation charges of less than 1,000 grams are not too popular in industrial laboratories is the need to use a smaller mill so as to approximate the product size distribution to that observed in the flotation mill. Wet grinding more than one float test quantity, and dividing the pulp into various portions, is definitely not recommended, as surface aging will vitiate the comparison between different test conditions.

The ball charge used varies with the operator. Usually it is a mixed charge of 1 1/4 inch, down to 1/2 inch ball bearings with a total weight of 7 and 10 kg. A typical size distribution of the balls used in a standard charge for a 7 x 8.8 inch mill is:

ball size	number	weight g*
1.25"	9	1,037
1.125"	16	1,530
1.00"	19	1,310
0.75"	103	2,304
0.50"	135	980
		7,161

*the void volume of this charge is approximately 870 ml.

Batch flotation tests

Field laboratory practice in batch flotation tests: the most straightforward flotation testing procedure is that used in the quality control laboratory of an operating mill. Generally the grind, flotation time, ph, and reagent addition points are all set by current mill practice. The test procedure consists of weighing out 1,000 g of a representative mill feed, crushed to 100% -10 mesh, and making up a laboratory ball mill pulp having around 70% solids using industrial process water from the plant. Usually the grinding time is set to give the same plus 100 mesh as observed in the flotation plant feed, and the lime required to adjust the flotation ph is added at the start of the grind. The ground pulp is separated from the grinding balls using a suitable coarse screen, and the pulp is diluted to 33% solids with mill process water. The flotation is carried out for a time that experience indicates properly simulates the rougher banks in the mill (frequently this is less than the theoretical pulp residence time in the plant). The concentrate is collected as a whole, dried, weighed, and riffled for an assay sample. The same is done with the tails. The heads are not assayed. The data is processed to give a calculated head grade, which can be compared to the bulk sample assay, and thus provide an index of the accuracy of the physical data, and the calculated recoveries for the different tests run.

A very important caveat which must be kept in mind when evaluating competitive reagents is that mill process water will have residual quantities of the standard mill reagents. These frequently will interact with new reagents in an unpredictable way, improving or impairing recoveries. Also, a bulk concentrate sample gives very little information on the effect of new reagents on the rate of flotation.

In a more sophisticated rougher flotation test programme, the grinding time and ball charge would be determined by a series of standard flotations, and the grinding conditions then chosen from the conditions that most closely matched the plant performance on tails, recovery and concentrate grade with the standard ore. Flotation tests required to evaluate new reagents are run at the standard grind time, and the flotation concentrate sampled over fixed time periods, with each sample weighed and assayed independently. The overall flotation time can be set to that available in the mill, or until the concentrate grade from the last sample drops to a specified level below the bulk concentrate value. This procedure provides kinetic data, as well as flagging differential flotation of the different mineral species in the ore.

Collector consumption, and the addition sequence in the test, usually correspond to the amounts employed in the mill. Frother addition on the other hand does not correlate well with the point of addition or the dosage used in the concentrator. In the laboratory tests it is added during the course of the flotation as a function of the changes observed in the character of the froth. Normally the test frother used is the same as that in use in the mill, or one that experience shows to be effective with the gangue and minerals that are predominant in the ore. The frother type must be optimised for the laboratory test programme, despite the fact that the optimum structure or mixture and the consumption does not scale up reliably, so that most definitive testing when a change in frother is involved is done on a test row in the operating mill, if one exists, or on the full mill after pilot planting. The frother is a key to controlling the rate of flotation of the different minerals, and therefore strongly affects concentrator grades, so that for unknown ores, chemistry and dosage should be optimised in the laboratory testing, but unfortunately frothers tend to act differently in the batch laboratory tests as well as being extremely sensitive to the water analysis, so the process of optimisation must be repeated at full scale.

Cycle tests are performed when the circuit called for in a mill includes returning a middling or cleaner tail to the rougher. This can be imitated in the lab by flotation of the concentrate to simulate a cleaning step, returning the tails from this flotation step to another rougher flotation of the same ore, and repeating the cycle until the cleaner tail assay becomes constant.

Mill testing is essential when studying the possible replacement of a frother, as batch lab tests are normally too insensitive to differentiate between frother types. Generally, if the mill has large ball mills feeding independent flotation banks, the test is run on one bank, comparing the recoveries and operation of the test bank with a similar bank with pulp fed from the same ball mill. If at all possible it is desirable during the test to use fresh water only in the test section, to avoid the effect of recycled reagents from the other flotation banks. If this is not possible, testing can become quite involved as it requires operating enough back to back tests, lasting at least a week each, to allow for a steady state amount of the reagent under test to return in the flotation water recovered in the tailings thickeners. Recycled water effects can be very significant not only on how a frother performs, but also on collector performance, and should always be considered in the experimental design.

At this point it is appropriate to re-emphasise the shape of the recovery response surface obtained when frother and collector dosages are varied (Fig. 73 page 267), and to note that pulp pH also interacts with the reagents; and that it is a common mistake to set both frother and collector dosage higher than required, as these dosages tend never to be lowered to check if recoveries might not actually increase with a lower reagent feed rate. Any mill that is using more than 30 to 40 gm/t of collector should at least run a 2 or 3 variable experimental design in the laboratory to check this point. For testing at full mill scale, an EVOP programme is recommended. For a description of this type of experimental design, see G.E.P. Box, "Evolutionary Operation - A method of Increasing Industrial Productivity", Applied Statistics (1957).

Designing batch tests: the simplest experimental study routinely faced by a copper mill metallurgist is the evaluation of the reagent dosage rates at his plant when the miners have noted a small change in the ore that is being delivered to the mill. His ore samples are obtained from the routine sampling points for the head grade determination. The experimental programme is a three variable experiment in which the collector, frother and lime dosages are varied. The most efficient and economic experimental design is cubic, with a symmetrical change in the dosages which can be plotted as the apices of a cube (8 conditions) plus a centre point which has been replicated to check experimental error. This is a ten flotation experiment, which should require one day's service from his laboratory flotation machine.

To minimise costs, the assays required can be limited to two or three head assays to verify the uniformity of his overall sample, and ten concentrate and tailings assays. Elements assayed would be total copper, non-sulphide copper, and iron. The in-house incremental cost of this series would probably run about US$1,000, including technician wages.

If the study required is to determine the processability and the optimum flotation circuits for a new ore, one still starts with an arbitrarily fixed grinding time, and then has to design a more comprehensive flotation sequence to try and determine the exact circuit which will give optimum recoveries for the particular ore. This can rarely be simulated in the lab.

The flotation conditions then depend on the bias of the particular float technician and his standard laboratory procedures. Most labs try to maintain a reasonably constant flotation environment by stage-adding reagents while adding fresh water to keep a constant cell level; or water containing the standard reagents, fed from an overhead container above the flotation machine, can be used so as to keep a standard cell level and thus approximately maintain a constant reagent concentration in the pulp.

Most tests include collecting timed concentrate samples during the test, rather than a single composite. Assays, and screen analysis, of these partial samples will provide flotation rate information and an idea of the floatability of coarse fractions.

Normally the minimum assay done is to weigh the concentrate obtained and assay only the tails, calculating the grade of the concentrate and recovery from these data. Better practice is to weigh and assay both the tails and concentrate and back-calculate the head assay to check the accuracy of the test.

From this brief description of the test procedure there are a number of points of non-duplication of the operating plant that should be obvious sources of possible errors. First, the ball milling cannot produce heads with a comparable size distribution to that obtained in the mill because there is no oversize fed back to the mill during grinding. Secondly, in the full scale mill, the rougher flotation is continuous and conditions in any one rougher cell are steady state and in no way comparable to a decreasing pulp density and overall grade that occurs in the batch flotation test. Thirdly, the environment is not identical to that in the mill as the dissolved air in the pulp and the age distribution of mineral particles, etc., cannot be duplicated in a batch test. Operator-affected factors include the rate of removal of concentrate, the rate of addition of water to maintain cell level, and the visual froth control through manual adjustment of air flows to the cell impeller. In practice the difference between a really good float technician and a poor one is the intuitive uniformity of flotation conditions maintained by the good technician just by observing changes in the froth and other operating characteristics. Most labs have tried mechanising the concentrate removal by installing rotating paddles, but as the level must also be controlled, most labs have found that their good technicians can obtain more reproducible results on the standard test than can be obtained by the mechanical paddles.

One problem that can be solved by a very simple trick is the reliable addition of fixed amounts of collectors and frothers to the float test. Most labs use hypodermic syringes. The operator changes needles for the different reagents, usually choosing a thinner needle for the less viscous liquids, trying to maintain the same volume drops. A procedure that has been found more accurate is the use of home made droppers made by drawing capillaries between pieces of 6 mm glass tubing over a Bunsen flame and then cutting the capillary at a diameter for each reagent that provides drops having the same weight for the different reagents. If the reagent is a non-volatile oil, at least a week's tests can be handled per filling of a particular dropper. Water soluble reagents should be prepared fresh every day. Generally xanthate solutions are fed as a 0.5% concentration solution, and conditioning agents such as sodium cyanide, sodium sulphate, copper sulphate, etc. are fed as 5% solutions.

Concentrate grade or copper recovery for a test series tend to fluctuate more erratically than the metal content of the tails, so that experienced metallurgists believe trends signalled by changes in the tail assays, more than changes in calculated recoveries. It is also considered a good idea to slip in at least one blind duplicate into each series of tests with a given ore for an error estimate, plus routinely duplicate or triplicate tests with identified samples. The reason for blind standards is not distrust of the tester, but rather the human impossibility to avoid all subconscious bias in flotation testing. Because of this human trait, and the lack of reliable automatic flotation testing equipment, one research lab has had the habit of keeping the flotation collectors and frothers unlabelled, but experienced technicians with a good sense of smell can usually identify most common frothers and collectors, so the use of unidentified reagents in a flotation lab is probably a waste of time. The steps required in testing an ore for floatability will explain the reason for the human bias problem involved.

Standard test procedure used by Sherex on coal tests

There is an extensive (40 page) standard for the flotation of hard coal issued by the International Organisation for Standardisation, which includes detailed examples of how to calculate all results. A similar standard is under study by the ASTM in the U.S.

Sherex's test procedures, detailed below, exemplify one company's approach to testing in this field.

Flotation apparatus required:
Denver subaeration D-12 machine
2,500 ml stainless steel cell
Pulp level - 2,400 ml
Impeller type - receded disk
Impeller diameter - 2 3/4 inch (70 mm)
Impeller speed - 1,250 rpm
Peripheral speed - 901.3 ft/minute (4.6 m/sec)
Air flow - 3.0 - 6.0 litres per minute

A.- *Rougher flotation conditioning time*
For setting dry samples - 5 minutes with 1,500 ml of water. Then with the impeller stopped, add water to the desired 2,400 ml level.

Before reagent addition - without air - 10 seconds (to redisperse solids).

After reagent addition - without air - 40 seconds
with air - 20 seconds

Flotation time
Three minutes, unless plant flotation time is known, in which case, plant flotation time is used.

Paddling intensity
Three two-handed strokes every 15 seconds. Paddle during the first five seconds and allow froth to build up during next ten.

Washing down cell sides
After the first 30 seconds (after the first three paddling intervals) start washing down the coal from the sides of the cell and the centre well shaft, while maintaining the original 2,400 ml pulp level.

Percent solids
Simulate plant conditions. If not known, use 200 grams dry weight, or 209 grams "moist weight", or 8% solids.

Reagent consumption
Use the same ratio of standard reagents (frother/collector) as used in the plants, and, for a better comparison, maintain this ratio to fuel oil or kerosene with Shur-Coal reagents. Testing other ratios can be beneficial, but the final decision on this will be made in the plant, where the effect of the hydrocarbon oils on the froth structure and volume can be taken into consideration in determining the optimum ratio.

When planning the test work, add the reagents in the ratio of their specific density. In compiling data for consumption curves, attempt to hit the same level of reagent feed rate as used in the plant, and at least two other levels, one of which should be twice the standard to determine overdose effects.

Visual observations
Make notes on froth structure, abundant froth volume, tenacity of froth, end of mineralisation, etc., because these often can determine the effectiveness of a reagent under plant conditions.

Lab flotation test sheets
The data is recorded in lab books and in lab flotation test sheets.

B.- *Stage reagent addition*

If the effect of reagent stage addition is to be determined, or if such a practice is used in the plant, the following procedure is to be used:

For wetting dry samples - 5 minutes with 1,500 ml of water. Then with the impeller stopped, add water to the desired 2,400 ml level.

Initial conditioning time
Before reagent addition - without air - 10 seconds (to disperse solids).

Add 60% of the total reagents with a microsyringe. The reagent should be injected below the pulp surface.

After reagent addition - without air - 40 seconds
with air - 20 seconds

Initial flotation time - 1 minute
Stop impeller

Bring water level back to 2,400 ml level by washing down the cell sides and the impeller shaft.

Turn on the impeller and condition for 10 seconds. Add remaining 40% of the total reagents with the impeller going.

Secondary conditioning time
5 seconds without air
15 seconds with air (to be subtracted from the flotation time)

Secondary flotation time
1 minute 45 seconds (plus 15 seconds of initial froth buildup in secondary conditioning step)

C.- *Cleaner flotation* (if applicable)
Transfer the froth product from the rougher flotation into 2,500 ml stainless steel cell. Add water till pulp is at 2,400 ml level with impeller-diffuser centre shaft submerged.

Impeller speed - 1,250 rpm

Conditioning time
Before reagent addition - 10 seconds
After reagent addition - without air - 5 seconds
with air - 15 seconds
Flotation time
1 minute 30 seconds

Note: to better assess the desired cleaner flotation time it is recommended to take incremental cleaner concentrates at 0 - 15 seconds, 15 - 30 seconds, and 30 - 90 seconds.

Instructions for paddling intensity and washing down all sides remain the same as for the rougher flotation.

Additional reagents to the cleaner flotation
Tests should be run with and without additional reagents. Recommended cleaner flotation reagent consumption - 0, 12.5%, and 25% of the total rougher reagent dosages.

Column testing

Flotation columns: The concept of flotation columns has been around for nearly 30 years, but it has become a hot topic since the copper mining crisis of the early eighties forced companies to drastically reduce capital expenditures and operating costs. A schematic of a classical flotation column is shown in Fig. 86. The obvious operating difference with flotation cells is the lack of an impeller, or any other agitation mechanism, generating a saving in energy and maintenance costs. Turbulence and mixing is deleterious to column operation, and is usually avoided in large diameter columns by incorporating internal baffles to maintain plug flow. The other major difference in the operation of a column, when compared to standard flotation cells, is that in most ore processing applications wash water is sprayed into the froth at the top of the column, impossible to do in a mechanical cell as it would kill the flotation froth. The amount of wash water added is a major factor in flotation selectivity, recovery and column operating stability. The ore pulp is fed into the column via a distributor located at about 2/3 of the height of the column; the tails are removed from the bottom, concentrate overflows at the top, and air bubbles are generated at

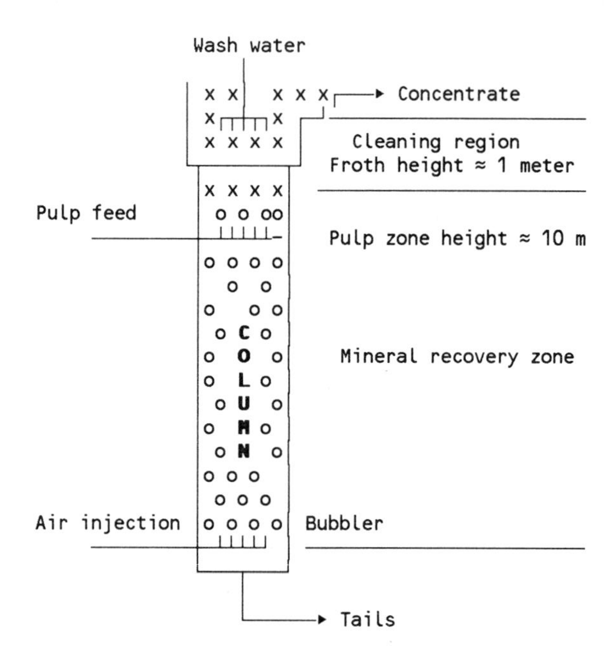

Fig. 86.- Schematic of a flotation column

the bottom of the column by a porous sparger. The design of this sparger is critical to efficient operation, as, with most process waters available at mine sites, plugging of the pores of the bubble generating device is a significant maintenance problem. The sparger is also critical in setting up a laboratory test programme, as bubble sizes and size distribution may affect plant operating efficiencies, so must be evaluated during a test programme.

Industrial columns have a height of 9-14 m, and a diameter of not more than 2 m, without baffling. Usually they are operated with sufficient overhead wash water to provide a net downward flow of water, a condition known as a positive bias. Positive bias has been the norm in column operation because the froth layer in a column is stabilised by the wash water. The greater the flow of water down the column, the greater the selectivity and the thicker the froth layer. The usual froth depth in stable operation is somewhat over a meter. A negative bias will eliminate the froth altogether, very deleterious for a process where the concentrate is the desired product, but a condition to consider for coal cleaning when the reagent selection results in the lower volume product (ash) going to the tails.

A major factor that must be kept in mind in the design of dressing experiments with columns is that the rate controlling flotation mechanism is always bubble capture, and usually with mineral that has been precoated with collectors in a prior flotation.

It is customary to describe the operating conditions of flotation columns in terms of superficial velocities (J) of the fluids, to normalise the information for different size columns. Typical values are as follows:

J_g = gas velocity - 0.5 to 3.0 cm/second
J_p = pulp feed velocity - 0.7 to 2.0 cm/second
J_w = wash water velocity - 0.1 to 0.8 cm/second
J_b = bias water velocity - 0.07 to 0.3 cm/second

In addition, in scale-up equations it is also customary to normalise the gas velocity for different height columns by the pressure correction:

$$J_g = \frac{(P_c)(J_g^*)(\ln(P_T/P_c)}{P_T - P_c}$$

where:

J_g^* = gas velocity at standard conditions at the top of the column
P_c = absolute pressure at the top of the column
P_T = absolute pressure at the sparger

For a 10 m column P_T is approximately twice P_c, so that $J_g \sim 0.69 J_g^*$. The effect of gas velocity on recovery and grade is dominated by the bubble size, which is affected by both the absolute amount of gas fed to the column as well as the pore size of the sparger. In practice, bubble sizes can range from 0.2 to 3.0 mm; industrially, a target range for normal pulp mesh sizes is 0.4 - 0.8 mm. Because bubble size is a function of the porosity of the sparger, the type and quantity of frother present, and the gas rate, it is not possible to generalise on the effect of the gas rate on recovery, although for a specific system there is usually a reduction in mineral carrying capacity with an increase in gas rate.

An increase in wash water rate will increase the height of the froth layer, increase the concentrate grade, and reduce recovery. The optimum water rate is an important variable to determine experimentally. Other variables, such as column height, have a logical effect on flotation; i.e., the higher the column, the greater the selectivity. A serious mill operating problem is detecting and maintaining constant the height of the pulp/froth interface. In the laboratory, where glass columns are prevalent, this level can be controlled visually. The simplest arrangement in a plant is to put flow controllers on the feed and the tails, and to vary the wash water as a function of the interface height, which is approximated by a differential pressure measurement at the bottom of the column. As this control method is inherently unstable because there is a relation between the froth height and the wash water rate, this is not the ideal operating procedure.

Flotation column pilot plant at Chile's CIMM: Chile's Mining and Metallurgical Research Centre (CIMM) purchased a Deister/Flotaire 8" diameter, 24 foot tall column and a Column Flotation Company of Canada Ltd. (CFCC) 2" diameter by 26 feet tall cleaning unit. These columns can be operated independently or as a rougher cleaner unit as shown in Fig. 87.

The Deister/Flotaire column consists of two 8" diameter by 12' metal sections. When assembled, feed pulp is pumped by a Galigher centrifugal slurry pump to a pipe at the top of the 11.34 meter column which ends in a distributor located 2.5 m below the column lip. Tails are withdrawn from the column bottom and pumped to concrete lined sloped side pools, and the concentrate overflows to a stirred tank, from which it can be transferred to the cleaner column by a Moyno pump. Air is supplied at the bottom of the column using one of the following: a) a porous diffuser capable of producing uniform micro-bubbles, b) an external bubbler, such as the standard Deister air aspirator system supplied with the column, or c) the Turbo Air system patented by the US Bureau of Mines, and currently sold by ARMCO Chile, S.A.

Under normal conditions the column operation can be stabilised after about three hours of operation, at which point representative composite samples of the streams can be taken. Usually for each run six or seven samples of each stream are taken at 10 minute intervals, which means each test requires about 4 hours running time, including feed preparation. Total feed required per run is about 4 m^3. If additional reagents are required for each run, these are added to the feed conditioning tank.

The 2" column consists of four glass tubes, each about 6.5' long, which when assembled form a 26' (12.3 m) flotation column with a pulp feed about 2 m below the concentrate lip. Operation is very similar to that of the 8" column.

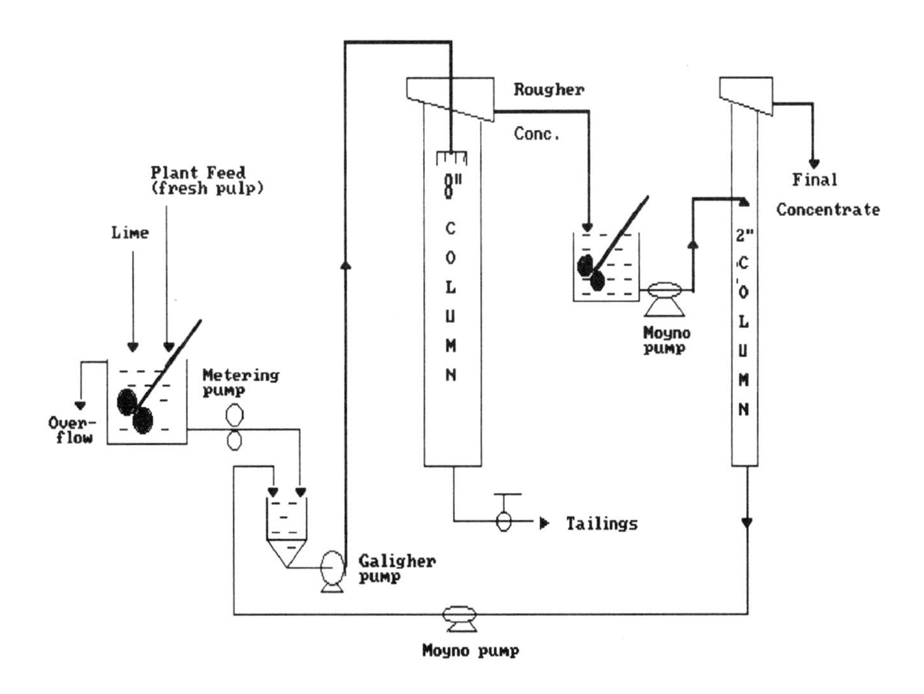

Fig. 87.- Chile's Mining and Metallurgical Research Centre column flotation pilot plant

CIMM has successfully completed a number of laboratory scale, as well as pilot plant, studies for CODELCO and the major privately owned copper and silver mines, including EXXON's Disputada de las Condes concentrator at San Francisco. A number of typical cases will be described, sourced from an unpublished report by Mr. Leonel Gutierrez, Project Manager, Metallurgy Division, CIMM.

Flotation column testing examples

Reground concentrate cleaning: the following cleaner test programme, on a reground rougher concentrate (RRC), was performed in 1987. The objective of the study was to determine if column flotation would be capable of replacing two cleaner stages in a copper/moly recovery circuit. The RRC sample was collected from the ball mill cyclone discharge, during normal operation of the concentrator, at 17% solids, 96% minus 200 mesh; assaying 21.5% total copper, 0.43% Mo_T and 13.9% Fe_T, with a screen analysis of 96% < 74 micron and 32% < 9 micron. The tests were carried out in the 2" and 8" CIMM columns. The two inch column testing is detailed below:

Target - determine the effect on the recovery and grade of the cleaner concentrate of the following variables:

Feed rate	:	F (l/h)
Froth bed height	:	H_e (cm)
Air feed rate	:	Q_A (l/h)
Bias (B = T-F > 0	:	B (l/h)

- Perform identical standard laboratory flotation tests, if necessary, as a comparison with the results of the column tests.
- Determine empirical mathematical model of the data to be able to recommend an optimum circuit for the plant, and optimum operating conditions, based on a computer simulation.
- Size the column(s) recommended as a replacement for the existing cleaner circuits in the plant.

Sampling - about 600 l/day of pulp was collected by means of a pump. This was sufficient to perform 3 - 4 daily tests, which were carried out for a total of 7 days (a total of 21 complete tests). To avoid changes in the sample it was kept in a continuously stirred feed tank during all the testing period.

Table 96.- COLUMN FLOTATION VARIABLES USED IN REGROUND PULP CLEANER STUDY.

Operating Variable	Level	Minimum (-)	Middle (0)	Maximum (+)	Typical Values
Feed Volume	F (l/h)	43	54	56	
	J_F (cm/sec)	0.59	0.74	0.89	0.7-2.0
Tailings volume	T (l/h)	46	60	74	
	J_T (cm/sec)	0.63	0.82	1.01	
Bias B = T - F	B (l/h)	3	6	9	
	J_B (cm/sec)	0.04	0.08	0.12	0.07-0.30
Air Volume	Q_A (l/h)	74	106	138	
	J_A (cm/sec)	1.01	1.45	1.89	0.5-3.0
Froth height	H_e (cm)	17	30	60	

Assays - were done using a fast X-ray fluorescence equipment to allow changing the experimental design on the run, and the final assays were duplicated at the in-house laboratory using atomic absorption. All samples were assayed for Cu_T, Mo_T, Fe_T, and insolubles. In addition, selective assays were made on given screen fractions. The full experiment was repeated during a period where the ore had changed.

Experimental design - the range of values of the variables chosen for this study are shown in Table 96.

Later, a centre point was selected as a reference point, shifting the other operating conditions to the extreme values, as shown in Table 97. These decisions were taken on site as the assay results were obtained, and taking into account how stable the column operation was at the previous conditions, and evaluating visually the froth quality.

Supplementary laboratory batch flotation tests - batch flotation tests, measuring the rate of flotation, were performed to simulate the cleaner, scavenger and/or rougher flotations in the plant, assuming that the column tails were sent back into the existing plant circuit, using the tails generated in RRC - 14 and RRC - 17. These results were included in the concurrent computer simulation made of the overall plant operations.

Review of the experimental results - the best results were obtained from tests RRC - 09, RRC - 10, RRC - 11, and RRC - 12, with the following results:

	Concentrate grade, %	Average recovery, %
Total copper	39.5%	97.4%
Total moly	0.52	84.4
Insolubles	5.6	-

These results were equivalent to those obtained in the plant with two cleaner stages, with conventional recleaner sections. It would result in a significant circuit simplification if an industrial size column were installed at the plant. It is interesting to note that the best copper recoveries were obtained for the -74 micron/+ 18 micron size range, and the best molybdenite recoveries occurred in the -53 micron/+34 micron fraction.

As the pulp characteristics varied somewhat, during the testing procedure, a formal statistical analysis was made of the following variables' influence on recovery:

x_1 = % by weight solids in the pulp feed to the column
x_2 = % copper grade (or molybdenum) in the feed
x_3 = apparent average residence time of the pulp in the column;
minutes = $1.216(792.5-H_e)/(B+F)$
x_4 = apparent average residence time of air in the column;
minutes = $V_A/Q_A = 1.216\ (H_e/Q_A)$

While the dependent variables (Y) were:

- % copper (or molybdenum) grade in the concentrate,
- % recovery of copper (or molybdenum) in the concentrate

Table 97.- OVERALL EXPERIMENTAL PLAN FOR THE RRC SAMPLES IN THE 2" COLUMN

Test	Operating level for each variable						
	F		B		H_e		Q_A
RRC - 01	(-)		(-)		(-)		(-)
RRC - 02	(0)		(0)		(-)		(0)
RRC - 03	(+)		(+)		(-)		(+)
RRC - 04	(+)		(-)		(-)		(0)
RRC - 05	(-)	*	(0)	*	(-)		(0)
RRC - 06	(0)		(+)		(-)		(0)
RRC - 07	(0)		(-)		(-)		(0)
RRC - 08	(0)	(	(0)	1	(0)	)	(0)
RRC - 09	(0)	(	(0)	3	(+)	)	(0)
RRC - 10	(0)		(0)		(-)		(-)
RRC - 11	(0)		(0)		(-)		(+)
RRC - 12	(-)		(0)		(-)		(-)
RRC - 13	(+)		(0)		(-)		(+)
RRC - 14	(0)	(	(0)	3	(+)	)	(0)
RRC - 15	(0)		(-)		(+)		(0)
RRC - 16	(0)		(-)		(-)		(0)
RRC - 17	(0)	(	(-)	2	(0)	)	(0)
RRC - 18	(0)		(-)		(0)		(+)
RRC - 18'	(0)	(	(-)	2	(0)	)	(0)
RRC - 21'	(0)	(	(0)	1	(0)	)	(0)

** Column operation could not be stabilised
(1),(2), (3) = error determining replications

Multiple lineal polynomial regressions for each experimental condition yielded $r^2 > 0.95$ for most of the equations, so level curves of copper and moly recovery, versus copper or moly grade in the feed, were obtained for the different experimental ranges studied. These curves yielded the optimum operating condition for the 2" column shown, in Table 98.

Computer simulation and final design of the proposed flotation circuit: Column flotation can replace the two cleaner circuits used in the plant to upgrade the rougher reground concentrate with one full size column, as demonstrated by the fact that the test 2" column produced a concentrate grade of 39.3% Cu and 0.69% Mo, as compared to an average of 35.3% Cu and 0.5% Mo in the plant operations.

Table 98.- OPTIMUM FLOTATION CONDITIONS FOR THE 2" COLUMN WITH THE RRC SAMPLES

1.-Average specifications of the test pulp feed
Weight % solids = 17.5; Copper grade = 21.5%;
Moly grade = 0.43%; Screen size = 96% -200 # Tyler

2.-Optimum column operating conditions:

F = Pulp feed rate	= 54 - 55 l/h
T = Tailings volume	= 60 - 62 l/h
B = Bias = T-F	= 6 l/h
H_e= Froth height	= 15 - 20 l/h
τ_p= Average pulp residence time	= 15.7 minutes

3.-Performance projected

	Concentrate Grade, %	Recovery, %
Copper	38.5 - 40.0%	97.0 - 98.0%
Molybdenum	0.6 - 0.7	84.0 - 85.0

Based on these results CIMM proposed the flowsheet shown in Fig. 88. Computer simulation of this circuit indicates that it is feasible to obtain a final copper/moly concentrate with 38.6% Cu and 0.7% Mo, with overall recoveries of 87.6% on Cu and 61.3% on moly. This performance would be very similar to that obtained in the current plant. They recommended that, prior to implementing their design, a further test be performed on-site with a semi-commercial size column.

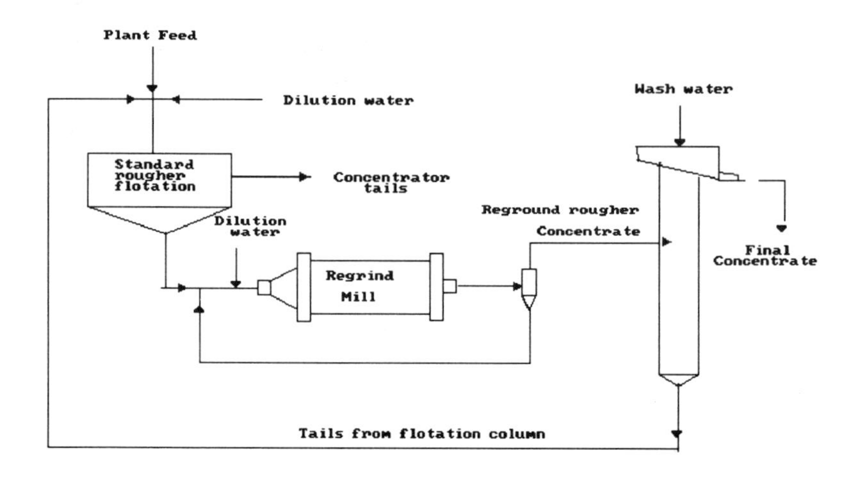

Fig. 88.- CIMM recommended flowsheet for cleaner column flotation

Design of the industrial column: Based on the experimental results, the key design data used in the calculation of the recommended column flotation flowsheet were:

τ_p = 15.7 minutes = apparent average pulp residence time in 2" column
$Fe(\tau)$ = 1.5 = scale-up factor for τ_p, then
$(\tau_p)_{industrial}$ = 23.6 minutes

F_s = RRC fed to cleaners in the plant = 9,788 t/day.
F = pulp feed in the plant = 2033 m^3/h = current feed to the first cleaner stage.
T = plant tailings volume from the cleaners = 2,251 m^3/h of pulp

V_{ep} = $(\tau_p)_{industrial} \times T/60$ = 885.4 m^3 of pulp = actual pulp volume of
7 the flotation column
V_c = volume of a standard industrial 12 m high column, with a cross section of 72"x72" = 40.1 m^3
H_e = design height of froth = 1 m
V_{pc} = effective pulp volume of the above industrial column = 36.8 m^3
$Fe(Q_A)$ = air consumption scale-up factor = 0.048 t conc./m^3 of air
J_A = air superficial velocity in the column = 1.44 cm/second

Number of columns required = V_{ep}/V_{pc} = 88.4/36.8 = 24 columns

C_s/F_s = Split-factor or concentration factor for plant, calculated from the enrichment of the final concentrate as compared to the feed to the cleaners in the experimental programme = 0.5148

C_s = tonnage of final concentrate = 0.5148 x F_s = 5,039 t/day
Q_A = 5039/Fe(Q_A) = 5039/0.048 = 4,374 m^3 of air/hour
air per column = 4,374/24 = 183 m^3 air/hour/column

Based on the superficial air velocity, J_A, the calculated value is 173 m^3/h/column.

Therefore the recommended design for the final concentrate cleaners to produce 1,800,000 t of concentrate per year requires 24 columns, 12 m in height, and a square 1.83 m x 1.83 m cross-section. Each column will operate at:

F = 85 m^3/h pulp feed rate
T = 94 m^3/h tailings discharge rate
B = 9 m^3/h bias flow of water
Q_{AC} = 183 m^3/h air feed rate

and these rates will result in a froth bed 1 m high, and a pulp residence time, t_p = 24 minutes.

BIBLIOGRAPHY

Abramov, A.A. (1969) Role of sulphidising agent in flotation of oxidised minerals, *Tsvetn. Metall.* 12(5), 7-13.

Agar, G.E. (1985) The optimisation of flotation circuit design from laboratory rate data, *XVth Int. Min. Proc. Congress*, Cannes, June 2-9, 1985, vol. II, pp.100-111.

Alferova, L.A., and Titova, G.A. (1969) Study of the reaction rates and mechanisms of oxidation of hydrogen sulphide, sodium hydrosulphide and sulphides of sodium, iron and copper in aqueous solutions by atmospheric oxygen, (in Russian), *Zh. Prikl. Khim.* 42, 192-196.

Arbiter, N., Young & Baker, (1951) U.S. Patent N° 2,559,104.

Arbiter, N., Young & Baker, (1952) N° 2,608,298.

Arbiter, N., Young & Baker, (1953) N° 2,664,199.

Baleshta, T.M., and Dibbs, H.P. (1970), An introduction to the theory, measurement and application of semiconductor transport properties of minerals, *Mines branch technical bulletin TB 106*, Canada.

Ball, B., and Rickard, R.S. (1976) The chemistry of pyrite flotation and depression, M.C.Fuerstenau, Ed. *Flotation*, AIME, pp.458-484

Beas-Bustos, E., and Crozier, R.D. (1991) Moly/copper separation from concentrate of the combined acid and basic circuits at El Teniente, *Reagents in Minerals Engineering*, Camborne School of Mines, September 18-20.

Bickerman, J. (1953) *Foams: theory and industrial applications*, Reinhold, New York.

Bogdanov, O.S., Emlyanov, M.F., Maximov, I.I., and Otrozhdennova. L.A. (1980) Influence of some factors on fine particle flotation, P. Somasundaran Ed. Fine Particle Processing, vol.1 , *Proc. of Int. Symp. AIME*, Las Vegas February 24-28, 1980, pp706-720, (impeller speed)

Bolth, F., Crozier, R.D., and Strow, L.A. (1975) Dialkyl thionocarbamate catalytic method, US Patent No. 3,907,854.

Booth, R.B., & Freyberger, W.L. (1962) Froth and frothing agents, *Froth flotation 50th anniversary volume*, D.W. Fuerstenau, ed., SME of AIME, New York, pp. 258-276.

Booth, R.B. (1954) U.S. Patent 2,675,101.

Born, C.A., Bender, F.N., and Kiehn, O.A. (1976) Molybdenite flotation reagent development at Climax, CO, *Flotation*, Ed. M.C. Fuerstenau, SME-AIME, Littleton, CO, pp.1147-1184.

Box, G.E.P., and Draper, N.R. (1969) *Evolutionary Operation*, John Wiley & Son, New York.

Bulatovic, S.M., and Salter, R.S. (1989) High intensity conditioning, a new approach to improving flotation of mineral slimes, *Processing of complex ores*, Proc. Int. Symp, CIMM, Halifax, August 20-24, 1989, pp169-181, Pergamon Press, New York.

Cambron, Adrien, and Whitby, G.S. (1930) The oxidation of xanthates and some new dialkyl sulphur- and disulphur-dicarbothionates, *Can. J. Research*, 2, 144-52.

Cambron, Adrien (1930) The mechanism of formation of thiuram and xanthogen monosulphides, and observations on thiocarbamyl thiocyanates, *Canadian J. Research*, 2, 341

Castleman Jr., A. W., Yang, X, and Shi, Z. (1991) *J. Chem. Phys*. Vol. 94, p. 3268.

Castleman Jr., A. W., and Yang, X. (1989) *J. Am. Chem. Soc.* Vol. 111, p. 6845.

Castro, S., Goldfarb, J., and Laskowski, J. (1974) Sulphidizing reactions in the flotation of oxidized copper minerals, Part I, *Int. J. Miner. Process*. 1(2), 141-149; Castro, S., Soto, H., Goldfarb, J., and Laskowski, J. (1974), Part II, *Int. J. Miner. Process*. 1(2), 151-161.

Castro, S., Gaytan, H., and Goldfarb, J. (1976) The stabilizing effect of Na_2S on the collector coating of chrysocolla, *Int. J. Miner. Process*. 3(1), 71-82.

Castro, S., and Pavez, O. (1977) Mechanisms of action of Anamol-D, Nokes reagent and sodium hydrosulphide in MoS_2 flotation, *Avances en flotacion*, vol 3, pp. 111 -130, Univ. of Concepcion, Third Conference on surface chemistry and flotation, 21 to 24 June, 1977, Concepcion, Chile, (in Spanish).

Cea, Zocimo, and Castro, S. (1975) Depressant effect of the calcium ion on pyrite flotation, *Avances en flotacion*, Vol. 1, p.97-105 (in Spanish).

Chander, S.(1988) Electrochemistry of sulfide mineral flotation, *Min. & Met. Processing*, p 104, August 1988.

Chander, S., and Fuerstenau, D.W. (1972) On the natural floatability of molybdenite, *AIME-SME Trans.*, 252, pp. 62-69.

Chanturiya, V.A., and Vigdergauz, V.E. (1989) Electrochemical modification of sulphide flotation, *Processing of complex ores, Proc. Int. Symp,*

CIMM, Halifax, August 20-24, 1989, pp.71-85, Pergamon Press, New York.

Chudacek, M.W. (1990) A new quantitative test-tube flotability test, *Minerals Engineering*, Vol. 3, pp. 461-472.

Chudacek, M.W., and Fichera, M.A. (1991) The relationship between the test-tube flotability test and batch cell flotation, *Minerals Engineering*, Vol. 4, pp. 25-35.

Clark, Alfred, (1974), *The chemisorptive bond*, Academic Press, New York, NY.

Clingan, B.V. and McGregor, D.R. (1987) Column flotation experience at Magma Copper Co., with related experience of other mineral processors, *SME annual meeting*, Denver, CO, Feb. 24-27, Preprint No. 87-91.

Crozier, R.D. (1957) *Froth stratification and liquid mixing in a bubble tray column*, Ph.D. Dissert. Univ. of Michigan, Ann Arbor, Mich. 1956; University Microfilms, Ann Arbor, Mich.

Crozier, R.D. (1977) Reactivos de flotacion en recuperacion de minerales sulfurados, *Avances en flotacion*, 3, Univ. Concepcion, Chile, pp. 301-344, Summarized in: Processing of Copper Sulphide Ores: Froth Flotation reagents - A Review, Min. Mag. London, April, 1978.

Crozier, R.D. (1978) Reactivos de flotacion aplicados a la recuperacion de molibdenita con minerales de cobre porfiricos, Primer Congreso de Ingenieros del Cobre, Chuquicamata, Chile, 4-8, December, 1978. Summarized in: Flotation Reagent Practice in Primary and By-Product Molybdenum Recovery, *Mining Mag*. London, February 1979.

Crozier, R.D. (1979) La funcion del espumante en la flotacion de minerales sulfurados, *Avances en flotacion*, 4, Univ. Concepcion, Chile, pp. 19-56, Summarized in: Frother function in sulphide flotation, *Mining Magazine*, January 1980, pp.26 -36.

Crozier, R.D. (1984) Changing patterns in the supply of flotation reagents, *Mining Magazine*, London, June, 1984.

Crozier, R.D., (1991) Sulphide collector mineral bonding and the mechanism of flotation, *Minerals Engineering*, vol 4, Nos 7-11, pp.839-858

Crozier, R.D. and Klimpel, R. (1989) Frothers: plant practice, *Mineral processing and extractive metallurgy review*, 5, pp 257-279, Gordon and Breach Science Publishers, London.

Crozier, R.D., and Strow, L.A. (1976) Process for the flotation of copper sulfide ores with improved thionocarbamates, U.S. Patent No. 3,975,264.

Cyanamid, (1989) *Mining chemicals handbook*, Mineral dressing notes No. 26-1, American Cyanamid, Wayne, NJ,

Davis, Farlow C. (1976) San Manuel's process for molybdenite recovery, Local Chapter Meeting, AIME, San Manuel AZ., October, 1976.

Dobby, G.S., and Rao, S.R., Editors (1989) *Processing Complex Ores*, Proc. Int. Symp. on processing of complex ores, Halifax, Aug. 20-24, 1989, Pergamon Press, New York

Douglass, W. (1927) US Patent No. 1,625,099.

Douglass, W. (1928) US Patent No. 1,659,396.

Down, R.F., and Turner, J. (1970) Concentration of oxide ores at Tynagh, *Metallurgy of lead and zinc*, Eds. Rausch, D.O., and Mariacher, B.C., Vol. 1, pp.710-731, TMS of AIME, New York, NY.

Drew, M.G.B. (1977) Seven-coordination chemistry, *Progress in inorganic chemistry*, Vol. 23, John Wiley & Sons, New York, pp. 67-210.

Drzmala, J., and Lekki, J. (1988) Application of flotometry for characterising flotation in the presence of particle aggregation, *Minerals Engineering*, Vol. 1, pp.327-336.

Dudenkov, S.V., and Abramov, A.A., Discussion of Lekki & Laskowski (1975).

Espinosa-Gomez, R., Finch, J.A., and Jonson, N.W. (1988) Column flotation of very fine particles, *Minerals Engineering*, Vol. 1, 3-18

Ewers, W.E., and Sutherland, K.L. (1952) The role of surface transport in the stability and breakdown of foams, *Austr. J. Sci. Res.* A5, pp.697.

Filyk, A. (1976) A review of plant practice at Molycorp Inc., Questa Division, *Flotation*, Ed. M.C. Fuerstenau, SME-AIME, Littleton, CO, pp.1185-1199.

Finkelstein, N.P. and Allison, S.A. (1976) The chemistry of activation and depression in the flotation of zinc sulfide: A review, *Flotation*, Ed. M.C. Fuerstenau, SME-AIME, Littleton, CO, pp. 414-457.

Finkelstein, N.P., Allison, S.A., Lovell, V.M., and Stewart, B.V. (1975) Natural and induced hydrophobicity in sulfide mineral systems, advances in interfacial phenomena of particulate-solution-gas systems; applications to flotation research, Somasundaran, P., and Grieves, R.B., ed., *AICHE, CEP Symposium Series No. 150*, Vol. 71, pp.165-175, AICHE, New York.

Finkelstein, N.P., & Poling, G.W. (1977) The role of dithiolates in the flotation of sulphide minerals, *Mineral Sci. Engng.*, 9, pp.177-97.

Fischer, A.H. (1928) US Patent No. 1,684,536

Fleming, M.G. (1953) Effects of soluble sulphide in the flotation of lead minerals, in Recent Developments in Mineral Dressing, Proc. *First Int. Mineral Proc. Congress*, London, 1952, IMM, London. pp. 521-528.

Forssberg, K.S.E., and Hallin, M.I. (1989) Process water recirculation in a lead-zinc plant and other sulphide flotation plants, *Challenges in mineral processing*, Eds., K.V.S. Sastry and M.C. Fuerstenau, SME, Littleton, CO, pp.452-466.

Fuerstenau, D.W. Editor (1962) *Froth Flotation.* 50th anniversary volume, SME of AIME, New York.

Fuerstenau, M.C., Editor (1976) *Flotation*, A.M.Gaudin memorial volume, Vol. 1 and 2, SME of AIME, New York, NY.

Gardner, J.R., and Woods, R. (1973) The use of a particulate bed electrode for the electrochemical investigation of metal and sulphide flotation, *Australian Journal of Chemistry*, vol 26, pp. 1635-1644,

Gardner, J.R., and Woods, R. (1974) An electrochemical investigation of contact angle and of flotation in the presence of alkylxanthates, I. Platinum and Gold surfaces, *Australian Journal of Chemistry*, vol 27, pp. 2139-2148.

Gardner, J.R., and Woods, R. (1977) An electrochemical investigation of contact angle and of flotation in the presence of alkylxanthates. II. Pyrite and Galena surfaces, *Australian Journal of Chemistry*, vol 30, pp. 981-

Gaudin, A.M. (1939) *Principles of Mineral Dressing*, McGraw-Hill Book Co. New York and London.

Gaudin, A.M. (1957) *Flotation*, Second Edition, McGraw-Hill Book Co., New York.

Glembotskii, V.A., Klassen, V.I., & Plaksin, I.N. (1963) *Flotation*, Engl. transl., Primary Sources, New York.

Glembotskii, V.A., Klassen, V.I., & Plaksin, I.N. (1972) *Flotation*, Engl. transl., Primary Sources, New York.

Glembotskii, V.A., et al (1963) Quoting *Concentration of non-metallic minerals by flotation*, Izd. AN SSSR, 1952.

Grainger-Allen, T.J.N. (1970) Bubble generation in froth flotation machines, *Trans. I.M.M.*,79, pp. C15-C22.

Granville, A., Finkelstein, N.P., and Allison, S.A. (1972) Review of reactions in the flotation system galena-xanthate-oxygen, *Trans. I.M.M.*, vol. 81, pp. C1-C30.

Gray, H.B. (1973) *Chemical bonds: an introduction to atomic and molecular structure*, W.A.Benjamin, Inc., Menlo Park, C.A.

Groppo, J.G., and Yoon, R.H. (1982) Selective flotation of sulfide ores using alkyl pyridinium collectors, Yarar,B. and Spottiswood, J.A. Editors, (1982), *Interfacial phenomena in mineral processing*. Proceedings of the Engineering Foundation Conference, Rindge, New Hampshire, Aug. 2-7, 1981. Engineering Foundation, New York, pp. 271-285.

Gutierrez, G., & Sanhueza, J. (1977) Efecto de carbon activado en la selectividad del proceso de separacion molibdenita-cobre, *Avances en flotacion*, 3, Univ. de Concepcion, Chile, pp.223-236.

Hansen, R.D., and Klimpel, R.R. (1986) The influence of frothers on particle size and selectivity in coal/sulfide mineral flotation, *Trans. AIME*, 280, pp. 1804-1811.

Hansen, R.D., Klimpel, R.R., and Bergman, R. (1986) U.S. Patent 4,582,496.

Harris, C.C. (1976) Flotation Machines, in *Flotation, A.M.Gaudin memorial volume*, Vol. 2, M.C.Fuerstenau, ed., AIME, New York, pp. 753-815.

Harris, Guy, (1954) Alkyl thionocarbamate collectors, US Patent No. 2,691,635.

Hayes, R.A., Price, D.M., Ralston, J., and Smith, R.W. (1987) Collectorless flotation of sulphide minerals, *Mineral processing and extractive metallurgy review*, vol. 2, pp.1-20.

Hayes, R.A., and Ralston, J. (1988) Collectorless flotation and separation of sulphide minerals by Eh control, *Int. J. Min. Processing*, vol. 45, pp.55-84.

Heyes, G.W., & Trahar, W.J. (1953) The natural floatability of chalcopyrite, *Int. J. Mineral Dressing* (London, Inst. Min. Metall.).

Heyes, G.W., & Trahar, W.J. (1977) The natural floatability of chalcopyrite, *Int. J. Min. Proc.*, vol. 4, pp.317-344.

Heyes, G.W., & Trahar, W.J. (1979) Oxidation-reduction effects in the flotation of chalcocite and cuprite. *Int. J. Min. Proc.*, vol. 4, pp.317-344.

Hoover, R.M., and Malhotra, D. (1976) Emulsion flotation of molybdenite, M.C.Fuerstenau, Ed. *Flotation*, AIME, pp.485-505

Hosten, Ç., and Tezcan, A. (1990) The influence of frother type on the flotation kinetics of a massive copper sulphide ore, *Minerals Engineering*, Vol. 3, pp. 637-640.

Houben-Weyl, (1973) *Methoden zur organischen chemie*, 12/2, 685 George Thieme Verlag.

Huggins, D.A., Wesley, R.J. and Jomoto, K. (1987) Column flotation, its status and potential for Escondida, SME annual meeting, Denver, CO, Feb. 24-27, Preprint No. 87-116.

Iwasaki, I., Malicsi, A.S., Li Xiaoxei, and Weiblen, P.W. (1989) Insights into beneficiation losses of platinum group metals from gabbroic rocks, *Challenges in mineral processing*, Eds., K.V.S. Sastry and M.C. Fuerstenau, SME, Littleton, CO, pp.437-451.

Jones, M.H., and Woodcock, J.T. (1979) Control of laboratory sulphidisation with a sulphide ion-selective electrode before flotation of oxidized lead-zinc-silver dump material, *Int. J. Miner. Process*, 6, 17-30.

Jones, M.H., and Woodcock, J.T. (1979) Perxanthates - a new factor in the theory and practice of flotation, *Int. J. Miner. Process.*, 5, pp.285-296.

Jones, M.J., and Oblatt, R., Editors (1984) *Reagents in the mineral industry*, Proc. Symp. Rome, Sept. 18-21, 1984, IMM, London.

Kapur, P.C., and Mehrotra, S.P. (1989) Modeling of flotation kinetics and design of optimum flotation circuits, *Challenges in mineral processing*, Eds., K.V.S. Sastry and M.C. Fuerstenau, SME, Littleton, CO, pp.300-322.

Kepert, D.L. (1977) Aspects of the stereochemistry of Six-coordination, *Progress in Inorganic Chemistry*, Vol. 23, John Wiley & Sons, New York, pp.1-66.

King, R.P. Editor (1982) *Principles of Flotation*, South African Institute of Mining and Metallurgy, Monograph Series No. 3, Johannesburg.

King, R.P. (1976) The use of simulation in the design and modification of flotation plants, *Flotation*, Ed. M.C. Fuerstenau, SME-AIME, Littleton, CO, pp. 937-962.

Klassen, V.I., and Mokrousov, V.A. (1963) *An Introduction to the Theory of Flotation*, Engl. transl. by J. Leja and G.W. Poling, Butterworths, London.

Klimpel, Richard (1977) An engineering analysis of the effects of chemical reagents in the laboratory evaluation of mineral flotation, *Avances en flotacion*, 3, Univ. de Concepcion, Chile.

Klimpel, R.R. (1987) The industrial practice of sulfide mineral collectors - Chapter 21, *Reagents in the mineral industry*, P. Somasundaran and B. Moudgil, eds., Marcel Dekker, New York, pp. 663-682.

Klimpel, R.R. (1987) The use of mathematical modeling to analyze collector dosage effects in froth flotation, *Symp. on Mathematical Modeling of Metals Processing Operations*, Palm Springs, Calif., The Metallurgical Society.

Klimpel, R.R., and Hansen, R.D. (1987) Chemistry of fine coal flotation - Chapter 4, *Fine Coal Processing*, S. Mishra and R. Klimpel, eds., Noyes Publishing, Park Ridge, New Jersey, 78-106.

Klimpel, R.R., and Hansen, R.D. (1987) Frothers - Chapter 12, *Reagents in the mineral industry*, P. Somasundaran and B. Moudgil, eds., Marcel Dekker, New York, pp. 385-409.

Lager, T., and Forssberg, F.S.E. (1989) Current processing technology for Antimony bearing ores, A review, part 2, *Minerals Engineering*, vol. 2, pp. 543-556.

Lager, T., and Forssberg, F.S.E. (1989) Beneficiation characteristics of Antimony minerals, A review, part 1, *Minerals Engineering*, vol.2, pp. 321-336.

Laskowski, J. (1974) Particle-bubble attachment in flotation, *Minerals Sci. Engng*, vol 6, no.4, pp. 223-235.

Laskowski, Janusz (1974) *Fundamentos Fìsico Quìmicos de la Mineralurgia*, Translated to Spanish from the original Polish by Mariano Rawicz and reviewed and up-dated by the author. University of Concepciòn, Chile, p. 503.

Laskowski, J. (1982) Redox conditions in flotation: oxides, Yarar,B. and Spottiswood, J.A. Editors, (1982), Interfacial phenomena in mineral processing. *Proceedings of the Engineering Foundation Conference*, Rindge, New Hampshire, Aug. 2-7, 1981. Engineering Foundation, New York,pp.189-206.

Lefond, S.J., Editor (1983) *Industrial Minerals and Rocks*, Vol 1 and 2, SME of the AIME, New York.

Leja, J. (1957) Mechanism of collector adsorption and dynamic attachment of particles to air bubbles as derived from surface- chemical studies, *Bulletin of IMM*, 607, pp. 425-437.

Leja, J. (1982) *Surface Chemistry of Froth Flotation*, Plenum Press, New York.

Leja, J., & Schulman, J.H. (1954) Flotation Theory: molecular interaction between frothers and collectors at solid-liquid-air interfaces, *Trans. AIME.*, 16, pp.221-8.

Lekki, J., & Laskowski, J. (1971) On the dynamic effect of frother-collector joint action in flotation, *Trans. I.M.M.*, Section C,80, C174-80.

Lekki, J., & Laskowski, J. (1975) A new concept of frothing in flotation systems and a general classification of flotation frothers, *Proceedings of the 11th IMP Congress, Cagliari*, M. Carta, ed., Instituto di Arte Mineraria, Universitá di Cagliari, Cagliari, Italy, pp. 427-448.

Leroux, M., Rao, S.R., Finch, J.A., Gervais, V., and Labontè, G. (1989) Collectorless flotation in the processing of complex sulphide ores, *Adv. in Coal and mineral processing using Flotation*, Chander S., and Klimpel. R.R. Eds., AIME, Littleton,

Li Guongming and Zhang Hingen (1989) The chemical principles of flotation, activation and depression of arsenopyrite, *Processing of complex ores*, Proc. Int. Symp, CIMM, Halifax, August 20-24, 1989, pp.61-70, Pergamon Press.

Luttrell, G.H., Adel, G.T. and Yoon, R.H. (1987) Modeling of column flotation, SME annual meeting, Denver, CO, Feb. 24-27, Preprint No. 87-130.

Malhotra, D., Hoover, R.M., and Bender, F.N. (1977) An analysis of the effect of some operating variables on flotation of molybdenite and its implication on control, Preprint No 77-B-89, AIME annual meeting, Atlanta, March 6-10, 1977.

Malhotra, D., Hoover, R.M., and Bender, F.N. (1980) Effect of agitation on flotation of molybdenite, *Mining Engineering*, p. 1392, September 1980.

Mathieu, G.I., and Sirois, L.L. (1984) New processes to float feldspathic and ferrous minerals from quartz, *Reagents in the Minerals Industry*, M.J.Jones and R.Oblatt, Eds, (London: IMM), pp. 57-67.

McKay, J.D., Foot Jr, D.G. and Shirts, M.B. (1987) Parameters affecting column flotation of fluorite, *SME annual meeting*, Denver, CO, Feb. 24-27, Preprint No. 87-122.

Moys, M.H. (1978) A study of a plug-flow model for flotation froth behaviour, *Int. J. Miner. Process.*, 5, pp. 21-38.

Mular, A.L. (1989) Modelling, simulation and optimization of mineral processing circuits, *Challenges in mineral processing*, Eds., K.V.S. Sastry and M.C. Fuerstenau, SME-AIME, Littleton, CO, pp.323-349.

Mular, A.L. (1976) Optimization in flotation plants, *Flotation*, Ed. M.C. Fuerstenau, SME-AIME, Littleton, CO, pp. 895-936.

Mumme W.G.& Winter, G. (1971) *Inorg. Nucl. Chem. Lett.*,*7*, 505.

Nagaraj, D.R. (1988) The chemistry and application of chelating or complexing agents in mineral separations, *Reagents in Mineral Technology, Surface Science Series*, vol. 27, P. Sumasundaran, Ed. Marcel Dekker, New York, pp.257-333.

Nagaraj, D.R., and Gorken, A. (1989) Potential controlled flotation and depression of copper sulphides and-xanthate systems, *Processing complex ores, Proc. Int. Symp. CIM*, Halifax, Aug. 20-24, 1989,pp.203-213 Pergamon Press, New York.

Nagaraj, D.R., Basilio, and Yoon, R.H. (1989) The chemistry and structure-activity relationships for new sulphide collectors, *Processing complex ores, Proc. Int. Symp. CIM*, Halifax, Aug. 20-24, 1989,pp.157-166, Pergamon Press, New York.

Nokes, C.M., & Quigley, C.G., (1948), Differential froth flotation of sulphide ores, U.S. Patent N° 2,492,936.

O'Connor, C.T., Bradshaw, D.J., and Upton, A.E. (1990) The use of dithiophosphates and dithiocarbamates for the flotation of Arsenopyrite, *Minerals Engineering*, Vol. 3, pp. 447-459.

O'Connor, C.T., and Mills, P.J.T. (1990) The effect of temperature on the pulp and froth phases in flotation of pyrite, *Minerals Engineering*, Vol. 3, pp. 615-624.

Ozbayoglu, G., Atalay, M.U., and Basaran, B. (1985) Barite flotation: a statistical experimental design approach, *XVth Int. Min. Proc. Congress*, Cannes, June 2-9, 1985, vol. II, pp.337-345.

Palagi, C.G., and Stillar, S.S. (1976) The Anaconda C.E. Weed concentrator, *Flotation*, Ed. M.C. Fuerstenau, SME-AIME, Littleton, CO, pp. 1029-1042.

Pauling, Linus (1970) Crystallography and chemical bonding of sulfide minerals, *Mineral. Soc. Amer. Spec. Pap.* 3, pp. 125-131.

Pauling, Linus (1967) *The chemical bond*, Cornell University Press, Ithaca, N.Y.

Pauling, Linus (1960) *The Nature of the chemical bond*, Cornell University Press, Ithaca, N.Y.

Poling, G.W. (1976) Reactions between thiol reagents and sulphide minerals, *Flotation*, Ed. M.C. Fuerstenau, SME-AIME, Littleton, CO, pp. 334-363.

Poling, G.W., and Vreugde, M.I.A. (1985) *Principles of mineral flotation, The Wark Symposium*, Australasian IMM, Parksville, Australia.

Potgieter van Aardt, J.H. (1951) U.S. Patent 2,561,251.

Powell, R.F. et al (1951) U.S Patent 2,591,289.

Ramsey, T. (1976) The Cyprus Pima concentrator, *Flotation*, Ed. M.C. Fuerstenau, SME-AIME, Littleton, CO, pp.1079-1099.

Ranney, M.N. Editor (1980) *Flotation agents and process: technology and application,* Noyes Data Corporation, Park Ridge, New Jersey.

Rao, S.R., and Finch, J.A. (1989) A review of water re-use in flotation, *Minerals Engineering*, Vol. 2, pp. 65-85

Rausch, D.O., and Mariacher, B.C. Editors (1970) *Metallurgy of lead and zinc*, Vol. 1 and 2, TMS of AIME, New York, NY.

Redeker, Immo H. and Bentzen, E.H. (1986) Plant and laboratory practice in non-metallic flotation, *Chemical reagents in the mineral processing industry*, Editors, Deepak Malhotra and W.F. Riggs. SME of AIME, Littleton, CO

Reid, E. Emmet (1962) *Organic chemistry of bivalent sulfur*, vol IV, Thiocarbonic Acid and Derivatives, Chemical Publishing Co. New York, N.Y.

Seidell-Linke, *Solubilities of inorganic and metal-organic compounds*, vol.2, American Chem. Soc. Washington, DC.

Sharp, F.H. (1976) Lead-zinc-copper separation and current practice at Magmont Hill, *Flotation*, Ed. M.C. Fuerstenau, SME-AIME, Littleton, CO, pp.1215-1231.

Shashtry, K.V.S., and Fuerstenau, M.C. Editors (1989) *Challenges in mineral processing*, SME of AIME, Littleton, CO.

Shimoiizaka, J. (1972) Fundamental studies on the flotation of complex ores, Joint meeting MMIJ, AIME, Tokyo.

Shirley, J.F. (1979) By-product Mo, in A. Sutulov, Ed., *International Molybdenum Encyclopaedia*, Vol. II, Intermet Publications, Santiago, Chile, pp. 37-56.

Smit, F.J., and Foth, H.C. (1970) The 5000 TPD concentrator for milling the complex oxidized lead-zinc ore at Tintic Division, , *Metallurgy of lead and zinc*, Eds. Rausch, D.O., and Mariacher, B.C., Vol. 1, pp. 751-769, TMS of AIME, New York, NY.

Smith P.R. (1976) Phosphate flotation, *Flotation*, Ed. M.C. Fuerstenau, SME-AIME, Littleton, CO, pp.1265-1284.

Spendly, W., et al., (1962) Sequential application of simplex designs in optimization and EVOP, *Technometrics*, 4, No. 4, p.441

Steininger, J. (1968) The depression of sphalerite and pyrite by basic complexes of copper and sulphhydryl flotation collectors, *Trans. SME/AIME*, vol. 24, No. 1, March 1968, p.34-42.

Strydom, P.J., Spitzer, D.P. and Goodman, R.M. (1983) Fine coal flotation with alcohol: dialkyl sulfosuccinate, *Colloids and surfaces*, 8, pp. 175-185.

Sulman, H.L. and Edser, E. (1924) U.S. Pat. 1492904.

Sutherland, K.L. and Wark, I.W. (1955) *Principles of flotation* Australasian Institute of Mining and Metallurgy, Melbourne.

Sutulov, A. (1963) *El proceso de lixiviaciòn, precipitaciòn y flotaciòn*, Universidad de Concepciòn, Chile.

Sutulov, A. (1963) *Flotaciòn de minerales*, Universidad de Concepciòn, Chile.

Sutulov, Alexander (1965) *Molybdenum extractive metallurgy*, Univ. de Concepcion, Chile.

Sutulov, A. (1974) *Copper porphyries*, University of Utah Printing Services, Salt Lake City, Utah.

Sutulov, Alexander (1977) Flotation recovery of molybdenite, Paper presented at 16th Ann. Conf. of Metallurgists, Vancouver, Canada, August 1977.

Sutulov, Alexander (1978) *Molybdenum and Rhenium, 1778-1977*, Univ. de Concepcion, Chile.

Taggart, A.F. (1945, 1947) *Handbook of mineral dressing*, J. Wiley & Sons, New York.

Tsai, M., Matsouka, I, and Shomoiizaka, J. (1971) The role of lime in xanthate flotation of pyrite, *J. Mining and Met. Inst. of Japan*, 87, pp. 1053-57, (in Japanese).

Tuwiner, S.B., & Korman, S. (1950) Effect of conditioning on flotation of chalcocite, *Trans. I.M.M.E.*, 187, pp.226-37, Min. Engng., Feb, 1950.

Tveter, E.C. (1952) U.S. Patent 2,611,485.

Tveter, E.C., & McQuiston Jr., F.W. (1962) Plant practice in sulphide mineral flotation, *Froth flotation*, 50th anniversary volume, AIME, New York.

Vartanian, K.T., & Gomelauri, I.V. (1941) Depression of sulphides with sodium sulphide,' U.S.S.R. Patent N° 48,010 (1936) and 63,803.

Weiss, A., Editor (1985) *SME Mineral processing handbook*, Soc. of Mining Engineers, New York, N.Y.

Wills, B.A. (1988) *Mineral processing technology*, 4th Edition, Pergamon Press, Oxford.

Winter, G. (1975) Xanthates of sulfur: their possible role in flotation, *Inorganic and nuclear chemistry letters*, Vol. 11, pp.113-118.

Woods, R. (1971) The oxidation of ethyl xanthate on platinum, gold, copper, and galena electrodes. Relation to the mechanisms of mineral flotation, *J. Phys. Chem.*, vol.75, pp. 354-362.

Woods, R. (1972) Electrochemistry of sulfide flotation, *Proc. Austr. IMM*, No 241, pp. 53-62.

Woods, R. (1972) The anodic oxidation of ethylxanthate on metal and galena electrodes, *Austr. J. Chem*. vol. 25, pp.2329-2335.

Woods, R. (1976) Electrochemistry of sulfide flotation, *Flotation*, A.M. Gaudin memorial volume, AIME, New York, N.Y.

Woods, Ronald (1977) Mixed potential mechanism in metallurgical systems, *Avances en flotacion*, 3, Univ. de Concepcion, Chile.

Woods, R. (1981) Mineral flotation, Chapter 11, *Comprehensive treatise of electrochemistry*, Bockris, J. O'M., Conway, B.E., Yeager, E., and White, R.E., Eds., vol. 2, Plenum Publishing Corp. pp. 571-595.

Wrobel, S.A. (1953) Power and stability of flotation frothers, governing factors, *Mine and Quarry Eng.*, 19, pp.363-7.

Yarar, B. and Spottiswood, J.A. Editors (1982) Interfacial phenomena in mineral processing. *Proceedings of the engineering foundation conference*, Rindge, New Hampshire, Aug. 2-7, 1981. Engineering Foundation, New York.

Yehia, Ahmed (1991) Adsorption of Aerosol OT on some synthetic carbonate apatite, *Minerals Engineering*, In press.

Yianatos, J.B., Espinosa, R.G., Finch, J.A. and Laplante, A.R. (1989) Effect of column height on flotation column performance, SME annual meeting, Denver, CO, Feb. 24-27, Preprint No. 87-24.

Yoon, R.H., Luttrell, G.H., and Adel, G.T. (1989) Advances in fine particle flotation, *Challenges in mineral processing*, Eds., K.V.S. Sastry and M.C. Fuerstenau, SME, Littleton, CO, pp. 487-506.

SUBJECT INDEX

Activating agents 9, 102
Alaskite
  typical composition of 235
Alcoholic frothers
  properties 94
Alkali metal alkyl xanthates
  see also, xanthates
  manufacture 49
  properties 51, 52, 55
  stability 55
  suppliers and trade names 67
Allied Colloids (UK) 71
Am. Cyanamid (USA) 57-72, 94, 98-100, 115, 123, 125, 207, 209
Amines
  as cationic collectors 226
  critical miscelle concentration 226
  ionization constant 226
  solubility 228
Ammonium hydrosulphide 116
Anamol D, or arsenic Nokes 116
  chemistry 121
Anionic collector chemistry
  miscelle concentration 224
  structure 223
Arsenic Nokes 116
  chemistry 121
Arsenopyrite
  lime as depressant 163
Barite 231
Batch tests 295-313
Bauxite 231
Beryl 231
Beryllium 231
Borates 232
Boron, important minerals 232
Box-Wilson design layout 271, 272
  five-level 272
  three-variable 272, 273
Bubble formation
  cavitation mechanism 18
Bubble-particle
  bond 40
  contact 40-43
  interpenetration theory 43
Bulk flotation 9
CANDINA (Italy) 67-72
Carbonate ores 214
Carboxylates, see fatty acids
Case histories
  CODELCO Andina 136-147
  CODELCO El Teniente 147-173
  San Manuel 127-135
Cationic collectors 226
  chemistry 225, 227
  miscellization 226
  pK_a 228
  solubility 228

Cement rock 234
Chalcocite
 depression of 132
Iso-recovery curves 88, 89
Chalcopyrite
 depression of 132, 133
 structure 39
CIMM column flotation
 pilot plant 317
Chromic acid 104
Chromates
 depressant for galena 104
Chromite 233
CODELCO Andina 136-147
 lab results sand/slime test 138-143
 mill test sand/slime circuit 144-148
 rougher circuit 136, 137
CODELCO El Teniente 147-173
 column tests moly plant 172
 copper cleaner 158-161, 166-168
 dissolved Mo in tailings 164
 effect of pH on Cu/Mo/pyrite 164
 lime, depressant effect on moly 162
 mine sample floatability profile 157
 mill material balance 151, 152
 moly recovery 1985/91 149, 169,170
 moly recovery, acid vs alcaline 163
 Nokes depressant studies 171
 ore mineralogy 153-155
 sand/slime split 168
Colemanite 232
Collectors for non-metallics 220
Collector suppliers and usage
 alkyl xanthates 67, 74-77
 dialkyl thionocarbamates 70, 80
 dialkyl thiophos. 70, 71, 72, 81-83
 dialkyl xanthogen formates 69, 79
 miscellaneous collectors 72, 84
 xanthic esters 68, 78
Collectors 9
 consumption 176, 198
 popularity 195
 structures 7
Collector-mineral
 bonding 40-46
 interface 35
Column testing 313
 two inch laboratory column 321
 variables in cleaner study 318
Complex ores
 flotation circuit 204, 209
 minerals contained in 206, 210
Concentrate steaming 113, 116
Contact angle 13
 definition 37
 values for xanthates 16
Copper sulphate 102
Copper-molybdenite separation 128-135
Copper mill flotation practice 179-191
Cresylic acid
 frothing properties 97
Critical miscelle concentration
 anionic collectors 224
 cationic collectors 226
Critical pH for flotation 163
Cyanamid, see Am. Cyanamid
Cyanide, depressant 103
Cycle tests 307
Cyclic alcohols as frothers 96
Depressants 103-105
 quebracho 105
 starch 105
Depressants, copper moly circuits 116
 ammonium sulphide 116
 Nokes reagents 116, 120-122
 sodium cyanide 103, 116
 sodium ferri, ferro cyanide 104, 116, 132
 sodium sulphide and sulphhydrate 107, 116
 sodium zinc cyanide 116
Dialkyl thionocarbamates
 manufacture 62-64
 suppliers and trade names 69
Dialkyl xanthogen formates
 manufacture 60
 properties 61
 suppliers and trade names 69

Differential flotation 9
Dithiolate 35
chemistry 29
nomenclature 33
reagents 73, 108
Dithiolate/Thiol couples
reduction potentials 34
Dithiophosphoric acids
boiling points 56
manufacture 57
Dixanthogen 29
adsorption 105
chemical properties 31
galena 20
Double layer, 17
potential, 20
frothers 99
Dow Chemical Co (USA) 49-55,62-70, 73, 91, 98-100, 115, 134, 150, 177, 285
Electrical double layer theory 21
Electrochemical properties
phase diag., see potential-pH diagram
minerals 38
potential of sulphides, 103
Electron transfer to min. surface 19
Enargite
structure 39
collectors for 73, 205
Equipping the flotation lab. 278
apparatus required 311
ball mill 301, 302, 305
charge 306
flotation machines 279
Exfoam '636' 134
Experimental design
false optima one variable 274
Fatty acid collectors 221
see also anionic collectors
composition 222
formulae 223
Feldspar 234
mica-quartz-lithia, reagents 237
Ferro and ferricyanides 104, 132
Float technician selection 284
Flotation
coursing bubble 12
mill operating norms 193
nascent bubble 17
process design 200
Flotation cells
impeller and diffuser 283
standard 5.5 liter (2 kg) 281, 282
Flotation column 314
design of the indus. columns 315
pilot plant at Chile's CIMM 316
testing examples 317-323
Flotation machines, laboratory
comparison of machines 285
Flotation of complex ores 209
Flotation mechanism 13, 16
cavitation 18
Flotation reagents
collectors 66-73
depressants 101-106
frothers 94-100
modifiers 101-106
Fluorspar 238
grade specifications 240
reagents for flotation 240
zinc heavy media concentration 239
Foote spudomene flotation 254
Froth flotation
definition of terms 5
in the U.S.A. 175-191
Froth hydrodynamics
effect of collectors 86
stability, dynamic 85
Frothers 9
alcoholic frother 94
alkoxy paraffins 98
chemical properties 94-100
consumption 177
cresols 97

Frothers (continued)
  Cyanamid 99
  Dowfroths 99
  effect of, on chalcocite rec. 88
  ketone 96
  popularity 197
  power 86
  stability 85
  structures 8
  terpenes 96
  US consumption 177
  use in copper mills 185-190
  volume versus alcohol chain 87
  water soluble 99
Frother-collector
  interaction complex 44
  surface complexes 43-46
Fuel oil 134
Glass sands 241
  flotation results 245, 246
  lock cycle tests 244
  reagents, flotation 243
  specifications 242
Glue as depressant 105
Graphite flotation 245
Gums 106
Gypsum 247
Hallimond tube 13, 86, 212, 266, 287
Hoechst (Germany) 67-72
Hydrofluoric acid 104
Hydrogen peroxide 130
Hydrogen sulphide 103
Ice, structure of 22
Interpenetration theory 43
Iodine flotation 246
IQM (Mexico) 67, 68
Isorecovery curves
  chalcocite 89
Kaolin flotation 247
Kerley Chemical (USA) 67, 68
Kyanite flotation 247
Laboratory ball mills 310, 302, 305
Laboratory flot. machine 285-287
  comparison of performance 285
Lead nitrate or acetate 103
  activator for NaCl 215
Leeds Autofloat, reproducibility 28
Lime, see also depressants 104
  depressant effect on moly 162
LPF or
Leach Precipitation Float 124, 125, 213
LR-744, see phosphate Nokes
  chemistry 120
Lubrizol (USA) 71
Manufacture of
  thionocarbamates 62-64
  thiophosphates 57
  xanthates 49-54
  xanthogen formates 60
Mechanism of bubble formation 18
Mercaptans 5, 36, 62, 64, 72, 73, 20
Mercaptobenzothiazole 65, 72
  surface properties 146
Methyl isobutyl carbinol see MIBC
MIBC 134
Microscopic analysis moly plant 135
Minerals, electrochemical properties 38
Minerec (USA) 1, 31, 69-71
Modifiers 9, 10, 101-106
Molybdenite plants 112-116
  depressants 129-134
  mineralogy 135
  reagent consumption 135
  recovery circuit, San Manuel 128
  recovery study El Teniente 147-173
  the sulphide process 113, 119
Molybdenum conc. cleaning 111-126
Nascent bubble flotation 17
Natural oils 95
Natural gums 106
Nitrogenous cationic agents 226
Non-metallic minerals
  collectors for 230
  fatty acid collectors 221
  flotation mechanism 217

Nokes reagents, see LR-744, Anamol
chemistry of 120-122
Oils and Fats
fatty acid content of 222
Oily collector flotation 213
ORP-Oxidation reduction potential 129
Over oiling 16, 92
Oxide minerals 229
flotation model 213
Oxygen role in collector adsorption 19
Particle size
effect on flotation 168
froth stabilization 85
Pegmatite processing 236
ores 256
Pennwalt (USA) 64
Perxanthates 20
pH
critical pH for flotation 101
effect of collectors 102
effect of cyanide 102
effect on molecular adsorption, 57
effect on sulphide ion 120
Phillips 66 (USA) 64, 70, 72
Phosphate Nokes 116
chemistry 120
Phosphate rock process 248
reagents 250
Pine oil 96
Pilot plant flowsheet for sands 224
Polar minerals
classification of 6
Polyglycol glycerol ethers 100
Polypropylene glycol ethers 99
Potash flotation 250
reagents used 252, 253
Potential-pH diag. moly-sulphur-H_2O 165
Properties of
alcoholic frothers 94
alkyl xanthates 51, 52, 55
alkyl xanthogen ethyl formates 61
dithiophospahtes 58
terpene and ketone frothers 96
water soluble frothers 99
Prospect Chemical (Canada) 68, 69
Pyrite
depression 133, 162, 164
Pyrrhotite
activation by copper sulphate 102
Quartz and feldspars, specification 242
Quebracho 105
Rate of flotation
copper ores 90
effect of collector 90
effect of frother 88
Klimpel curve 93
RENASA (Peru) 67
Sampling and sample preparation 287-294
minimum representative size 289
Sampling working concentrators 291
San Manuel
copper and moly assays 129
molybdenite recovery circuit 127-135
Sand/slimes study - Andina 136-147
Scheelite flotation 253
reagents 253
Screen size equivalents 201
Semi-soluble salt flotation 215
Shellflot (Chile) 67, 69, 70
Sherex Chemical Co,
coal tests 310-313
standard test procedures 303
Sherrit-Gordon (Canada) 67, 68
Slimes
flotation recovery 139-146
sulphidisation sand/slime separation 136

Sodium cyanide 103
Sodium hydrosulphide 109
Sodium hydroxide 104
Sodium hypochlorite 133, 134
Sodium silicate 104
Sodium sulphide, 107-111
  chemical properties 108
  concentration effect on pH 111
  ionization as a function of pH 109
  ionization idem temperature 111
  phases in equilibrium 108
  solubility in water 108
Sodium sulphites and hyposulphites 104
Sodium thioglycolate 116
Sodium zinc cyanide 129
Sphalerite
  structure 39
Spudomene 254
  mineralogical analysis 255
  and co-products (pegmatite ores) 256
  and feldspars-quartz-mica flotation 235
SSDEVOP: a response surface 276
Starch 105
Stereochemistry 25-30
  dithiophosphates 29
  metal xanthates 25
  $[M(bidentate)_3]^{x\pm}$ 26
  octahedron 27
  pentagonal bipyramid 27
  seven-coordination 26
  six-coordination 25
Structure of water 23-25
Structures, suppliers and mineral use of
  alkyl xanthates 67, 74-77
  dialkyl thionocarbamates 70, 80
  dialkyl thiophosphate 70, 71, 72, 81-83
  dialkyl xanthogen formates 69, 79, 136
  miscellaneous collectors 72, 84
  xanthic esters 68, 78
Sulphide minerals
  band-gap energies 40
Sulphidisation
Non-sulphide copper ores 208
  of refractory ores 122-126
Sulphonates 5
Sulphur dioxide 104
Sulphuric acid 104, 134
Talc 257
Tannic acid 105
TEB 98
Tecnomin (Chile) 69
Testing
  design 260
  variables 259
Thiocarbanilide 65
Thiol collectors
  bonding with sulphide minerals 36
  stereochemistry of xanthates 25-30
  idem dithiophosphates 29
Thiolated surfaces
  products extracted from 35
Thiophosphates 55-58
  O,O-dialkyl dithiophosphates 55
  O,O-diaryl dithiophosphoric acids 57
  manufacture 57
  properties 58
  suppliers and trade names 70-72
UCB (Belgium) 67
U.S.A. flotation statistics 175-191
Vermiculite 257
  suppliers and trade names 69
Water
  at mineral surface 17
  clathrate 218
  dilution 134
  structure 23, 218
Wollastonite 257
Wurtzite
  structure 39

Xanthates, see also thiols
analysis 52
analitical method 55
handling and safety 54
manufacture 49, 53
properties of 47-55
solubility 51
solutions stability 51, 55
suppliers and trade names 67
Xanthic acids, see xanthates
Xanthic anhydrides
properties 31
Xanthic esters
allyl ester 64
suppliers and trade names 68
Xanthogen formates 58
manufacture of 60
properties 61
Zinc sulphate 104
ZUPA (Yugoslavia) 67, 70

AUTHOR INDEX

Baleshta, T.M. 40
Beas-Bustos, E. 147
Booth, R.B. 98, 196
Cambron, Adrien 29
Castleman Jr., A. W. 217
Castro, S. 129, 162, 171
Cea, Zocimo 162
Chander, S. 162
Clark, Alfred 36
Crozier, R.D. 23, 44, 63, 66, 116, 176
Davis, Farlow C. 127, 128
Douglass, W. 32, 62, 73
Drew, M.G.B. 24, 27
Ewers, W.E. 85
Finkelstein, N.P. 5, 29, 32, 33, 34
Fischer, A.H. 32, 73, 153
Fuerstenau, D.W. 162
Gardner, J.R. 14
Gaudin, A.M. 12
Glembotskii, V.A. 41, 162
Grainger-Allen, T.J.N. 18
Granville, A. 33
Gutierrez, G. 121
Hansen, R.D. 89
Harris, Guy 49, 50, 62
Houben-Weyl 56
Li Guongming 162
Jones, M.H. 19, 122
Kepert, D.L. 25, 28
Klassen, V.I. 12, 14, 15, 17, 40, 117
Klimpel, Richard 44, 89, 90, 92
Laskowski, Janusz 14, 23, 40, 87, 88, 266
Leja, J. 14, 24, 30, 43, 46, 86, 117, 217, 228
Lekki, J. 87, 88, 266
Malhotra, D. 169
Mathieu, G.I. 242
Mular, A.L. 274, 275, 276
Mumme, W.G. 27
Nokes, C.M. 120
O'Connor, C.T. 285, 291
Palagi, C.G. 125
Pauling, Linus 22, 37, 39, 45
Poling, G.W. 5, 29, 34
Powell, R.F. 98
Redeker, Immo H. 230, 234, 241, 257
Reid, E. Emmet 29
Shimoiizaka, J. 166
Shirley, J.F. 119, 120, 162
Smith, R. W. 228
Sutherland, K.L. 13, 38, 85, 101
Taggart, A.F. 15, 17, 31, 68
Tsai, M., 166
Tveter, E.C. 98
Wills, B.A. 5, 6
Winter, G. 15, 27, 36
Woods, R. 14, 19, 33, 34, 36
Wrobel, S.A. 86
Yehia, Ahmed 218